Stochastic Modelling of Reaction–Diffusion Processes

This practical introduction to stochastic reaction–diffusion modelling is based on courses taught at the University of Oxford. The authors discuss the essence of mathematical methods that appear (under different names) in a number of interdisciplinary scientific fields bridging mathematics and computations with biology and chemistry.

The book can be used both for self-study and as a supporting text for advanced undergraduate or beginning graduate-level courses in applied mathematics. New mathematical approaches are explained using simple examples of biological models, which range in size from simulations of small biomolecules to groups of animals. The book starts with stochastic modelling of chemical reactions, introducing stochastic simulation algorithms and mathematical methods for analysis of stochastic models. Different stochastic spatio-temporal models are then studied, including models of diffusion and stochastic reaction–diffusion modelling. The methods covered include molecular dynamics, Brownian dynamics, velocity-jump processes and compartment-based (lattice-based) models.

RADEK ERBAN is Professor of Mathematics at the University of Oxford, a Fellow of Merton College, Oxford, and a Royal Society University Research Fellow.

S. JONATHAN CHAPMAN is Professor of Mathematics and its Applications at the University of Oxford, and a Fellow of Mansfield College, Oxford.

T0201455

Cambridge Texts in Applied Mathematics

All titles listed below can be obtained from good booksellers or from Cambridge University Press. For a complete series listing, visit www.cambridge.org/mathematics.

Stochastic Modelling of Reaction–Diffusion Processes

RADEK ERBAN

University of Oxford

S. JONATHAN CHAPMAN

University of Oxford

CAMBRIDGE
UNIVERSITY PRESS

CAMBRIDGE
UNIVERSITY PRESS

University Printing House, Cambridge CB2 8BS, United Kingdom

One Liberty Plaza, 20th Floor, New York, NY 10006, USA

477 Williamstown Road, Port Melbourne, VIC 3207, Australia

314–321, 3rd Floor, Plot 3, Splendor Forum, Jasola District Centre,
New Delhi – 110025, India

79 Anson Road, #06–04/06, Singapore 079906

Cambridge University Press is part of the University of Cambridge.

It furthers the University's mission by disseminating knowledge in the pursuit of
education, learning, and research at the highest international levels of excellence.

www.cambridge.org
Information on this title: www.cambridge.org/9781108498128
DOI: 10.1017/9781108628389

First published 2020

Printed in Singapore by Markono Print Media Pte Ltd

A catalogue record for this publication is available from the British Library.

Library of Congress Cataloging-in-Publication Data
Names: Erban, Radek, author. | Chapman, Jon, author.
Title: Stochastic modelling of reaction–diffusion processes / Radek Erban (University of
Oxford), S. Jonathan Chapman (University of Oxford).
Description: Cambridge ; New York, NY : Cambridge University Press, [2020] | Series:
Cambridge texts in applied mathematics | Includes bibliographical references and index.
Identifiers: LCCN 2019019436 | ISBN 9781108498128 (alk. paper)
Subjects: LCSH: Stochastic processes – Textbooks. | Reaction–diffusion
equations – Textbooks.
Classification: LCC QA274 .E73 2020 | DDC 515/.3534–dc23
LC record available at https://lccn.loc.gov/2019019436

ISBN 978-1-108-49812-8 Hardback
ISBN 978-1-108-70300-0 Paperback

Additional resources for this publication at www.cambridge.org/9781108498128

Contents

Preface

In this textbook, we provide an introduction to stochastic modelling of reaction–diffusion processes presented in a way that is aimed at advanced undergraduate or beginning graduate students. We assume that the reader has a basic understanding of differential equations, but we do not assume any prior knowledge of advanced probability theory or stochastic analysis. We introduce and explain some common stochastic simulation methods using illustrative examples. At the same time we present some basic theoretical tools that are used to analyse the methods. New theory is introduced whenever it provides a better insight into a particular example. In our experience, such an example-based approach is more accessible to students than a theory-first approach. To make this textbook self-contained, we also summarize some basic facts about deterministic modelling (based on differential equations) and probability distributions in the Appendix.

The material in this textbook has been developed (and used in teaching students in Oxford) over a period of 12 years. It started as a short 30-page-long set of lecture notes on stochastic modelling of reaction–diffusion processes written by Erban, Chapman and Maini (Erban et al., 2007), where we explained simple algorithms on simple (linear) models. Many important topics had to be omitted due to lack of space. Over the next 12 years, we significantly extended that material to the present form. The resulting book covers stochastic models of chemical reactions, stochastic differential equations (SDEs), and stochastic models of diffusion, reaction–diffusion and reaction–diffusion–advection processes. We explain how stochastic models can be derived from more detailed molecular dynamics approaches and provide an introduction to methods designed for modelling processes on multiple spatial and temporal scales. We introduce methods for analysis of stochastic models (for example, the chemical master equation and the Fokker–Planck equation) and efficient stochastic simulation algorithms (SSAs). We also discuss connections and differences between stochastic and deterministic modelling approaches.

Chapters of this textbook have been used as reading material for three types of courses at the Mathematical Institute, University of Oxford. The most advanced topics were initially used in lectures designed for graduate students, including students on a one-year taught master programme in applied mathematics. During the last seven years, we have also taught an undergraduate course offered to fourth- and later third-year students, which is based on the material in this textbook. Variants of practice problems and exam questions used for assessment of both undergraduate courses have been included in Exercises supporting each chapter.

Although some simple stochastic models can be analysed using only a pen-and-paper approach, one cannot apply stochastic modelling to more complicated examples without using computers. Thus, our discussion is built not only around mathematical equations and their analysis, but also around the corresponding algorithms. However, we do not assume prior knowledge of a particular computer language: the algorithms we will present are all written in a general pseudocode. The simulations we show were computed using Matlab or Fortran. In the Oxford courses, we encouraged students to write their own computer codes to reproduce examples from the textbook, which they did using a variety of other programming languages, including C/C++ and Python. The computer codes that compute illustrative simulations and figures in this book are available online at

http://people.maths.ox.ac.uk/erban/cupbook/

Our illustrative computational examples help students to increase their confidence in writing relatively simple computer codes that can verify the presented theory. They can also be viewed as building blocks of complicated computational models of complex biological systems. Since theoretical results for such models are difficult to obtain, one can gain confidence in the accuracy of the computed results by analysing each simple building block separately using the techniques from the book. If each building block exactly matches the corresponding theory, we can better trust that a complex computer program correctly implements a complex biological system.

Although we do not focus on biological details, our textbook is written in such a way that students are well equipped to begin research projects in the area. To support this, we include physical units in some of our examples. They are not necessary for understanding the underlying mathematics, but they are an important reminder for students who want to apply the presented methods to practical problems. Physical units also help us to highlight the situations where some terms in equations are large or small compared to others. This is useful in some considered limiting processes or in our discussions of multi-scale approaches. Since physical units do not play the most important role, we denote a second as "sec" in this book (rather than the SI standard "s"). This is to avoid confusion in some parts of the book where letter s denotes a variable (for example, speed).

Finally, for many years of grant support and assistance, we would like to thank the Royal Society, Leverhulme Trust, European Research Council, Isaac Newton Institute, University of Oxford, including its Mathematical Institute, and a number of colleges at Oxford (Brasenose, Linacre, Merton, Mansfield, Somerville and St John's) and Cambridge (Peterhouse). We would like to thank our students and colleagues for their comments and suggestions, which improved the presentation of the material. Our special thanks goes to Professor Philip Maini, our co-author of the original 2007 lecture notes, who has made valuable comments during the preparation of this book.

Oxford, August 2019 Radek Erban and Jon Chapman

1

Stochastic Simulation of Chemical Reactions

In this chapter we will introduce stochastic methods for modelling spatially homogeneous systems of chemical reactions, including the chemical master equation and the Gillespie SSA, where the acronym SSA stands for the term "stochastic simulation algorithm" in this book. We will also discuss the connection between stochastic and deterministic modelling approaches.

1.1 Stochastic Simulation of Degradation

We start with the simplest possible example, which is the single chemical reaction

$$A \xrightarrow{k} \emptyset, \tag{1.1}$$

where A is the chemical species of interest and k is the rate constant of the reaction. The symbol $\emptyset$ denotes chemical species that are of no further interest. The rate constant k in (1.1) is defined so that $k\,dt$ gives the probability that a randomly chosen molecule of the chemical species A reacts (is degraded) during the time interval $[t, t + dt)$ where t is time and dt an (infinitesimally) small time step.

Let us denote the number of molecules of chemical species A at time t by $A(t)$ (a convention that will be used throughout the book). Then, in the time interval $[t, t + dt)$, a number of things may happen: none of the molecules may react, exactly one may react, or more than one may react. Assuming that each molecule acts independently, we may combine the individual probabilities of reaction to deduce that

no reactions occur	with probability	$1 - A(t)k\,dt + O(dt^2)$,
exactly one reaction occurs	with probability	$A(t)k\,dt + O(dt^2)$,
two or more reactions occur	with probability	$O(dt^2)$,

where $O(dt^2)$ signifies terms proportional to dt^2. If dt is small enough then the quadratic terms in dt are much smaller than the linear terms, and they may be safely neglected. Thus, in a small enough time interval, the chance of two or more reactions happening is negligible.

Let us assume that we have n_0 molecules of A in the system at time $t = 0$, i.e. $A(0) = n_0$. Our first goal is to compute the number of molecules $A(t)$ for times $t > 0$. The mathematical definition of the chemical reaction (1.1) can be directly used to design a "naive" stochastic simulation algorithm (SSA) for simulating it. We choose a small time step Δt, and compute the number of molecules $A(t)$ at times $t = i\Delta t$, $i = 1, 2, 3, \ldots$, by testing to see if a reaction occurs in each time interval and updating $A(t)$ accordingly.

To do that, we need a computer routine generating random numbers. Most modern programming languages contain a routine for generating random numbers uniformly distributed in the interval $(0, 1)$ (e.g. the function `rand` in Matlab). The routine will generate a number $r \in (0, 1)$, such that the probability that r is in a subinterval $(a, b) \subset (0, 1)$ is equal to $b - a$ for any $a, b \in (0, 1)$ with $a < b$. Using this routine we compute the number of molecules of $A(t)$ as follows. Starting with $t = 0$ and $A(0) = n_0$, we perform two steps at time t:

(a1) Generate a random number r uniformly distributed in the interval $(0, 1)$.

(b1) If $r < A(t)k\,\Delta t$, then put $A(t + \Delta t) = A(t) - 1$; otherwise, put $A(t + \Delta t) = A(t)$.
Then continue with step (a1) for time $t + \Delta t$.

Since r is a random number uniformly distributed in the interval $(0, 1)$, the probability that $r < A(t)k\,\Delta t$ is equal to $A(t)k\,\Delta t$. Consequently, step (b1) says that the probability that the chemical reaction (1.1) occurs in the time interval $[t, t+\Delta t)$ is equal to $A(t)k\,\Delta t$. Thus step (b1) correctly implements the definition of (1.1) provided that Δt is small. The time evolution of A obtained by the "naive" SSA (a1)–(b1) is given in Figure 1.1(a) for[*] $k = 0.1\,\text{sec}^{-1}$, $A(0) = 20$ and $\Delta t = 0.005\,\text{sec}$. We repeated the stochastic simulation twice and we plotted two realizations of the SSA (a1)–(b1). We see in Figure 1.1(a) that these two realizations of the SSA (a1)–(b1) give two different evolutions. Each time we run the algorithm, we obtain different results. This is generally true for any SSA. Therefore, one might reasonably ask what useful and reproducible

---

[*] We use "sec" rather than "s" to denote seconds throughout the whole book. This is to avoid possible confusion with the use of lower case letter s to denote speed or other variables in later chapters of this book.

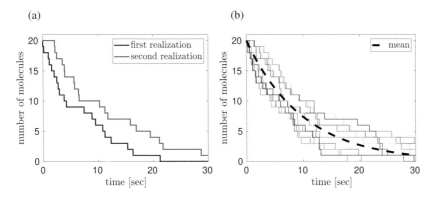

Figure 1.1 *Stochastic simulation of chemical reaction* (1.1) *for* $k = 0.1 \, sec^{-1}$ *and* $A(0) = 20$. (a) *Number of molecules of A as a function of time for two realizations of the "naive" SSA (a1)–(b1) for* $\Delta t = 0.005$ *sec.* (b) *Results of ten realizations of the SSA (a2)–(c2) (solid lines; different colours show different realizations) and stochastic mean* (1.8) *plotted by the dashed line.*

information can be obtained from stochastic simulations? We will come back to this question later in this section.

The probability that exactly one reaction (1.1) occurs during the infinitesimal time interval $[t, t+dt)$ is equal to $A(t)k \, dt$. To design the SSA (a1)–(b1), we replaced dt by the finite time step Δt. In order to get reasonably accurate results, we must ensure that $A(t)k \, \Delta t \ll 1$. For our simulations we used $k = 0.1 \, sec^{-1}$ and $\Delta t = 0.005$ sec. Since $A(t) \leq A(0) = 20$ for any $t \geq 0$, this gives $A(t)k \, \Delta t \in [0, 0.01]$ for any $t \geq 0$. Consequently, the condition $A(t)k \, \Delta t \ll 1$ is reasonably satisfied during the simulation. We might further increase the accuracy of the SSA (a1)–(b1) by decreasing Δt. However, decreasing Δt increases the computational intensity of the algorithm. The probability that the reaction (1.1) occurs during the time interval $[t, t+\Delta t)$ is less than or equal to 1% for our parameter values. Thus during most time steps we generate a random number r in step (a1) only to find out that no reaction occurs in step (b1): we need to generate a lot of random numbers before the reaction takes place. This naive SSA is extremely inefficient: we can do a lot better.

The key to improving the algorithm is a change in viewpoint. Instead of focusing on time we focus on events: rather than stepping forward in time and asking did a reaction take place, we ask at what time will the next reaction occur?

If the time is t now, our goal is to compute the time $t + \tau$ when the next reaction (1.1) takes place. Of course, τ is a random variable, so that what we need to calculate is its probability distribution function. Let us denote by

$f(A(t), s)\,\mathrm{d}s$ the probability that, given $A(t)$ molecules at time t in the system, the next reaction occurs during the time interval $[t + s, t + s + \mathrm{d}s)$ where $\mathrm{d}s$ is an (infinitesimally) small time step. In order for this to happen, there must have been no reaction during the interval $[t, t + s)$, and then a reaction must have occurred during the interval $[t + s, t + s + \mathrm{d}s)$. Thus, if we let $g(A(t), s)$ be the probability that no reaction occurs in interval $[t, t + s)$, the probability $f(A(t), s)\,\mathrm{d}s$ can be computed as a product of $g(A(t), s)$ and $A(t + s)k\,\mathrm{d}s$:

$$f(A(t), s)\,\mathrm{d}s = g(A(t), s)A(t + s)k\,\mathrm{d}s.$$

Since no reaction occurs in $[t, t + s)$, we have $A(t + s) = A(t)$, so that in fact

$$f(A(t), s)\,\mathrm{d}s = g(A(t), s)A(t)k\,\mathrm{d}s. \tag{1.2}$$

It remains for us to calculate $g(A(t), s)$. For any $\sigma > 0$, the probability that no reaction occurs in the interval $[t, t + \sigma + \mathrm{d}\sigma)$ can be computed as the product of the probability that no reaction occurs in the interval $[t, t + \sigma)$ and the probability that no reaction occurs in the interval $[t + \sigma, t + \sigma + \mathrm{d}\sigma)$. Hence

$$g(A(t), \sigma + \mathrm{d}\sigma) = g(A(t), \sigma)[1 - A(t + \sigma)k\,\mathrm{d}\sigma].$$

Since no reaction occurs in the interval $[t, t + \sigma)$, we have $A(t + \sigma) = A(t)$. Consequently, after some rearrangement,

$$\frac{g(A(t), \sigma + \mathrm{d}\sigma) - g(A(t), \sigma)}{\mathrm{d}\sigma} = -A(t)k\,g(A(t), \sigma).$$

Passing to the limit $\mathrm{d}\sigma \to 0$, we obtain the ordinary differential equation (in the σ variable)

$$\frac{\mathrm{d}g(A(t), \sigma)}{\mathrm{d}\sigma} = -A(t)k\,g(A(t), \sigma).$$

Solving this equation with initial condition $g(A(t), 0) = 1$, we obtain

$$g(A(t), \sigma) = \exp[-A(t)k\sigma].$$

Now (1.2) can be written as

$$f(A(t), s)\,\mathrm{d}s = A(t)k\exp[-A(t)ks]\,\mathrm{d}s. \tag{1.3}$$

Thus we have found that the time interval to the next reaction, τ, is distributed according to the exponential distribution with mean $(A(t)k)^{-1}$. The exponential distribution is, together with other useful distributions, defined in the Appendix.

To use this in our simulation algorithm we need to generate random numbers τ distributed according to (1.3). The easiest way to accomplish this is to use the following auxiliary function:

$$F(\tau) = \exp[-A(t)k\tau] = \int_{\tau}^{\infty} f(A(t), s) \, ds. \tag{1.4}$$

The function $F(\tau)$ represents the probability that the time to the next reaction is greater than τ, and is monotone decreasing for $A(t) > 0$. For our present purposes what is important is that, if τ is a random number in the interval $(0, \infty)$, then $F(\tau)$ is a random number in the interval $(0, 1)$. Moreover, if τ is a random number distributed according to the probability density function (1.3), then $F(\tau)$ is a random number *uniformly* distributed in the interval $(0, 1)$. To show this let $0 < a < b < 1$ be chosen arbitrarily. The probability that $F(\tau) \in (a, b)$ is equal to the probability that $\tau \in (F^{-1}(b), F^{-1}(a))$, which is given by the integral of $f(A(t), s)$ over s from $F^{-1}(b)$ to $F^{-1}(a)$. Using (1.3) and (1.4) we obtain

$$\int_{F^{-1}(b)}^{F^{-1}(a)} f(A(t), s) \, ds = \int_{F^{-1}(b)}^{F^{-1}(a)} A(t)k \exp[-A(t)ks] \, ds$$

$$= -\int_{F^{-1}(b)}^{F^{-1}(a)} \frac{dF}{ds} \, ds = -F[F^{-1}(a)] + F[F^{-1}(b)] = b - a.$$

Thus, if we have an algorithm that generates a random number r uniformly distributed on $(0, 1)$, we can generate the time of the next reaction by setting

$$r = F(\tau) = \exp[-A(t)k\tau].$$

Solving for τ, we obtain the formula

$$\tau = \frac{1}{A(t)k} \ln\left[\frac{1}{r}\right]. \tag{1.5}$$

Consequently, the improved SSA for the chemical reaction (1.1) can be written as follows. Starting with $t = 0$ and $A(0) = n_0$, we perform three steps at time t:

(a2) Generate a random number r uniformly distributed in the interval $(0, 1)$.

(b2) Compute the time when the next reaction (1.1) occurs as $t + \tau$ where τ is given by (1.5).

(c2) Compute the number of molecules at time $t + \tau$ by $A(t + \tau) = A(t) - 1$. Then continue with step (a2) for time $t + \tau$.

Steps (a2)–(c2) are repeated until we reach the time when there is no molecule of A in the system, i.e. $A = 0$. The SSA (a2)–(c2) computes the time of the next reaction $t + \tau$ using formula (1.5) in step (b2) with the help of one random number only. Then the reaction is performed in step (c2) by decreasing the number of molecules of chemical species A by 1. The time evolution of A obtained by the SSA (a2)–(c2) is given in Figure 1.1(b). We plot ten realizations of the SSA (a2)–(c2) for $k = 0.1 \sec^{-1}$ and $A(0) = 20$. Since the function $A(t)$ has only integer values $\{0, 1, 2, \ldots, 20\}$, it is not surprising that some of the computed curves $A(t)$ partially overlap. On the other hand, all ten realizations yield different functions $A(t)$. Even if we made millions of stochastic realizations, it would be very unlikely (with probability zero) that there would be two realizations giving exactly the same results. Therefore, the details of one realization $A(t)$ are of no special interest (they depend on the sequence of random numbers obtained from the random number generator). However, averaging values of A at time t over many realizations (that is, computing the stochastic mean of A), we obtain a reproducible characteristic of the system – see the dashed line in Figure 1.1(b). The mean of $A(t)$ over (infinitely) many realizations can be also computed theoretically as follows.

Let us denote by $p_n(t)$ the probability that there are n molecules of A at time t in the system, i.e. $A(t) = n$. Let us consider an (infinitesimally) small time step dt chosen such that the probability that two molecules are degraded during $[t, t + dt)$ is negligible compared to the probability that only one molecule is degraded during $[t, t+dt)$. Then there are two possible ways for $A(t+dt)$ to take the value n: either $A(t) = n$ and no reaction occurred in $[t, t+dt)$, or $A(t) = n+1$ and one molecule was degraded in $[t, t + dt)$, i.e.

$$p_n(t + dt) = p_n(t) \times (1 - kn \, dt) + p_{n+1}(t) \times k(n + 1) \, dt.$$

A simple algebraic manipulation yields

$$\frac{p_n(t + dt) - p_n(t)}{dt} = k(n + 1) \, p_{n+1}(t) - kn \, p_n(t).$$

Passing to the limit $dt \to 0$, we obtain the so-called *chemical master equation* in the form

$$\frac{dp_n}{dt} = k(n + 1) \, p_{n+1} - kn \, p_n. \tag{1.6}$$

Equation (1.6) looks like an infinite system of ordinary differential equations (ODEs) for p_n, $n = 0, 1, 2, 3, \ldots$. However, our initial condition $A(0) = n_0$ implies that there are never more than n_0 molecules in the system, so that $p_n \equiv 0$ for $n > n_0$ and the system (1.6) reduces to a system of $(n_0 + 1)$ ODEs for $p_0, p_1, \ldots, p_{n_0}$. The equation for $n = n_0$ reads

$$\frac{dp_{n_0}}{dt} = -kn_0 \, p_{n_0}.$$

Solving this equation with the initial condition $p_{n_0}(0) = 1$ (since we know with certainty that there are n_0 molecules of A at time $t = 0$), we find

$$p_{n_0}(t) = \exp[-kn_0 t].$$

Using this formula in the chemical master equation (1.6) for $p_{n_0-1}(t)$, we obtain

$$\frac{d}{dt} p_{n_0-1}(t) = kn_0 \exp[-kn_0 t] - k(n_0 - 1) p_{n_0-1}(t).$$

Solving this equation with initial condition $p_{n_0-1}(0) = 0$, we obtain

$$p_{n_0-1}(t) = \exp[-k(n_0 - 1)t] n_0 (1 - \exp[-kt]).$$

Using mathematical induction, it is possible to show that in general

$$p_n(t) = \exp[-knt] \binom{n_0}{n} \{1 - \exp[-kt]\}^{n_0-n}, \tag{1.7}$$

where

$$\binom{n_0}{n} = \frac{n_0!}{n!(n_0 - n)!}$$

is the binomial coefficient. Looking at the Appendix, we can observe that the formula (1.7) is the probability mass function of the binomial distribution with parameter $\exp[-kt]$. It provides complete information about the stochastic process defined by (1.1) with the initial condition $A(0) = n_0$. We can never say for sure that $A(t) = n$; we can only say that $A(t) = n$ with probability $p_n(t)$. In particular, we can use (1.7) to derive a formula for the mean value of $A(t)$ over (infinitely) many realizations, which is defined by

$$M(t) = \sum_{n=0}^{n_0} n \, p_n(t).$$

Using (1.7), we deduce that

$$
\begin{aligned}
M(t) &= \sum_{n=0}^{n_0} n \, p_n(t) \\
&= \sum_{n=0}^{n_0} n \exp[-knt] \binom{n_0}{n} \{1 - \exp[-kt]\}^{n_0-n} \\
&= n_0 \exp[-kt] \sum_{n=1}^{n_0} \binom{n_0 - 1}{n - 1} \{1 - \exp[-kt]\}^{(n_0-1)-(n-1)} \{\exp[-kt]\}^{n-1} \\
&= n_0 \exp[-kt], \tag{1.8}
\end{aligned}
$$

where the last step follows from the binomial theorem applied to the identity $(\{1 - \exp[-kt]\} + \exp[-kt])^{n_0-1} = 1$. The chemical master equation (1.6) and its solution (1.7) can be also used to quantify the stochastic fluctuations around the mean value (1.8), i.e. how much an individual realization of the SSA (a2)–(c2) can differ from the mean value given by (1.8). We will present the corresponding theory and results on a more complicated illustrative example in Section 1.2. Finally, let us note that a classic deterministic description of the chemical reaction (1.1) is given by the ODE (see the Appendix)

$$\frac{da}{dt} = -ka,$$

where $a(t) = A(t)/v$ is the concentration of chemical species A in a container of volume v. Solving this equation with initial condition $a(0) = n_0/v$, we obtain the function (1.8) divided by volume v, i.e. $a(t) = M(t)/v$. In other words, the stochastic mean can be obtained by solving the corresponding deterministic ODE. However, we should emphasize that this is not true for general systems of chemical reactions, as we will see in Section 1.4, Section 1.5 and Chapter 2.

1.2 Stochastic Simulation of Production and Degradation

The reaction (1.1) eventually leads to the elimination of all molecules of A. We now make it more interesting by adding also some production of A. Thus let us suppose that we have a chemical species A in a container of volume v which is subject to the following two chemical reactions:

$$A \xrightarrow{k_1} \emptyset, \qquad \emptyset \xrightarrow{k_2} A. \qquad (1.9)$$

The first reaction describes the degradation of chemical A with the rate constant k_1 previously studied. We couple it with the second reaction, which represents the production of chemical A with the rate constant k_2 per unit volume. The exact meaning of the second chemical reaction in (1.9) is that one molecule of A is created during the time interval $[t, t + dt)$ with probability $k_2 v\, dt$ where v is the system volume. As before, the symbol $\emptyset$ denotes chemical species that are of no special interest to us, i.e. the second reaction does not mean that chemical A would be produced from empty space. Indeed, in Section 2.4, we revisit this example and present a slightly larger chemical system which has similar dynamics as (1.9) without using notation $\emptyset$ in its production reaction. In this section, the impact of other chemical species on the rate of production of A is assumed to be time-independent and is already incorporated in the rate constant k_2.

The rate constants k_1 and k_2 have different physical units. The rate constant k_1 is expressed in the units of $[\text{sec}^{-1}]$. The units of the rate constant k_2 are $[\text{m}^{-3}\,\text{sec}^{-1}]$. It is the production rate per unit of volume and per unit of time, so that the probability that one molecule of A is created during the time interval $[t, t+dt)$ is equal to $k_2 v\, dt$. The scaling with the volume v is natural: if we divide the container into two equal parts, the production rate in each part will be half of the production rate in the whole container. In this section, the scaling with the system volume v is not crucial: to simulate the production of molecules in a container of volume v, we do not need to specify k_2 and v individually but only the product k_2v, which is the global production rate (with units $[\sec^{-1}]$). The scaling of the reaction rates with the volume v will be more important in later chapters, when we consider spatially inhomogeneous systems.

To simulate the system of chemical reactions (1.9) we want again to think in terms of events by jumping forwards to the time that the next reaction happens. We can do this by performing the following four steps at time t (starting with $A(0) = n_0$ at time $t = 0$):

(a3) Generate two random numbers r_1, r_2 uniformly distributed in $(0, 1)$.

(b3) Compute $\alpha_0 = A(t)k_1 + k_2v$.

(c3) Compute the time when the next chemical reaction takes place as
$t + \tau$ where

$$\tau = \frac{1}{\alpha_0} \ln\left[\frac{1}{r_1}\right]. \qquad (1.10)$$

(d3) Compute the number of molecules at time $t + \tau$ by

$$A(t + \tau) = \begin{cases} A(t) + 1 & \text{if } r_2 < k_2v/\alpha_0, \\ A(t) - 1 & \text{if } r_2 \geq k_2v/\alpha_0. \end{cases}$$

Then continue with step (a3) for time $t + \tau$.

Let us examine this algorithm, to see why it correctly simulates (1.9). First we observe that the probability that both reactions in (1.9) occur in the time interval $[t, t+dt)$ is quadratic in dt, and so is negligible as before. Thus the probability that one of the reactions takes place in the time interval $[t, t+dt)$ is equal to the probability that the first reaction occurs, $A(t)k_1\,dt$, plus the probability that the second reaction occurs, $k_2v\,dt$. We label this combined probability of some reaction occuring $\alpha_0\,dt$, and calculate it in step (b3).

The formula (1.10) in step (c3) gives the time $t + \tau$ when the next reaction takes place; it can be justified using the same arguments as for the formula (1.5). Now that we know a reaction has taken place, the final step is to decide which reaction it was. Whether a molecule is produced or degraded depends on the relative probabilities of the two reactions: a molecule is produced with probability k_2v/α_0; if a molecule is not produced then one must have been degraded. The decision as to which reaction takes place is given in step (d3) with the help of the second uniformly distributed random number r_2.

Five realizations of the SSA (a3)–(d3) are presented in Figure 1.2(a) as solid lines. We plot the number of molecules of A as a function of time for $A(0) = 0$, $k_1 = 0.1 \sec^{-1}$ and $k_2v = 1 \sec^{-1}$. We see that, after an initial transient, the

(a)   (b)

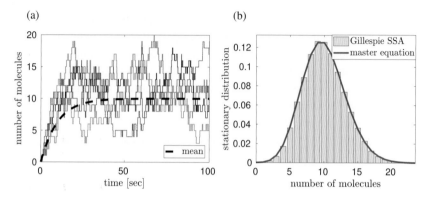

Figure 1.2 *Stochastic simulation of the system of chemical reactions (1.9) for*
$A(0) = 0$, $k_1 = 0.1 \, sec^{-1}$ *and* $k_2 \nu = 1 \, sec^{-1}$. *(a)* $A(t)$ *given by five realizations*
of the SSA (a3)–(d3) (solid lines) and stochastic mean (dashed line). (b) Station-
ary distribution $\phi(n)$ obtained by long-time simulation of the SSA (a3)–(d3) (grey
histogram) and by formulae (1.20)–(1.21) (red solid line).

number of molecules $A(t)$ fluctuates around its mean value. To compute the
mean and quantify the stochastic fluctuations, we again use the chemical master equation. As before, let $p_n(t)$ denote the probability that $A(t) = n$ for
$n = 0, 1, 2, 3, \ldots$. This time there are three ways we can arrive at n molecules
at time $t + dt$: there could have been n molecules at time t and no reactions
happened, or there could have been $n + 1$ molecules and one was degraded, or
there could have been $n - 1$ molecules and one was produced. Thus

$$p_n(t + dt) = p_n(t) \times (1 - k_1 n \, dt - k_2 \nu \, dt)$$
$$+ p_{n+1}(t) \times k_1(n + 1) \, dt + p_{n-1}(t) \times k_2 \nu \, dt.$$

Rearranging and passing to the limit $dt \to 0$ gives

$$\frac{dp_n}{dt} = k_1(n + 1) \, p_{n+1} - k_1 n \, p_n + k_2 \nu \, p_{n-1} - k_2 \nu \, p_n. \qquad (1.11)$$

Equation (1.11) needs to be slightly modified in the case $n = 0$: it is not possible to arrive at zero molecules at time $t + dt$ by having -1 molecules at time t
and producing one molecule! This means that the third term on the right-hand
side of (1.11) is missing for $n = 0$. To save ourselves the bother of writing a
separate equation for the $n = 0$ case we can continue to use (1.11) if we adopt
the convention that $p_{-1} \equiv 0$.

Since one of the reactions (1.9) involves production of A, there is no maximum possible number of molecules of A, as there was in Section 1.1. Thus
chemical master equation (1.11) does this time describe an infinite set of ODEs

for the variables p_n, $n = 0, 1, \ldots$. To solve this system numerically in practice it needs to be truncated, which is possible since the probability of a very large number of molecules is very small, that is, $p_n \to 0$ as $n \to \infty$.

The mean $M(t)$ and variance $V(t)$ are defined by

$$M(t) = \sum_{n=0}^{\infty} n \, p_n(t), \qquad V(t) = \sum_{n=0}^{\infty} (n - M(t))^2 \, p_n(t). \qquad (1.12)$$

The mean $M(t)$ gives the average number of molecules of A at time t, while the variance $V(t)$ describes the fluctuations. In Section 1.1, we first solved the chemical master equation (1.6) and then we used its solution (1.7) to compute $M(t)$. An alternative approach is to try to use the chemical master equation to derive an evolution equation for $M(t)$; if we can do this we could find $M(t)$ without having to first solve the chemical master equation. Let us see how such an approach might be carried out.

Multiplying (1.11) by n and summing over n, we obtain

$$\frac{d}{dt} \sum_{n=0}^{\infty} n p_n = k_1 \sum_{n=0}^{\infty} n(n + 1) \, p_{n+1} - k_1 \sum_{n=0}^{\infty} n^2 \, p_n$$

$$+ k_2 v \sum_{n=1}^{\infty} n \, p_{n-1} - k_2 v \sum_{n=0}^{\infty} n \, p_n.$$

Using definition (1.12) on the left-hand side and changing indices $n + 1 \to n$ (resp. $n - 1 \to n$) in the first (resp. third) sum on the right-hand side, we obtain

$$\frac{dM}{dt} = k_1 \sum_{n=0}^{\infty} (n - 1)n \, p_n - k_1 \sum_{n=0}^{\infty} n^2 \, p_n + k_2 v \sum_{n=0}^{\infty} (n + 1) \, p_n - k_2 v \sum_{n=0}^{\infty} n \, p_n.$$

Now combining the first and the second sums, and the third and the fourth sums, gives

$$\frac{dM}{dt} = -k_1 \sum_{n=0}^{\infty} n \, p_n + k_2 v \sum_{n=0}^{\infty} p_n. \qquad (1.13)$$

Since $p_n(t)$ is the probability that $A(t) = n$ and $A(t)$ is equal to a non-negative integer with probability 1, we have

$$\sum_{n=0}^{\infty} p_n(t) = 1. \qquad (1.14)$$

Using this fact together with the definition of $M(t)$, equation (1.13) implies an evolution equation for $M(t)$ in the form

$$\frac{dM}{dt} = -k_1 M + k_2 v. \qquad (1.15)$$

The solution of (1.15) depends on the initial condition, $M(0)$, as follows:

$$M(t) = \frac{k_2 v}{k_1} + \left(M(0) - \frac{k_2 v}{k_1}\right) \exp(-k_1 t).$$

Using $M(0) = 0$, we plot it as a dashed line in Figure 1.2(a). To derive the evolution equation for the variance $V(t)$, let us first observe that definition (1.12) implies

$$\sum_{n=0}^{\infty} n^2 p_n(t) = V(t) + M(t)^2. \tag{1.16}$$

Multiplying (1.11) by n^2 and summing over n, we obtain

$$\frac{d}{dt} \sum_{n=0}^{\infty} n^2 p_n = k_1 \sum_{n=0}^{\infty} n^2(n+1) p_{n+1} - k_1 \sum_{n=0}^{\infty} n^3 p_n$$

$$+ k_2 v \sum_{n=1}^{\infty} n^2 p_{n-1} - k_2 v \sum_{n=0}^{\infty} n^2 p_n.$$

Changing indices $n + 1 \to n$ (resp. $n - 1 \to n$) in the first (resp. third) sum on the right-hand side and combining the first and the second sums (resp. the third and the fourth sums) on the right-hand side gives

$$\frac{d}{dt} \sum_{n=0}^{\infty} n^2 p_n = k_1 \sum_{n=0}^{\infty} (-2n^2 + n) p_n + k_2 v \sum_{n=0}^{\infty} (2n+1) p_n.$$

Using (1.16), (1.14) and (1.12), we obtain

$$\frac{dV}{dt} + 2M \frac{dM}{dt} = -2k_1 [V + M^2] + k_1 M + 2k_2 v M + k_2 v.$$

Substituting (1.15) for dM/dt, we derive the evolution equation for the variance $V(t)$ in the following form:

$$\frac{dV}{dt} = -2k_1 V + k_1 M + k_2 v. \tag{1.17}$$

Often we are interested not so much in the initial transient behaviour, but more in the long-time behaviour of the system. We can see from (1.15) and (1.17) that as $t \to \infty$ both M and V approach stationary values, defined formally by

$$M_s = \lim_{t\to\infty} M(t), \qquad V_s = \lim_{t\to\infty} V(t). \tag{1.18}$$

The values of M_s and V_s can be computed using the steady-state versions of equations (1.15) and (1.17), namely by solving

$$0 = -k_1 M_s + k_2 v \qquad \text{and} \qquad 0 = -2k_1 V_s + k_1 M_s + k_2 v.$$

Consequently,

$$M_s = V_s = \frac{k_2 v}{k_1}.$$

For our parameter values $k_1 = 0.1 \sec^{-1}$ and $k_2 v = 1 \sec^{-1}$, we obtain $M_s = V_s = 10$. We see in Figure 1.2(a) that $A(t)$ fluctuates after a sufficiently long time around the mean value $M_s = 10$. To quantify the fluctuations, one often uses the square root of V_s, the so-called standard deviation, which in this case is equal to $\sqrt{10}$.

More detailed information about the fluctuations is given by the so-called *stationary distribution* $\phi(n)$, $n = 0, 1, 2, 3, \ldots$, which is defined as

$$\phi(n) = \lim_{t \to \infty} p_n(t). \tag{1.19}$$

This means that $\phi(n)$ is the probability that $A(t) = n$ after an (infinitely) long time. One way to compute $\phi(n)$ is to run the SSA (a3)–(d3) for a long time and create a histogram of values of $A(t)$ at given time intervals. Using $k_1 = 0.1 \sec^{-1}$ and $k_2 v = 1 \sec^{-1}$, the results of such a long-time computation are presented in Figure 1.2(b) as a grey histogram. To compute it, we ran the SSA (a3)–(d3) for 10^5 seconds, recording the value of $A(t)$ every second, and then divided the whole histogram by the number of recordings, i.e. by 10^5. An alternative way to compute $\phi(n)$ is to use the steady-state version of the chemical master equation (1.11), namely

$$0 = k_1 \phi(1) - k_2 v \phi(0)$$
$$0 = k_1(n + 1) \phi(n + 1) - k_1 n \phi(n) + k_2 v \phi(n - 1) - k_2 v \phi(n), \quad \text{for } n \geq 1,$$

which implies

$$\phi(1) = \frac{k_2 v}{k_1} \phi(0), \tag{1.20}$$

$$\phi(n + 1) = \frac{1}{k_1(n + 1)} [k_1 n \phi(n) + k_2 v \phi(n) - k_2 v \phi(n - 1)], \quad \text{for } n \geq 1. \tag{1.21}$$

By iteratively using (1.21) we can express $\phi(n)$ for any $n \geq 1$ in terms of $\phi(0)$. This remaining constant is then fixed by applying the normalization condition

$$\sum_{n=0}^{\infty} \phi(n) = 1, \tag{1.22}$$

which follows from (1.14) and (1.19). This enables us to compute $\phi(n)$, as follows. First set $\phi(0) = 1$ and compute $\phi(n)$, for sufficiently many n. Then normalize by dividing each $\phi(n)$ by $\sum \phi(n)$. The results obtained by (1.20)–(1.21) are plotted in Figure 1.2(b) as a (red) solid line. As expected, the results compare well with the results obtained by the long-time stochastic simulation.

In fact, the recurrence relations (1.20)–(1.21) are sufficiently simple that we can find the formula for $\phi(n)$ directly. We leave it as an exercise to verify that the solution of the equations (1.20)–(1.21) can be written as

$$\phi(n) = \frac{C}{n!} \left(\frac{k_2 v}{k_1} \right)^n, \tag{1.23}$$

where C is a real constant. Substituting (1.23) into the normalization condition (1.22), we get

$$1 = \sum_{n=0}^{\infty} \frac{C}{n!} \left(\frac{k_2 v}{k_1} \right)^n = C \sum_{n=0}^{\infty} \frac{1}{n!} \left(\frac{k_2 v}{k_1} \right)^n = C \exp\left[\frac{k_2 v}{k_1} \right],$$

where we used the Taylor series for the exponential function to give the last equality. Consequently,

$$C = \exp\left[-\frac{k_2 v}{k_1} \right],$$

which, together with (1.23), implies that the stationary distribution $\phi(n)$ is the Poisson distribution,

$$\phi(n) = \frac{1}{n!} \left(\frac{k_2 v}{k_1} \right)^n \exp\left[-\frac{k_2 v}{k_1} \right]. \tag{1.24}$$

Thus the red solid line in Figure 1.2(b) which was obtained numerically by the recurrence formula (1.20)–(1.21) can be also viewed as the stationary distribution $\phi(n)$ given by the explicit exact formula (1.24).

Considering initial condition $A(0) = 0$ which is used in Figure 1.2(a), we can also show that the time-dependent solution, $p_n(t)$, of the corresponding chemical master equation (1.11) is given by the Poisson distribution with mean $M(t)$ for all times $t > 0$. This is left as Exercise 1.6. Indeed, if $A(0) = 0$, then we have $M(0) = V(0) = 0$ and equations (1.15) and (1.17) imply that $M(t) = V(t)$ for $t \geq 0$, which is a necessary condition for $A(t)$ to be Poisson distributed at time t. If we consider a general initial condition, where $M(0) \neq V(0)$, then an analytical formula for $p_n(t)$ could be found using the approach of Jahnke and Huisinga (2007), but it would no longer be a simple Poisson distribution.

1.3 Higher-Order Chemical Reactions

The chemical reaction (1.1) is an example of a *first-order* chemical reaction, which is such that only one molecule is needed for the reaction to take place. As a result the rate of the first-order reaction depends linearly on the number of molecules of that one reactant. Another important class of chemical reactions are the so-called *second-order* (or bimolecular) chemical reactions. A simple example can be written as follows:

$$A + B \overset{k}{\longrightarrow} C. \tag{1.25}$$

Here, one molecule of A and one molecule of B react, with the rate constant k, to produce a molecule of C.

Now we have to think carefully about how the rate of reaction should scale with the volume of the system. We would expect that a reaction between one molecule of A and one molecule of B is more likely in a small box than in a large box. In fact, we would expect the two molecules to collide twice as often in a box that is half the size. Thus for one molecule of A and one molecule of B the probability of a reaction occurring in the time interval $[t, t + dt)$ is $(k/v)\, dt$. In particular, this means that the units of the rate constant k of the reaction (1.25) are $[\text{m}^3\ \text{sec}^{-1}]$. Now the number of different pairs of molecules A and B is equal to the product $A(t)B(t)$. Consequently, the probability that a reaction (1.25) takes place in the time interval $[t, t + dt)$ between any of these pairs is equal to $A(t)B(t)(k/v)\, dt$.

The *propensity function* $\alpha(t)$ of a reaction is defined to be such that the probability that the reaction occurs in the infinitesimally small time interval $[t, t+dt)$ is $\alpha(t)\, dt$. For the reaction (1.25) we have $\alpha(t) = A(t)B(t)k/v$. Other examples of second-order chemical reactions are

$$A + B \overset{k}{\longrightarrow} C + D, \qquad\qquad A + A \overset{k}{\longrightarrow} C. \tag{1.26}$$

In the first reaction, one molecule of A and one molecule of B react to produce a molecule of C and a molecule of D. The propensity function of the first reaction is the same as in the case of the reaction (1.25), i.e. $\alpha(t) = A(t)B(t)k/v$. The number of product molecules does not influence the characterization of the order of the chemical reaction, and does not change its propensity function.

In the second chemical reaction in (1.26), two molecules of A react with the rate constant k to produce C. To derive the propensity function for this reaction we follow the same reasoning as before. The key difference is that now the number of different pairs of molecules A is equal to

$$\binom{A(t)}{2} = \frac{A(t)(A(t) - 1)}{2}.$$

It is common to absorb the factor of 2 into the rate constant, so that the propensity function is written as $\alpha(t) = A(t)(A(t) - 1)k/v$. Note in particular that α is zero if A is zero or one: it takes two molecules of A for the reaction to happen.

Table 1.1 *The propensity functions of the basic chemical reactions.*

chemical reaction	order	propensity function $\alpha(t)$	units of k
$\emptyset \xrightarrow{k} A$	zeroth-order	kv	$m^{-3}\,sec^{-1}$
$A \xrightarrow{k} \emptyset$	first-order	$A(t)k$	sec^{-1}
$A + B \xrightarrow{k} \emptyset$	second-order	$A(t)B(t)k/v$	$m^3\,sec^{-1}$
$A + A \xrightarrow{k} \emptyset$	second-order	$A(t)(A(t)-1)k/v$	$m^3\,sec^{-1}$

Table 1.2 *The propensity functions of third-order chemical reactions.*

chemical reaction	propensity function $\alpha(t)$
$A + B + C \xrightarrow{k} \emptyset$	$A(t)B(t)C(t)k/v^2$
$2A + B \xrightarrow{k} \emptyset$	$A(t)(A(t)-1)B(t)k/v^2$
$3A \xrightarrow{k} \emptyset$	$A(t)(A(t)-1)(A(t)-2)k/v^2$

The units of the rate constant of any second-order chemical reaction are equal to $[m^3\,sec^{-1}]$. In Table 1.1, we summarize the order of the basic chemical reactions and their propensity functions. Since the number of product molecules does not influence the propensities and the order of the chemical reaction, we denote the products by $\emptyset$ for simplicity. We also include the production reaction (1.9), which we call a zeroth-order chemical reaction.

In a similar way, we can define *third-order* chemical reactions as reactions that require a collision of three molecules to take place. In Table 1.2, we specify the propensity functions of three possible types of third-order reactions. The rate constants of third-order chemical reactions have units of $[m^6\,sec^{-1}]$.

1.4 Stochastic Simulation of Dimerization

Systems with second-order (or higher-order) chemical reactions lead to chemical master equations that are more difficult to analyse (for example, the method presented in Section 1.2 does not work). In this section we use the example of dimerization to introduce an alternative approach to solving the chemical master equation. It is based on the so-called probability-generating function.

We consider a chemical species A in a container of volume v which is subject to the following two chemical reactions:

$$A + A \xrightarrow{k_1} \emptyset, \qquad \emptyset \xrightarrow{k_2} A. \qquad (1.27)$$

The first reaction describes the dimerization of the chemical A with the rate constant k_1. We couple it with the second reaction, which represents the production of the chemical A with the rate constant k_2. As before, the symbol $\emptyset$ denotes chemical species that are of no special interest to the modeller. Using Table 1.1, we get the propensity functions of the first and second reactions as $\alpha_1(t) = A(t)(A(t) - 1)k_1/v$ and $\alpha_2(t) = k_2 v$ respectively where, as usual, $A(t)$ is the number of molecules of A at time t and v is the system volume. To simulate the system of chemical reactions (1.27), we perform the following four steps at time t (starting with $A(0) = n_0$ at time $t = 0$):

(a4) Generate two random numbers r_1, r_2 uniformly distributed in $(0, 1)$.

(b4) Compute the propensity functions of both reactions:
  $\alpha_1 = A(t)(A(t) - 1)k_1/v$ and $\alpha_2 = k_2 v$. Compute $\alpha_0 = \alpha_1 + \alpha_2$.

(c4) Compute the time when the next chemical reaction takes place as
  $t + \tau$ where τ is given by (1.10).

(d4) Compute the number of molecules at time $t + \tau$ by

$$A(t + \tau) = \begin{cases} A(t) - 2 & \text{if } r_2 < \alpha_1/\alpha_0, \\ A(t) + 1 & \text{if } r_2 \geq \alpha_1/\alpha_0. \end{cases}$$

Then continue with step (a4) for time $t + \tau$.

As before, the probability that any of the reactions in (1.27) takes place in the time interval $[t, t + dt)$ is equal to $\alpha_0 \, dt$ where α_0 is the sum of the propensity functions of both reactions. Formula (1.10) gives the time $t + \tau$ when the next reaction takes place. Once we know the time $t + \tau$, we have to decide which of the two reactions actually took place. The first reaction in (1.27) takes place with probability α_1/α_0; otherwise, the second reaction occurs. The decision as to which reaction it was is made in step (d4) with the help of the second uniformly distributed random number r_2. In Figure 1.3(a), five realizations of the SSA (a4)–(d4) are presented as solid lines. We plot the number of molecules of A as a function of time for $A(0) = 0$, $k_1/v = 0.005 \text{ sec}^{-1}$ and $k_2 v = 1 \text{ sec}^{-1}$. We see that, after an initial transient, the number of molecules $A(t)$ fluctuates around its mean value. To compute the mean and quantify the stochastic fluctuations, we will analyse the corresponding chemical master equation.

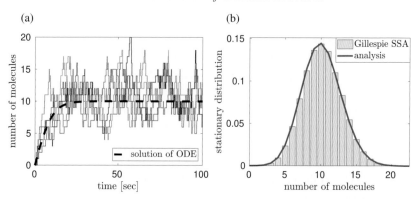

Figure 1.3 *Stochastic simulation of the system of chemical reactions* (1.27) *for* $A(0) = 0$, $k_1/\nu = 0.005\ sec^{-1}$ *and* $k_2\nu = 1\ sec^{-1}$. (a) $A(t)$ *given by five realizations of the SSA (a4)–(d4) (solid lines) and the solution of the ODE equation* (1.52) *(dashed line).* (b) *Stationary distribution* $\phi(n)$ *obtained by long-time simulation of the SSA (a4)–(d4) (grey histogram) and by formula* (1.50) *(red solid line).*

Let us again denote by $p_n(t)$ the probability that there are n molecules of A at time t in the system, i.e. that $A(t) = n$. As usual we consider an (infinitesimally) small time step dt chosen such that the probability that two reactions occur during $[t, t+dt)$ is negligible compared to the probability that only one reaction takes place during $[t, t + dt)$. Then there are three possible ways for $A(t + dt)$ to take the value n: either $A(t) = n$ and no reaction occurred in $[t, t + dt)$, or $A(t) = n+2$ and the first reaction in (1.27) occurred in $[t, t+dt)$, or $A(t) = n-1$ and one molecule was produced in $[t, t + dt)$ according to the second chemical reaction in (1.27). Hence

$$p_n(t + dt) = p_n(t) \times \left(1 - \frac{k_1}{\nu}n(n - 1)\,dt - k_2\nu\,dt\right)$$
$$+ p_{n+2}(t) \times \frac{k_1}{\nu}(n + 2)(n + 1)\,dt \ + \ p_{n-1}(t) \times k_2\nu\,dt.$$

A simple algebraic manipulation yields

$$\frac{p_n(t + dt) - p_n(t)}{dt} = \frac{k_1}{\nu}(n + 2)(n + 1)\,p_{n+2}(t) - \frac{k_1}{\nu}n(n - 1)\,p_n(t)$$
$$+ k_2\nu\,p_{n-1}(t) - k_2\nu\,p_n(t).$$

Passing to the limit $dt \to 0$, we obtain the chemical master equation for the chemical system (1.27) in the form

$$\frac{dp_n}{dt} = \frac{k_1}{\nu}(n + 2)(n + 1)\,p_{n+2} - \frac{k_1}{\nu}n(n - 1)\,p_n + k_2\nu\,p_{n-1} - k_2\nu\,p_n, \quad (1.28)$$

where the third term on the right-hand side is missing in (1.28) for $n = 0$; we again use the convention that $p_{-1} \equiv 0$.

The mean $M(t)$ and variance $V(t)$ are defined by (1.12). In Section 1.2, we were able to derive the evolution equations for $M(t)$ and $V(t)$. Unfortunately, the approach of Section 1.2 is not applicable to the chemical systems with second-order (or higher-order) reactions. If we try to follow the method from Section 1.2 (i.e. multiply the chemical master equation (1.28) by n and sum over n), we do not obtain a closed evolution equation for $M(t)$. However, we are mostly interested in the stationary values M_s and V_s, which are defined by (1.18), and the stationary distribution $\phi(n)$, which is defined by (1.19). Fortunately, it is possible to derive analytical formulae for M_s, V_s and $\phi(n)$. To do that, we define the *probability-generating function* $G: [-1, 1] \times (0, \infty) \to \mathbb{R}$ by

$$G(x, t) = \sum_{n=0}^{\infty} x^n p_n(t). \tag{1.29}$$

The function $G(x, t)$ encodes all the information about the solution of the chemical master equation (1.28). To find an equation for $G(x, t)$, and to see how the information it contains may be extracted, we first differentiate with respect to x to obtain

$$\frac{\partial G}{\partial x}(x, t) = \frac{\partial}{\partial x} \sum_{n=0}^{\infty} x^n p_n(t) = \sum_{n=1}^{\infty} n x^{n-1} p_n(t). \tag{1.30}$$

Differentiating with respect to x again, we deduce

$$\frac{\partial^2 G}{\partial x^2}(x, t) = \frac{\partial}{\partial x} \sum_{n=1}^{\infty} n x^{n-1} p_n(t) = \sum_{n=2}^{\infty} n(n-1) x^{n-2} p_n(t). \tag{1.31}$$

Substituting $x = 1$ into (1.30) and (1.31), we obtain the identities

$$\frac{\partial G}{\partial x}(1, t) = \sum_{n=0}^{\infty} n p_n(t), \qquad \frac{\partial^2 G}{\partial x^2}(1, t) = \sum_{n=0}^{\infty} n(n-1) p_n(t). \tag{1.32}$$

Using the definition of $M(t)$ given by (1.12), we immediately conclude that

$$M(t) = \frac{\partial G}{\partial x}(1, t). \tag{1.33}$$

Moreover, using the identity (1.16), we get

$$V(t) = \sum_{n=0}^{\infty} n^2 p_n(t) - M(t)^2 = \sum_{n=0}^{\infty} n(n-1) p_n(t) + M(t) - M(t)^2,$$

which together with (1.32) implies

$$V(t) = \frac{\partial^2 G}{\partial x^2}(1, t) + M(t) - M(t)^2 \tag{1.34}$$

$$= \frac{\partial^2 G}{\partial x^2}(1, t) + \frac{\partial G}{\partial x}(1, t) - \left(\frac{\partial G}{\partial x}(1, t)\right)^2.$$

Thus the mean $M(t)$ and variance $V(t)$ may be extracted from a knowledge of the x-derivatives of $G(x, t)$ at $x = 1$. The individual probabilities $p_n(t)$ may also be recovered easily from a knowledge of $G(x, t)$. Substituting $x = 0$ into (1.29), (1.30) and (1.31), we obtain

$$p_0(t) = G(0, t), \qquad p_1(t) = \frac{\partial G}{\partial x}(0, t), \qquad p_2(t) = \frac{1}{2}\frac{\partial^2 G}{\partial x^2}(0, t).$$

Using mathematical induction, we can prove the general formula

$$p_n(t) = \frac{1}{n!}\frac{\partial^n G}{\partial x^n}(0, t) \tag{1.35}$$

for any non-negative integer n. Thus, if we know the probability-generating function $G(x, t)$, we can use the formulae (1.33), (1.34) and (1.35) to find $M(t)$, $V(t)$ and $p_n(t)$.

To find $G(x, t)$, we multiply the chemical master equation (1.28) by x^n and sum over n to deduce

$$\frac{\partial}{\partial t}\sum_{n=0}^{\infty} x^n p_n = \frac{k_1}{v}\sum_{n=0}^{\infty} x^n(n + 2)(n + 1)\, p_{n+2} - \frac{k_1}{v}\sum_{n=2}^{\infty} x^n n(n - 1)\, p_n$$

$$+ k_2 v\sum_{n=1}^{\infty} x^n p_{n-1} - k_2 v\sum_{n=0}^{\infty} x^n p_n.$$

Changing indices in the first and third sums on the right-hand side, we obtain

$$\frac{\partial}{\partial t}\sum_{n=0}^{\infty} x^n p_n = \frac{k_1}{v}\sum_{n=2}^{\infty} x^{n-2}n(n - 1)\, p_n - \frac{k_1}{v}x^2\sum_{n=2}^{\infty} x^{n-2}n(n - 1)\, p_n$$

$$+ k_2 v x\sum_{n=0}^{\infty} x^n p_n - k_2 v\sum_{n-0}^{\infty} x^n p_n.$$

Using (1.29) and (1.31), we obtain a partial differential equation for G in the following form:

$$\frac{\partial G}{\partial t} = \frac{k_1}{v}(1 - x^2)\frac{\partial^2 G}{\partial x^2} + k_2 v(x - 1)\, G. \tag{1.36}$$

Solving this equation numerically we could find $G(x, t)$ and, with the help of formulae (1.33), (1.34) and (1.35), we could also find $M(t)$, $V(t)$ and $p_n(t)$.

Of course, to solve (1.36), numerically or otherwise, we need to supply some initial and boundary conditions. The initial condition is provided simply by using the initial values of p_n in the formula (1.29). Since (1.36) is second-order in x, we need two boundary conditions. One is given by (1.14), which gives

$$G(1,t) = \sum_{n=0}^{\infty} p_n(t) = 1. \tag{1.37}$$

The second is the condition that the second derivative $\partial^2 G/\partial x^2$ should be bounded at $x = -1$. This means that the second term on the right-hand side of (1.36) vanishes at $x = -1$. From this we deduce that

$$\frac{\partial}{\partial t} G(-1,t) = -2k_2 v\, G(-1,t),$$

so that

$$G(-1,t) = G(-1,0)\, \exp[-2k_2 v\, t]. \tag{1.38}$$

This relation can also be derived directly from the chemical master equation by subtracting the sum of (1.28) over odd n from the sum over even n.

Since we are mostly interested in the stationary values M_s, V_s and $\phi(n)$, which are defined by (1.18) and (1.19), we look for the stationary probability-generating function $G_s: [-1,1] \to \mathbb{R}$ which is defined as

$$G_s(x) = \lim_{t \to \infty} G(x,t) = \sum_{n=0}^{\infty} x^n \phi(n). \tag{1.39}$$

Then formulae (1.33), (1.34) and (1.35) imply that

$$M_s = \frac{dG_s}{dx}(1), \tag{1.40}$$

$$V_s = \frac{d^2 G_s}{dx^2}(1) + M_s - M_s^2, \tag{1.41}$$

$$\phi(n) = \frac{1}{n!} \frac{d^n G_s}{dx^n}(0), \qquad n = 0, 1, 2, 3, \ldots, \tag{1.42}$$

and thus M_s, V_s and $\phi(n)$ can be computed if $G_s(x)$ is known. The stationary probability-generating function satisfies the stationary version of equation (1.36), namely

$$0 = \frac{k_1}{v}(1 - x^2)\frac{d^2 G_s}{dx^2} + k_2 v\,(x - 1)\, G_s.$$

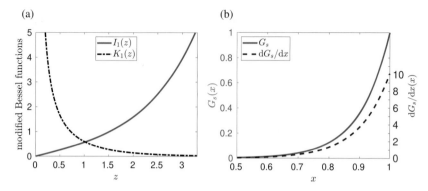

Figure 1.4 (a) *Modified Bessel functions I_1 and K_1. (b) The solution $G_s(x)$ of (1.43) given by (1.44) for $k_1/v = 0.005 \, sec^{-1}$, $k_2 v = 1 \, sec^{-1}$ and C_1 given by (1.47) (red line). The derivative $G'_s(x)$ is plotted as a blue dashed line.*

Using a simple algebraic manipulation, we obtain the ordinary differential equation

$$G''_s(x) = \frac{k_2 v^2}{k_1} \frac{1}{1+x} G_s(x), \tag{1.43}$$

where the prime denotes differentiation with respect to x. The general solution of (1.43) is

$$G_s(x) = C_1 \sqrt{1+x} \, I_1\left(2\sqrt{\frac{k_2 v^2(1+x)}{k_1}}\right) + C_2 \sqrt{1+x} \, K_1\left(2\sqrt{\frac{k_2 v^2(1+x)}{k_1}}\right), \tag{1.44}$$

where I_1 and K_1 are modified Bessel functions and C_1 and C_2 are arbitrary constants. The modified Bessel functions $I_n(z)$ and $K_n(z)$ are independent solutions of the ordinary differential equation

$$z^2 I''_n(z) + z I'_n(z) - (z^2 + n^2) I_n(z) = 0; \tag{1.45}$$

this can be used to verify that (1.44) is indeed a solution of (1.43). The evaluation of modified Bessel functions is part of any standard mathematical software (e.g. the functions `besseli` and `besselk` in Matlab). In Figure 1.4(a), we plot the modified Bessel functions I_1 and K_1.

To determine the coefficients C_1 and C_2 we impose the steady versions of the boundary conditions (1.37) and (1.38). Taking the limit as $t \to \infty$ in (1.38) we find $G_s(-1) = 0$. Since $I_1(z) \sim z/2$ and $K_1(z) \sim 1/z$ as $z \to 0$, this implies that $C_2 = 0$. Taking the limit as $t \to \infty$ in the normalization condition (1.37) gives

$$G_s(1) = 1, \qquad (1.46)$$

so that

$$C_1 = \left[\sqrt{2} \, I_1\left(2 \sqrt{\frac{2k_2 v^2}{k_1}} \right) \right]^{-1}. \qquad (1.47)$$

Thus we have been able to find an analytic expression for $G_s(x)$, which we can now use to compute M_s, V_s and $\phi(n)$. Differentiating (1.44) with respect to x, we obtain (note that $C_2 = 0$ in (1.44))

$$G'_s(x) = \frac{C_1}{2\sqrt{1+x}} I_1\left(2 \sqrt{\frac{k_2 v^2(1+x)}{k_1}} \right) + C_1 \sqrt{\frac{k_2 v^2}{k_1}} I'_1\left(2 \sqrt{\frac{k_2 v^2(1+x)}{k_1}} \right).$$

Substituting $x = 1$ and using (1.40) and (1.47) gives

$$M_s = G'_s(1) = \frac{1}{4} + \sqrt{\frac{k_2 v^2}{2k_1}} I'_1\left(2 \sqrt{\frac{2k_2 v^2}{k_1}} \right) \left[I_1\left(2 \sqrt{\frac{2k_2 v^2}{k_1}} \right) \right]^{-1}. \qquad (1.48)$$

Substituting $x = 1$ into (1.43) and using (1.46), we obtain

$$G''_s(1) = \frac{k_2 v^2}{2k_1} G_s(1) = \frac{k_2 v^2}{2k_1}.$$

Consequently, formula (1.41) implies

$$V_s = \frac{k_2 v^2}{2k_1} + M_s - M_s^2. \qquad (1.49)$$

Finally, using (1.42), one can verify that the stationary distribution is given by (see Exercise 1.3)

$$\phi(n) = C_1 \frac{1}{n!} \left(\frac{k_2 v^2}{k_1} \right)^n I_{n-1}\left(2 \sqrt{\frac{k_2 v^2}{k_1}} \right), \qquad n = 0, 1, 2, 3, \dots, \qquad (1.50)$$

where C_1 is given by (1.47). Using our parameter values $k_1/v = 0.005\,\text{sec}^{-1}$ and $k_2 v = 1\,\text{sec}^{-1}$, we plot $\phi(n)$ given by (1.50) in Figure 1.3(b) as the red solid line. Comparison with the results of stochastic simulation is excellent. The probability-generating function $G_s(x)$ together with its derivative $G'_s(x)$ is plotted in Figure 1.4(b). Formulae (1.48) and (1.49) give $M_s \doteq 10.13$ and $V_s \doteq 7.56$, which are also in excellent agreement with the results obtained by long-time stochastic simulation.

To close this section, let us consider a classic deterministic description of the chemical system (1.27). This would be written for the concentration $a(t) = A(t)/v$ as the following ODE (see the Appendix):

$$\frac{da}{dt} = -2k_1 a^2 + k_2. \tag{1.51}$$

The solution of this ODE is presented in Figure 1.3(a) for comparison. More precisely, we compare the results of the stochastic simulation with the function

$$\overline{A}(t) = a(t)v,$$

which gives the number of molecules in the volume v with concentration $a(t)$. Multiplying (1.51) by v, we obtain that $\overline{A}(t)$ satisfies the ODE

$$\frac{d\overline{A}}{dt} = -\frac{2k_1}{v}\overline{A}^2 + k_2 v. \tag{1.52}$$

The solution of (1.52) is plotted for $k_1/v = 0.005\ \mathrm{sec}^{-1}$, $k_2 v = 1\ \mathrm{sec}^{-1}$ and initial condition $\overline{A}(0) = 0$ in Figure 1.3(a) as the black dashed line.

We emphasize that, in contrast to the case of first-order reactions considered in Section 1.1 and Section 1.2, equation (1.52) does not give the time evolution of the stochastic mean $M(t)$, that is, $\overline{A}(t) \neq M(t)$. To see that, let us consider the stationary value of $\overline{A}(t)$, which is given by

$$0 = -(2k_1/v)\overline{A}_s^2 + k_2 v,$$

i.e.

$$\overline{A}_s = v\sqrt{\frac{k_2}{2k_1}}.$$

Using our parameter values, we conclude that $\overline{A}_s = 10$. On the other hand, we have already found (using formula (1.48)) that $M_s \doteq 10.13$. The difference between the exact value $M_s \doteq 10.13$ and the deterministic approximation $\overline{A}_s = 10$ is not large. However, even if we compute M_s not exactly, but as an average over many stochastic realizations, the difference can still be identified. Thus even in the case of the simple second-order chemical reaction presented here, the deterministic system of ODEs does not provide the exact description of the stochastic mean. Moreover, the deterministic approach does not give any description of stochastic fluctuations about the mean value. The size of fluctuations can be estimated as the mean standard deviation, which is the square root of the variance V_s. Using (1.49), we find that $\sqrt{V_s} \doteq 2.75$.

In Chapter 2, we will present examples of chemical systems where the difference between the results of stochastic simulation and the corresponding deterministic approximation is more significant.

1.5 Gillespie Algorithm

The SSAs (a2)–(c2), (a3)–(d3) and (a4)–(d4) are special forms of the so-called Gillespie SSA. To conclude this chapter, we formulate this algorithm in its full generality. It was presented in papers by Gillespie (1976, 1977). Although one can find similar ideas in earlier works of other researchers, and some authors even give different names to this algorithm, we call it the Gillespie SSA in this book, because that is its most common name in the literature. There are equivalent ways to formulate the Gillespie SSA or implement it on a computer. Some versions of this algorithm have been even given special names in the literature, because the corresponding computer implementations are more efficient. We discuss such issues in Chapter 5.

Suppose that we have a system of q chemical reactions. Let $\alpha_i(t)$ be the propensity function of the ith reaction, $i = 1, 2, \ldots, q$, at time t, that is, $\alpha_i(t)\,dt$ is the probability that the ith reaction occurs in the time interval $[t, t + dt)$ (the propensity functions for basic chemical reactions are given in Tables 1.1 and 1.2). Then the Gillespie SSA consists of the following four steps at time t:

(a5) Generate two random numbers r_1, r_2 uniformly distributed in $(0, 1)$.

(b5) Compute the propensity function $\alpha_i(t)$ of each reaction. Compute

$$\alpha_0 = \sum_{i=1}^{q} \alpha_i(t). \tag{1.53}$$

(c5) Compute the time when the next chemical reaction takes place as $t + \tau$, where

$$\tau = \frac{1}{\alpha_0} \ln\left[\frac{1}{r_1}\right]. \tag{1.54}$$

(d5) Compute which reaction occurs at time $t + \tau$. Find j such that

$$r_2 \geq \frac{1}{\alpha_0} \sum_{i=1}^{j-1} \alpha_i(t) \qquad \text{and} \qquad r_2 < \frac{1}{\alpha_0} \sum_{i=1}^{j} \alpha_i(t).$$

Then the jth reaction takes place, so update numbers of reactants and products of the jth reaction.

Continue with step (a5) for time $t + \tau$.

The Gillespie SSA (a5)–(d5) provides an exact method for the stochastic simulation of systems of chemical reactions. It was applied previously as the SSA (a2)–(c2) for the chemical reaction (1.1), as the SSA (a3)–(d3) for the chemical system (1.9) and as the SSA (a4)–(d4) for the chemical system (1.27).

At each time step, we first ask the question: When will the next reaction occur? The answer is given by formula (1.54), which can be justified using the same arguments as formulae (1.5) or (1.10). Then we ask the question: Which reaction takes place? The probability that the ith chemical reaction occurs is given by α_i/α_0 for $i = 1, 2, \ldots, q$. The decision on which reaction takes place is given in step (d5) with the help of the second uniformly distributed random number r_2. Then we update the number of reactants and products accordingly.

Our simple examples can be simulated quickly in Matlab (in less than a second on present-day computers). If one considers systems of many chemical reactions and many chemical species, then the SSA (a5)–(d5) might be computationally intensive. In such a case, there are ways to make the Gillespie SSA more efficient. For example, it would be a waste of time to recompute all the propensity functions at each time step (step (b5)). We simulate one reaction per one time step. Therefore, it makes sense to update only those propensity functions which are changed by the chemical reaction that was selected in step (d5) of the SSA (a5)–(d5). A more detailed discussion about the efficient computer implementation of the Gillespie SSA is given in Chapter 5.

In this section, we present the SSA (a5)–(d5) for a more complicated illustrative example, which will also involve the second-order chemical reactions (1.26). We consider two chemical species A and B in a container of volume v. We assume that A and B are subject to the following system of four chemical reactions:

$$A + A \xrightarrow{k_1} \emptyset, \qquad\qquad A + B \xrightarrow{k_2} \emptyset, \qquad\qquad (1.55)$$

$$\emptyset \xrightarrow{k_3} A, \qquad\qquad \emptyset \xrightarrow{k_4} B. \qquad\qquad (1.56)$$

We are not interested in the chemical species C and D (which were present as products in the illustrative example (1.26)), so we have replaced them by $\emptyset$, consistent with our previous notation of unimportant chemical species. To simulate the system of chemical reactions (1.55)–(1.56), we apply the SSA (a5)–(d5). To do that, we have to find the propensities of every reaction. Using Table 1.1, we find that the propensities are: $\alpha_1(t) = A(t)(A(t) - 1)k_1/v$, $\alpha_2(t) = A(t)B(t)k_2/v$, $\alpha_3 = k_3 v$ and $\alpha_4 = k_4 v$. In the step (b5), we compute $\alpha_0 = \alpha_1 + \alpha_2 + \alpha_3 + \alpha_4$, which is used in the step (c5) to determine the time when the next reaction occurs (formula (1.54)). The decision on which reaction should occur is done in the step (d5) with the help of the second uniformly distributed random number r_2. In the case of the chemical system (1.55)–(1.56), the step (d5) reads as follows:

$$A(t + \tau) = \begin{cases} A(t) - 2 & \text{if } 0 \le r_2 < \alpha_1/\alpha_0, \\ A(t) - 1 & \text{if } \alpha_1/\alpha_0 \le r_2 < (\alpha_1 + \alpha_2)/\alpha_0, \\ A(t) + 1 & \text{if } (\alpha_1 + \alpha_2)/\alpha_0 \le r_2 < (\alpha_1 + \alpha_2 + \alpha_3)/\alpha_0, \\ A(t) & \text{if } (\alpha_1 + \alpha_2 + \alpha_3)/\alpha_0 \le r_2 < 1, \end{cases}$$

$$B(t + \tau) = \begin{cases} B(t) & \text{if } 0 \le r_2 < \alpha_1/\alpha_0, \\ B(t) - 1 & \text{if } \alpha_1/\alpha_0 \le r_2 < (\alpha_1 + \alpha_2)/\alpha_0, \\ B(t) & \text{if } (\alpha_1 + \alpha_2)/\alpha_0 \le r_2 < (\alpha_1 + \alpha_2 + \alpha_3)/\alpha_0, \\ B(t) + 1 & \text{if } (\alpha_1 + \alpha_2 + \alpha_3)/\alpha_0 \le r_2 < 1. \end{cases}$$

Results of five realizations of the SSA (a5)–(d5) are plotted in Figure 1.5(a) and (b) as solid lines. We use $A(0) = 0$, $B(0) = 0$, $k_1/\nu = 10^{-3} \sec^{-1}$,

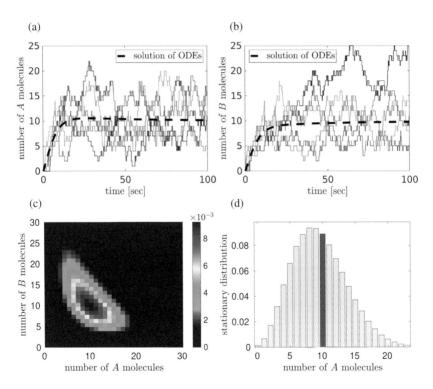

Figure 1.5 (a), (b) *Five realizations of the SSA (a5)–(d5). Number of molecules of chemical species A and B are plotted as functions of time as solid lines. Different colours correspond to different realizations. The solution of ODEs (1.60)–(1.61) is given by the dashed line. We use $A(0) = 0$, $B(0) = 0$, $k_1/\nu = 10^{-3} \sec^{-1}$, $k_2/\nu = 10^{-2} \sec^{-1}$, $k_3\nu = 1.2 \sec^{-1}$ and $k_4\nu = 1 \sec^{-1}$. (c) Stationary distribution $\phi(n, m)$ obtained by long-time simulation of the SSA (a5)–(d5). (d) Stationary distribution of A obtained by (1.62).*

$k_2/v = 10^{-2} \sec^{-1}$, $k_3 v = 1.2 \sec^{-1}$ and $k_4 v = 1 \sec^{-1}$. We plot the number of molecules of chemical species A and B as functions of time. We see that, after initial transients, $A(t)$ and $B(t)$ fluctuate around their average values. They can be estimated from long-time stochastic simulations as 9.6 for A and 12.2 for B.

To formulate the chemical master equation for this system of reactions we let $p_{n,m}(t)$ denote the probability that $A(t) = n$ and $B(t) = m$. Note that $p_{n,m}(t)$ is now parametrized by two indices, since there are two chemical species, A and B. By following the usual arguments the chemical master equation can be derived as

$$
\begin{aligned}
\frac{dp_{n,m}}{dt} &= \frac{k_1}{v}(n+2)(n+1)\,p_{n+2,m} - \frac{k_1}{v}n(n-1)\,p_{n,m} \\
&\quad + \frac{k_2}{v}(n+1)(m+1)\,p_{n+1,m+1} - \frac{k_2}{v}nm\,p_{n,m} \\
&\quad + k_3 v\,p_{n-1,m} - k_3 v\,p_{n,m} + k_4 v\,p_{n,m-1} - k_4 v\,p_{n,m}
\end{aligned}
\tag{1.57}
$$

for $n, m \geq 0$, with the convention that $p_{n,m} \equiv 0$ if $n < 0$ or $m < 0$. The probability $p_{n,m}(t)$ is sometimes denoted by $p(n, m, t)$, especially when we consider systems of many chemical species. We will adopt this convention in the following chapters to avoid long subscripts.

Since the system contains second-order chemical reactions, we cannot solve the chemical master equation (1.57) analytically as we did with (1.6) and the equation (1.57) does not lead to closed evolution equations for the stochastic mean and variance; so we cannot follow the same technique as in the case of equation (1.11). Although such methods are applicable to general zeroth-order and first-order chemical reaction networks (Gadgil et al., 2005; Jahnke and Huisinga, 2007), they are not suitable for the analysis of our system (1.55)–(1.56). Therefore we conclude this introductory chapter with a comparison with the deterministic model.

The classic deterministic description of the chemical system (1.55)–(1.56) is given for concentrations $a(t) = A(t)/v$ and $b(t) = B(t)/v$ by the system of ODEs (see the Appendix)

$$
\frac{da}{dt} = -2k_1 a^2 - k_2\,ab + k_3,
\tag{1.58}
$$

$$
\frac{db}{dt} = -k_2\,ab + k_4.
\tag{1.59}
$$

The results given by (1.58)–(1.59) are also plotted in Figure 1.5 for comparison. More precisely, we plot

$$
\overline{A}(t) = a(t)v, \qquad \overline{B}(t) = b(t)v,
$$

which give the numbers of molecules of chemical species A and B in the volume v with concentrations $a(t)$ and $b(t)$, respectively. Multiplying (1.58)–(1.59) by v, we obtain that $\overline{A}(t)$ and $\overline{B}(t)$ satisfy the system of ODEs

$$\frac{\mathrm{d}\overline{A}}{\mathrm{d}t} = -(2k_1/v)\,\overline{A}^2 - (k_2/v)\,\overline{A}\,\overline{B} + k_3 v, \qquad (1.60)$$

$$\frac{\mathrm{d}\overline{B}}{\mathrm{d}t} = -(k_2/v)\,\overline{A}\,\overline{B} + k_4 v. \qquad (1.61)$$

The solution of (1.60)–(1.61) with initial conditions $\overline{A}(0) = 0$ and $\overline{B}(0) = 0$ is plotted using dashed lines in panels (a) and (b) of Figure 1.5. As in Section 1.4, equations (1.60)–(1.61) do not describe the stochastic means of $A(t)$ and $B(t)$. For example, the steady-state values of (1.60)–(1.61) are (for the parameter values of Figure 1.5) equal to $\overline{A}_s = \overline{B}_s = 10$. On the other hand, the average values estimated from long-time stochastic simulations are 9.6 for A and 12.2 for B. The difference between the stochastic averages and the deterministic steady states is more significant for this example than for that of Section 1.4. We will see in Chapter 2 that the difference between the results of stochastic simulations and the corresponding ODEs can be even more significant.

The stationary distribution is defined by

$$\phi(n, m) = \lim_{t \to \infty} p_{n,m}(t).$$

This can be computed by long-time simulations of the SSA (a5)–(d5) and is plotted in Figure 1.5(c). We see that there is a correlation between the values of A and B: values of A below average correspond to values of B above average, and vice versa. This can also be observed in Figure 1.5(a) and (b). Looking at the darker blue realizations for example, we see that the values of $A(t)$ are below the average and the values of $B(t)$ are above the average. One can also define the stationary distribution of A only by averaging over the values of B:

$$\phi(n) = \sum_{m=0}^{\infty} \phi(n, m). \qquad (1.62)$$

Summing the results of Figure 1.5(c) over m, we obtain $\phi(n)$, which is plotted in Figure 1.5(d) as a grey histogram. The red bar highlights the steady-state value $\overline{A}_s = 10$ of the ODE system (1.60)–(1.61).

Exercises

1.1 Consider the chemical reaction

$$A \xrightarrow{k} \emptyset.$$

Let $p_n(t)$ denote the probability that there are n molecules of A at time t. Derive and solve ordinary differential equations for the mean

$$M(t) = \sum_{n=0}^{\infty} n p_n(t)$$

and variance

$$V(t) = \sum_{n=0}^{\infty} (n - M(t))^2 p_n(t).$$

1.2   Consider two chemical species A and B in a reactor of volume v which are subject to the following two chemical reactions:

$$A + A \xrightarrow{k_1} B, \qquad\qquad B \xrightarrow{k_2} A + A.$$

These two reactions can also be equivalently written as one reversible chemical reaction:

$$2A \underset{k_2}{\overset{k_1}{\rightleftarrows}} B.$$

Suppose there are initially 100 molecules of A and no molecules of B. Denote by $p(n, t)$ the probability that there are n molecules of B at time t, and by $\phi(n)$ the corresponding stationary probability distribution. Find $\phi(n)$ as a function of the rate constants k_1 and k_2.

1.3   By changing the independent variable to $z = 2\sqrt{k_2 v^2 (1 + x)/k_1}$, show that

$$G_s(x) = C_1 \sqrt{1 + x}\, I_1\!\left(2 \sqrt{\frac{k_2 v^2 (1 + x)}{k_1}}\right)$$

is a solution to

$$G_s''(x) = \frac{k_2 v^2}{k_1} \frac{1}{1 + x} G_s(x).$$

Using the derivative formulae for the modified Bessel functions, namely

$$I_n'(z) = I_{n-1}(z) - (n/z) I_n(z) = I_{n+1}(z) + (n/z) I_n(z),$$

show that

$$\frac{d^n G_s}{dx^n}(0) = C_1 \left(\frac{k_2 v^2}{k_1}\right)^n I_{n-1}\!\left(2 \sqrt{\frac{k_2 v^2}{k_1}}\right),$$

and thereby verify the identity (1.50).

1.4 A chemical species A is subject to the following two reactions:

$$A \xrightarrow{k_1} 2A, \qquad A \xrightarrow{k_2} \emptyset,$$

where $k_2 \neq k_1$. Let $p_n(t)$ denote the probability that there are n molecules of A at time t. Write down the chemical master equation for $p_n(t)$. Show that the probability-generating function $G(x, t)$ satisfies

$$\frac{\partial G}{\partial t} = (k_2 - k_1 x)(1 - x)\frac{\partial G}{\partial x}.$$

Suppose that there are n_0 molecules of A initially. Show that

$$G(x, t) = \left(\frac{(k_1 x - k_2)\exp[(k_1 - k_2)t] - k_2(x - 1)}{(k_1 x - k_2)\exp[(k_1 - k_2)t] - k_1(x - 1)}\right)^{n_0}.$$

Hence show that the mean number of molecules is

$$M(t) = n_0 \exp[(k_1 - k_2)t].$$

Verify this result by deriving an ordinary differential equation for $M(t)$ directly from the chemical master equation.

1.5 Consider two chemical species A_1 and A_2 in a reactor of volume v which are subject to the following system of four chemical reactions:

$$\emptyset \xrightarrow{k_1} A_1 \xrightarrow{k_2} \emptyset, \qquad A_1 \xrightarrow{k_3} A_2 \xrightarrow{k_4} \emptyset. \qquad (1.63)$$

Assume that there are initially no molecules of A_1 or A_2 in the system, i.e. $A_1(0) = A_2(0) = 0$, and also that $k_2 + k_3 \neq k_4$. Denote by $p(n_1, n_2, t)$ the probability that $A_1(t) = n_1$ and $A_2(t) = n_2$.

(a) Write down the chemical master equation for $p(n_1, n_2, t)$.

(b) The average number of molecules of A_i at time t is defined to be

$$M_i(t) = \sum_{n_1=0}^{\infty}\sum_{n_2=0}^{\infty} n_i\, p(n_1, n_2, t), \qquad i = 1, 2.$$

Find $M_2(t)$ as a function of rate constants k_1, k_2, k_3, k_4 and volume v.

(c) Use the chemical master equation to derive a system of ordinary differential equations for the second moments S_{11}, S_{12} and S_{22}, which are defined by

$$S_{ii}(t) = \sum_{n_1=0}^{\infty}\sum_{n_2=0}^{\infty} n_i^2\, p(n_1, n_2, t), \qquad i = 1, 2,$$

$$S_{12}(t) = \sum_{n_1=0}^{\infty} \sum_{n_2=0}^{\infty} n_1 n_2 \, p(n_1, n_2, t).$$

(d) Let the stationary distribution be denoted by $\phi(n_1, n_2)$, i.e.

$$\phi(n_1, n_2) = \lim_{t \to \infty} p(n_1, n_2, t),$$

and let $V_{2,s}$ be the stationary value of the variance of A_2, given by

$$V_{2,s} = \sum_{n_1=0}^{\infty} \sum_{n_2=0}^{\infty} (n_2 - M_{2,s})^2 \, \phi(n_1, n_2), \quad \text{where } M_{2,s} = \lim_{t \to \infty} M_2(t).$$

Find $V_{2,s}$ as a function of rate constants k_1, k_2, k_3, k_4 and volume v.

(e) Write a computer code which applies the Gillespie SSA (a5)–(d5) to the chemical system (1.63). Use initial conditions $A_1(0) = A_2(0) = 0$, and parameter values $k_1 v = 2 \sec^{-1}$, $k_2 = 0.2 \sec^{-1}$, $k_3 = 0.2 \sec^{-1}$ and $k_4 = 0.1 \sec^{-1}$.

Calculate $M_2(t)$, for times $t \leq 10$ sec, as an average over many realizations of the Gillespie SSA (a5)–(d5) and compare it with your result in part (b).

Estimate $V_{2,s}$ from long-time simulations of the Gillespie SSA and compare it with your result from part (d).

Calculate the stationary distribution $\phi(n_1, n_2)$ and compare it with the analytical result.

1.6 Consider the production–degradation system (1.9) which is described by the chemical master equation (1.11). Assume that there are initially no molecules in the system, i.e. $A(0) = 0$, as in Figure 1.2(a).

(a) By considering equation (1.15), equation (1.17) and $M(0) = V(0) = 0$, prove that $V(t) = M(t)$ for all time $t \geq 0$.

(b) Show that $p_n(t)$ is the Poisson distribution with mean $M(t)$.

2

Deterministic versus Stochastic Modelling

In the previous chapter we found that for zeroth-order and first-order chemical reactions the evolution equations for the stochastic mean are exactly the ordinary differential equations (ODEs) describing the corresponding deterministic system. On the other hand, when we considered higher-order chemical reactions, for which the deterministic description is nonlinear, we found that the deterministic ODEs do not provide an exact description of the stochastic mean. Nevertheless, the solution of these ODEs was still a reasonable approximation of the evolution of the stochastic mean: the results of the SSAs looked like "noisy solutions" of the corresponding deterministic ODE model. Thus for all the models presented so far one could use the deterministic ODEs to obtain a reasonable description of the average behaviour of the system.

In this chapter, we present examples where deterministic modelling fails and a stochastic approach is necessary; for the problems we discuss below, SSAs give results which cannot be obtained from the corresponding deterministic models. We start with a simple example of stochastic switching between favourable states of a system. Then we illustrate the fact that the stochastic model might have qualitatively different properties than its deterministic counterpart for some parameter regimes, that is, the stochastic model is not just "equal" to the "noisy solution" of the corresponding deterministic ODEs. We present a simple system of chemical reactions for which the deterministic description converges to a steady state. On the other hand, the stochastic model of the same system of chemical reactions has oscillatory solutions. In Section 2.3, we present the phenomenon of stochastic focusing by showing a simple system for which the stochastic model is more sensitive (in amplification of the signal) than its deterministic counterpart. Finally, in Section 2.4, we show that a chemical reaction system can be redesigned, by adding extra reactions, in such a way that its stochastic behaviour qualitatively changes, while its deterministic ODEs do not change at all. In particular, we show that

33

there exist many stochastic reaction networks that correspond to the same deterministic ODE model. Although some of them can be well described by ODEs, the others could be used as counterexamples to any general statements which one could attempt to make on the relevance of the deterministic ODE modelling to the actual behaviour of stochastic chemical reaction networks.

2.1 Systems with Multiple Favourable States

The system (1.60)–(1.61) which we considered in Section 1.5 had only one non-negative (stable) steady state for our parameter values, namely $\overline{A}_s = \overline{B}_s = 10$. Solutions of (1.60)–(1.61) converge to $\overline{A}_s$ and $\overline{B}_s$ as $t \to \infty$ for all non-negative initial conditions (see Figure 1.5). Similarly the results of SSAs show fluctuation about the means, which are roughly equal to $\overline{A}_s$ and $\overline{B}_s$ (they are 9.6 for A and 12.2 for B).

The situation becomes more interesting when a chemical system has two or more favourable states, so that the corresponding ODEs have more than one non-negative stable steady state. The following system, first considered by Schlögl (1972), has two favourable states. Consider the chemical A in a container of volume v which is subject to the four chemical reactions:

$$3A \underset{k_2}{\overset{k_1}{\rightleftarrows}} 2A, \qquad\qquad A \underset{k_4}{\overset{k_3}{\rightleftarrows}} \emptyset. \qquad (2.1)$$

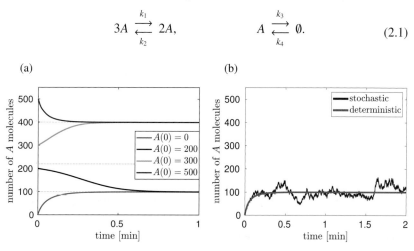

Figure 2.1 (a) *Solution of the ODE (2.3) for four different initial conditions. The steady states (2.5) are denoted as dotted lines. (b) The number of molecules of A as a function of time over the first 2 minutes of simulation of (2.1). One realization of the SSA (a5)–(d5) for the system of chemical reactions (2.1) (blue line) and the solution of the deterministic ODE (2.3) (red line).*

The (approximate) deterministic ODE model can be written for the concentration $a(t)$ as

$$\frac{da}{dt} = -k_1 a^3 + k_2 a^2 - k_3 a + k_4. \qquad (2.2)$$

We will compare results obtained by solving (2.2) with the results obtained by the Gillespie SSA (a5)–(d5). To do that, we rewrite equation (2.2) in terms of $\overline{A}(t) = a(t)v$, which is the average number of molecules of A in the volume v with concentration $a(t)$, to give

$$\frac{d\overline{A}}{dt} = -\frac{k_1}{v^2} \overline{A}^3 + \frac{k_2}{v} \overline{A}^2 - k_3 \overline{A} + k_4 v. \qquad (2.3)$$

The steady states of (2.3) can be found by solving the cubic equation

$$-(k_1/v^2) \overline{A}^3 + (k_2/v) \overline{A}^2 - k_3 \overline{A} + k_4 v = 0. \qquad (2.4)$$

We choose the rate constants to be $k_1/v^2 = 2.5 \times 10^{-4} \, \text{min}^{-1}$, $k_2/v = 0.18 \, \text{min}^{-1}$, $k_3 = 37.5 \, \text{min}^{-1}$ and $k_4 v = 2200 \, \text{min}^{-1}$. Then, solving (2.4), we find that the ODE (2.3) has the three steady states

$$\overline{A}_{s1} = 100, \qquad \overline{A}_u = 220 \quad \text{and} \quad \overline{A}_{s2} = 400. \qquad (2.5)$$

We leave it to the reader (Exercise 2.1) to verify that $\overline{A}_{s1}$ and $\overline{A}_{s2}$ are stable steady states and $\overline{A}_u$ is an unstable steady state. The solution of (2.3) converges to one of the steady states, with the choice of the steady state dependent on the initial condition. If the initial condition $\overline{A}(0)$ satisfies $\overline{A}(0) \in [0, \overline{A}_u)$, then the solution of (2.3) converges to the stable steady state $\overline{A}_{s1} = 100$. If $\overline{A}(0) > \overline{A}_u$, then the solution of (2.3) converges to the second stable steady state $\overline{A}_{s2} = 400$ (see Figure 2.1(a) where we plot the solution of (2.3) for four different initial conditions).

Now suppose that there are initially no molecules of A in the system, i.e. $\overline{A}(0) = 0$. The solution of (2.3) is plotted in Figure 2.1 as the red line. We see that the solution of (2.3) converges to the stable steady state $\overline{A}_{s1} = 100$. Next, we use the Gillespie SSA (a5)–(d5) to simulate the chemical system (2.1). Starting with no molecules of A in the system, we plot one realization of the SSA (a5)–(d5) in Figure 2.1(b) as a blue line. We see that the time evolution of A given by the SSA (a5)–(d5) initially (over the first 2 minutes) looks like the "noisy solution" of (2.3). However, we can find significant differences between the stochastic and deterministic models if we observe both models over sufficiently large times – see Figure 2.2(a) where we plot the time evolution of A over the first 100 minutes. As expected, the solution of the deterministic model (2.3) stays forever close to the stable steady state $\overline{A}_{s1} = 100$. The number of molecules given by the stochastic model initially fluctuates around one

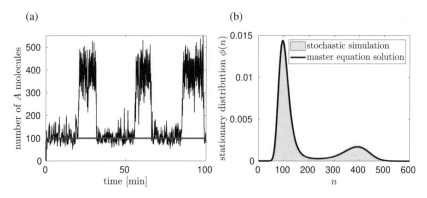

Figure 2.2 (a) *Simulation of (2.1) over a time of 100 minutes. One realization of the SSA (a5)–(d5) for the system of chemical reactions (2.1) (blue line) and the solution of the deterministic ODE (2.3) (red line). (b) Stationary distribution of (2.1) obtained by the long-time simulation of the SSA (a5)–(d5) (yellow histogram). Blue solid line is the solution of the stationary chemical master equation – formula (2.8).*

of the favourable states of the system (which is close to $\overline{A}_{s1} = 100$). However, the fluctuations are sometimes so strong that the system spontaneously switches to another steady state (which is close to $\overline{A}_{s2} = 400$). This random switching is missed by the deterministic description.

Let $p_n(t)$ be the probability that there are n molecules of A at time t in the system, i.e. that $A(t) = n$. The chemical master equation corresponding to (2.1) can be written as

$$\frac{dp_n}{dt} = (k_1/v^2)(n+1)n(n-1)\,p_{n+1} - (k_1/v^2)\,n(n-1)(n-2)\,p_n$$
$$+ (k_2/v)(n-1)(n-2)\,p_{n-1} - (k_2/v)\,n(n-1)\,p_n \qquad (2.6)$$
$$+ k_3(n+1)\,p_{n+1} - k_3 n\,p_n + k_4 v\,p_{n-1} - k_4 v\,p_n$$

for $n \geq 0$, where as usual we use the convention $p_{-1} = 0$. The stationary distribution $\phi(n)$ (which is defined by (1.19)) can be obtained as a solution of the equation $dp_n/dt = 0$, namely

$$0 = (k_1/v^2)(n+1)n(n-1)\,\phi(n+1) - (k_1/v^2)\,n(n-1)(n-2)\,\phi(n)$$
$$+ (k_2/v)(n-1)(n-2)\,\phi(n-1) - (k_2/v)\,n(n-1)\,\phi(n) \qquad (2.7)$$
$$+ k_3(n+1)\,\phi(n+1) - k_3 n\,\phi(n) + k_4 v\,\phi(n-1) - k_4 v\,\phi(n).$$

Using $n = 0$ in (2.7) and remembering that $\phi(-1) = 0$, we obtain

$$\phi(1) = \phi(0)\frac{k_4 v}{k_3}.$$

Now using $n = 1$ in (2.7), we obtain

$$\phi(2) = \phi(0) \frac{k_4 v}{k_3} \frac{k_4 v}{2k_3}.$$

Proceeding in this way we can express all $\phi(n)$ for $n \geq 1$ in terms of $\phi(0)$ and the rate constants. We leave it to the reader to prove the following general formula:

$$\phi(n) = \phi(0) \prod_{i=0}^{n-1} \frac{(k_2/v)\, i(i-1) + k_4 v}{(k_1/v^2)\,(i+1)i(i-1) + k_3(i+1)} \tag{2.8}$$

for $n \geq 1$. Using the normalization condition (1.22), we can determine the value of $\phi(0)$. To compute $\phi(n)$ in practice we can set $\phi(0) = 1$ and calculate $\phi(n)$, for sufficiently many n, by (2.8). Then we divide $\phi(n)$, $n \geq 0$, by $\sum \phi(n)$ so that the normalization condition (1.22) is satisfied. In Figure 2.2(b), we plot the resulting stationary distribution as the blue solid line. As expected, the results compare well with the results obtained by the long-time stochastic simulation, shown as the yellow histogram.

The most important characteristic of a system with two (or more) favourable states is the mean switching time between these states, that is, how long, on average, does the system spend on each visit to a favourable state. Such a quantity cannot be obtained from the deterministic ODE model because it depends crucially on the fluctuations.

By running an SSA for a long time, we can estimate how much time the system spends in each of its favourable states. However, it is also possible to analyse the model further analytically and estimate the mean switching time without doing long-time simulations. To do so, we will need some further theoretical concepts, which will be introduced in Chapter 3. In Section 3.8 we return to the model (2.1) and estimate the mean switching time analytically.

The model (2.1) is a pedagogical example, and is the simplest model (with one chemical species) which enables switching between multiple favourable states. However, we note that random switching between states has been found in more realistic models – see, for example, Kepler and Elston (2001) for examples of switching models of gene regulatory networks.

2.2 Self-Induced Stochastic Resonance

Our next example is a nonlinear system of chemical equations for which the stochastic model has a qualitatively different behaviour than its deterministic counterpart in some parameter regimes. The phenomenon we present is sometimes called self-induced stochastic resonance (Muratov et al., 2005; DeVille et al., 2006). We consider a system of two chemical species A and B in a reactor of volume v which are subject to the chemical reactions:

$$2A + B \xrightarrow{k_1} 3A, \qquad \emptyset \underset{k_3}{\overset{k_2}{\rightleftarrows}} A, \qquad \emptyset \xrightarrow{k_4} B. \qquad (2.9)$$

Such a system was first studied by Schnakenberg (1979). The approximate deterministic description is given by the system of ODEs

$$\frac{da}{dt} = k_1 a^2 b + k_2 - k_3 a, \qquad (2.10)$$

$$\frac{db}{dt} = -k_1 a^2 b + k_4, \qquad (2.11)$$

where $a(t)$ and $b(t)$ are concentrations of A and B. Translating this into the (approximate) number of molecules of A and B gives

$$\frac{d\overline{A}}{dt} = \frac{k_1}{v^2} \overline{A}^2 \overline{B} + k_2 v - k_3 \overline{A}, \qquad (2.12)$$

$$\frac{d\overline{B}}{dt} = -\frac{k_1}{v^2} \overline{A}^2 \overline{B} + k_4 v, \qquad (2.13)$$

where $\overline{A}(t) = a(t)v$ and $\overline{B}(t) = b(t)v$. We choose the rate constants

$$\frac{k_1}{v^2} = 4 \times 10^{-5} \text{ sec}^{-1}, \ k_2 v = 50 \text{ sec}^{-1}, \ k_3 = 10 \text{ sec}^{-1}, \ k_4 v = 25 \text{ sec}^{-1}. \quad (2.14)$$

We use the Gillespie SSA (a5)–(d5) to simulate the time evolution of this system. We also solve the deterministic system of ODEs (2.12)–(2.13). Using the same initial conditions $[A(0), B(0)] = [10, 10]$, we compare the results of the stochastic and deterministic models in Figure 2.3. We plot the time evolution of $A(t)$ in Figure 2.3(a). We see that the solution of the deterministic equations converges to a steady state, while the stochastic model has oscillatory solutions. Note that there is a log scale on the A-axis – numbers of A given by the (more precise) SSA vary between 0 and 10^4. In Figure 2.3(b), we use a linear scale on the A-axis. On this scale the low molecular fluctuations are invisible and the solution of the SSA looks as if there were "almost deterministic oscillations" (which are not present in the deterministic model). We also plot in Figure 2.3(b) the time evolution of the number of molecules of B.

The difference in behaviour between the stochastic and deterministic models of the chemical system (2.9) depends on the parameters. Changing the rate constants, we can find parameter regimes for which both systems have oscillatory solutions (although the periods of oscillations differ). To illustrate this point, we simulate the chemical system (2.9) with parameter values given by

$$\frac{k_1}{v^2} = 4 \times 10^{-5} \text{ sec}^{-1}, \ k_2 v = 50 \text{ sec}^{-1}, \ k_3 = 10 \text{ sec}^{-1}, \ k_4 v = 100 \text{ sec}^{-1}, \quad (2.15)$$

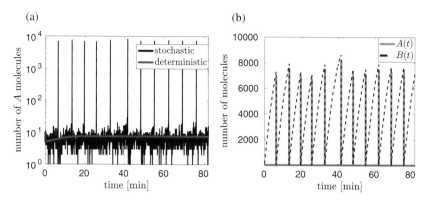

Figure 2.3 (a) *Time evolution of A(t) given by the Gillespie SSA (a5)–(d5) for the system of chemical reactions (2.9) with parameter values (2.14) (blue line). Time evolution of $\overline{A}(t)$ given by the system of ODEs (2.12)–(2.13) for parameter values (2.14) (red line). Log scale on the A-axis. (b) Time evolution of A(t) and B(t) given by the Gillespie SSA (a5)–(d5) for the system of chemical reactions (2.9) for the values of rate constants (2.14).*

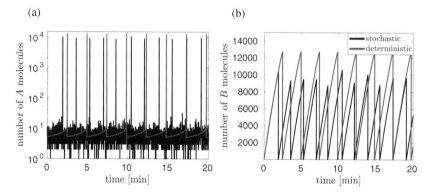

Figure 2.4 *One realization of SSA (a5)–(d5) for the system of chemical reactions (2.9) (blue lines) and the solution of the system of ODEs (2.12)–(2.13) (red lines). Parameter values are given by (2.15). (a) The time evolution of A(t). (b) The time evolution of B(t).*

i.e. we change the value of the rate constant $k_4\nu$, keeping all other rate constants the same as in (2.14). In Figure 2.4, we compare the results obtained by the Gillespie SSA (a5)–(d5) with the solution of the system of ODEs (2.12)–(2.13). We use the same initial conditions $[A(0), B(0)] = [10, 10]$ in both models. We see that both models have oscillatory solutions for the parameter values (2.15). However, the period of "stochastic oscillations" is shorter than the period of oscillations predicted by ODEs.

To understand this behaviour better, we plot the stochastic and deterministic trajectories in the A–B plane in Figure 2.5 (again we use a log scale on
the A-axis). We include the nullclines of the deterministic system of ODEs
(2.12)–(2.13) (green lines). In Figure 2.5(a), we replot the trajectory from
Figure 2.3, which was obtained for parameters (2.14). In Figure 2.5(b), we
replot the trajectory from Figure 2.4, which was obtained for parameters (2.15).
The $[d\overline{A}/dt = 0]$ nullcline is given by

$$\overline{B} = \frac{k_3 \overline{A} - k_2 v}{\overline{A}^2 k_1/v^2}; \tag{2.16}$$

we see that this is independent of $k_4 v$, so that it appears as the same green curve
in both panels of Figure 2.5. The maximum of this curve is attained at the point

$$a_m = \frac{2k_2 v}{k_3} = 10 \tag{2.17}$$

and it is equal to $b_m = 12\,500$.
   The $[d\overline{B}/dt = 0]$ nullcline is given by

$$\overline{B} = \frac{k_4 v}{\overline{A}^2 k_1/v^2}. \tag{2.18}$$

This nullcline depends on $k_4 v$. It intersects the nullcline (2.16) at the stable
steady state (red circle) for parameter values (2.14). Consequently, the deterministic system follows a stable nullcline into this steady state in Figure 2.5(a).
The stochastic model also initially "follows" the nullcline (2.16) (with some
noise). However, occasionally the noise is enough to move the trajectory to
the right of the right-hand branch of this nullcline. The deterministic trajectory starting from such a point exhibits a large excursion (sometimes called
an action potential), before finally returning to the steady state. The stochastic solution now follows this new trajectory (again with some noise), before
returning to the left-hand branch of the nullcline and repeating the procedure.
   In the situation of Figure 2.5(b), we have shifted the nullcline (2.18) so that
it intersects the other nullcline (2.16) on its unstable branch. The steady state
(red circle) is now unstable. The solution of ODEs (2.12)–(2.13) first follows
the stable branch of the nullcline (2.16) to the top point $[a_m, b_m]$ (where the
nullcline loses its stability) and then undergoes a large excursion before finally
returning to this nullcline at a lower value of $\overline{B}$ and repeating the procedure.
The stochastic simulation follows a similar trajectory, but may start its excursion sooner if the noise takes it to the right of the unstable branch of the
nullcline. Thus the (average) period of the stochastic oscillation is less than
the period of the deterministic oscillation.

(a)   (b)

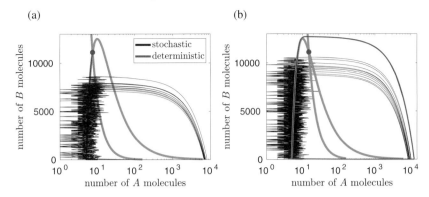

Figure 2.5 (a) *One realization of the SSA (a5)–(d5) for the system of chemical reactions (2.9) (blue line) and the solution of the system of ODEs (2.12)–(2.13) (red line). The parameters are given by (2.14). Nullclines of the deterministic ODEs are plotted as green lines.* (b) *The same plot for the parameters given by (2.15).*

The system of ODEs (2.12)–(2.13) changes dramatically its qualitative behaviour as the parameter $k_4 v$ is varied from $25 \sec^{-1}$ to $100 \sec^{-1}$. Such a change in qualitative behaviour is called a bifurcation. There exists a unique constant k_b (bifurcation value) in the interval $(25, 100) \sec^{-1}$ such that the solution of the ODEs converges to the steady state for $k_4 v < k_b$ and oscillates for $k_4 v > k_b$. The value of the constant k_b can be identified by investigating the stability of the steady state of the ODEs. We can also approximate k_b as the point for which the nullcline (2.18) intersects the other nullcline (2.16) at the top point $[a_m, b_m]$. Using (2.18), we obtain

$$b_m \doteq \frac{k_b}{a_m^2 k_1 / v^2}.$$

Thus the bifurcation value k_b can be approximated as

$$k_b \doteq a_m^2 \, b_m \, \frac{k_1}{v^2} = 50 \sec^{-1}.$$

The bifurcation of the deterministic system of ODEs (2.12)–(2.13) is a very sharp change in its qualitative behaviour. Choosing arbitrarily small ε, the solution of the ODEs will oscillate for $k_4 v = k_b + \varepsilon$ and converge to the stable steady state for $k_4 v = k_b - \varepsilon$. The noise in the stochastic model makes it oscillate even for parameter values less than k_b (as we observed in Figures 2.3 and 2.5), smoothing the transition between two qualitatively different states of the system. This is generally true for any bifurcation. Stochastic models will differ qualitatively from their ODE counterparts if the parameter values are close to the bifurcation values. Even for the case $k_4 v > k_b$ where both models have oscillatory solutions, the periods of oscillations differ. It is interesting to investigate how the period of "stochastic oscillations" depends on the parameters. We will return to this problem in Section 3.9 after we develop a mathematical theory suitable for such an analysis.

2.3 Stochastic Focusing

We consider now three chemical species A, B and C in a container of volume v. Following Paulsson et al. (2000), we will refer to A as a signal and to B as a product. We assume that A, B and C are subject to the following six chemical reactions. The product B is produced from the intermediate C (which is itself produced from a source) and degraded:

$$\emptyset \xrightarrow{k_1} C \xrightarrow{k_2} B \xrightarrow{k_3} \emptyset. \tag{2.19}$$

The coupling between the production of the product B and the signal A is indirect: the signal A catalyses the degradation of the intermediate C, so that

$$A + C \xrightarrow{k_4} A. \tag{2.20}$$

We will study how changes in the number of signal molecules A influence the changes in the number of product molecules B. We will show that the mean value of B depends not only on the mean value of A but also on the fluctuations of the number of signal molecules A. To that end, we let the signal A evolve according to the production–degradation chemical system (1.9) that was studied in Section 1.2, i.e.

$$\emptyset \xrightarrow{k_5} A \xrightarrow{k_6} \emptyset. \tag{2.21}$$

The approximate ODE description of the chemical system (2.19)–(2.21) can be written as follows:

$$\frac{da}{dt} = k_5 - k_6\, a, \tag{2.22}$$

$$\frac{db}{dt} = k_2\, c - k_3\, b, \tag{2.23}$$

$$\frac{dc}{dt} = k_1 - k_2\, c - k_4\, ac, \tag{2.24}$$

where $a(t)$, $b(t)$ and $c(t)$ are concentrations of A, B and C, respectively, in the reactor of volume v. The system (2.22)–(2.24) can be rewritten as the ODE system for approximate numbers of molecules in the reactor, which are given by $\overline{A}(t) = a(t)\nu$, $\overline{B}(t) = b(t)\nu$ and $\overline{C}(t) = c(t)\nu$. We obtain

$$\frac{d\overline{A}}{dt} = k_5 \nu - k_6 \overline{A}, \tag{2.25}$$

$$\frac{d\overline{B}}{dt} = k_2 \overline{C} - k_3 \overline{B}, \tag{2.26}$$

$$\frac{d\overline{C}}{dt} = k_1 \nu - k_2 \overline{C} - \frac{k_4}{\nu} \overline{A}\,\overline{C}. \tag{2.27}$$

The rate constants of reactions (2.19)–(2.20) are chosen as

$$k_1\nu = 10^2 \text{ sec}^{-1}, \ k_2 = 10^3 \text{ sec}^{-1}, \ k_3 = 10^{-2} \text{ sec}^{-1}, \ \frac{k_4}{\nu} = 9900 \text{ sec}^{-1}. \tag{2.28}$$

We assume that the value of the rate constant k_5 is abruptly halved at time 10 minutes and choose the rate constants of the chemical reaction (2.21) by

$$k_5\nu = \begin{cases} 10^3 \text{ sec}^{-1}, & \text{for } t < 10\,\text{min,} \\ 5 \times 10^2 \text{ sec}^{-1}, & \text{for } t \geq 10\,\text{min,} \end{cases} \qquad k_6 = 10^2 \text{ sec}^{-1}. \tag{2.29}$$

In Figure 2.6, we compare the time evolution of the system given by the (approximate) system of ODEs (2.25)–(2.27) with the results obtained by the Gillespie SSA (a5)–(d5). We use the same initial condition $[\overline{A}(0), \overline{B}(0), \overline{C}(0)]$ $= [10, 100, 0]$ for both models. We see that the average number of signal molecules A halves at the time 10 minutes because of the change of the production rate $k_5\nu$ given by (2.29). However, the response of the number of product

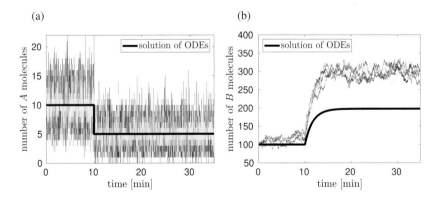

(a)   (b)

Figure 2.6 *Five realizations of the SSA (a5)–(d5) for the system of chemical reactions (2.19)–(2.21) (coloured lines) and the solution of the system of ODEs (2.25)–(2.27) (black line). The parameters are given by (2.28) and (2.29). (a) Time evolution of the signal A. (b) Time evolution of the product B.*

molecules B to the change in the signal is not correctly predicted by the ODE model (2.25)–(2.27).

The steady state of the ODE system (2.25)–(2.27) can be found by solving the algebraic system

$$0 = k_5 v - k_6 \overline{A}_s,$$
$$0 = k_2 \overline{C}_s - k_3 \overline{B}_s,$$
$$0 = k_1 v - k_2 \overline{C}_s - \overline{A}_s \overline{C}_s k_4 / v.$$

Consequently,

$$\overline{A}_s = \frac{k_5 v}{k_6}, \qquad \overline{B}_s = \frac{k_1 v k_2 k_6}{k_3 (k_2 k_6 + k_4 k_5)}, \qquad \overline{C}_s = \frac{k_1 v k_6}{k_2 k_6 + k_4 k_5}. \qquad (2.30)$$

Using the parameter values (2.28)–(2.29), we find that the steady-state values $\overline{A}_s$ and $\overline{B}_s$ are

$$\overline{A}_s = \begin{cases} 10, & \text{for } t < 10\,\text{min}, \\ 5, & \text{for } t \geq 10\,\text{min}, \end{cases} \qquad \overline{B}_s \doteq \begin{cases} 100, & \text{for } t < 10\,\text{min}, \\ 198, & \text{for } t \geq 10\,\text{min}. \end{cases} \qquad (2.31)$$

Thus the deterministic system predicts that the number of product molecules B should, roughly speaking, double, while the number of signal molecules A should be halved. The numerical solution of the approximate system of ODEs (2.25)–(2.27) (which is plotted in Figure 2.6 as the black line) confirms this observation. On the other hand, the stochastic model shows approximately a threefold increase in the number of the product molecules, i.e. it is more sensitive to the change in the signal – Figure 2.6(b).

To understand this behaviour, let $M_A(t)$ be the average number of A molecules at time t. The number of A molecules is only influenced by the chemical reactions (2.21). Such a process was studied in Section 1.2 as the chemical system (1.9). In particular, we were able to derive the evolution equation for the stochastic mean (1.15). Using the notation of this section, the evolution equation for $M_A(t)$ can be written as

$$\frac{dM_A}{dt} = k_5 v - k_6 M_A. \qquad (2.32)$$

The stationary value of the stochastic mean is $M_{A,s} = k_5 v / k_6$. Using the parameter values (2.28)–(2.29), we have

$$M_{A,s} = \begin{cases} 10, & \text{for } t < 10\,\text{min}, \\ 5, & \text{for } t \geq 10\,\text{min}. \end{cases} \qquad (2.33)$$

Thus the stationary values of the signal A are in agreement with the values (2.31) predicted by the ODE model (2.25)–(2.27).

We will show that it is the fluctuations in A that lead to the differences in B that we observe between the stochastic and deterministic models. To that end, we simulate the system of chemical reactions (2.19)–(2.20) with the rate constants given by (2.28). However, instead of simulating (2.21), we directly impose the time evolution of the signal $A(t)$ in the form

$$A(t) = \begin{cases} 10, & \text{for } t < 10\,\text{min}, \\ 5, & \text{for } t \geq 10\,\text{min}, \end{cases} \tag{2.34}$$

i.e. we ignore the fluctuations in A and postulate that A is equal to the corresponding steady-state values (2.33). Then the deterministic description is given by the system of ODEs

$$\frac{d\overline{B}}{dt} = k_2\,\overline{C} - k_3\,\overline{B}, \tag{2.35}$$

$$\frac{d\overline{C}}{dt} = k_1 v - k_2\,\overline{C} - \frac{k_4}{v}\,A(t)\overline{C}, \tag{2.36}$$

where $A(t)$ is given by (2.34). In Figure 2.7, we compare the time evolution of the system given by the ODEs (2.35)–(2.36) with the results obtained by the Gillespie SSA (a5)–(d5). We use the same initial condition $[\overline{B}(0), \overline{C}(0)] = [100, 0]$ for both models. We see that there are no significant differences between the evolution of the number of product molecules B computed by the deterministic and stochastic models. Moreover, one can show (Exercise 2.2) that the ODE system (2.35)–(2.36) is the exact description of the stochastic mean of B and C. In fact, the only higher-order (nonlinear) reaction in the

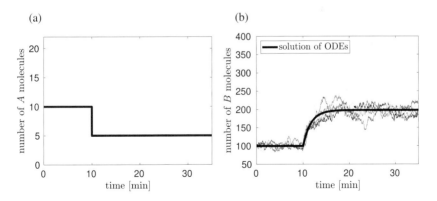

Figure 2.7 (a) *Time evolution of the signal A given by* (2.34). (b) *Five realizations of the SSA (a5)–(d5) for the system of chemical reactions (2.19)–(2.20) where A(t) is given by (2.34) (coloured lines) and the solution of the system of ODEs (2.35)–(2.36) (black line).*

chemical system (2.19)–(2.21) is the second-order chemical reaction (2.20). It can be rewritten as

$$C \xrightarrow{k_4 A(t)} \emptyset,$$
(2.37)

i.e. the degradation of C with the rate $k_4 A(t)$. If $A(t)$ is assumed to be constant (no fluctuations in A), then (2.37) is the first-order chemical reaction, degradation, which was studied in Section 1.1 as the chemical reaction (1.1). In particular, the linearity of the ODE model implies that it provides an exact description of the average behaviour of the chemical system. The same is true for the piecewise constant $A(t)$ given by (2.34).

The presence of fluctuations of A (or equivalently the second-order chemical reaction (2.20)) is not a sufficient condition to obtain such significant differences between the stochastic and deterministic descriptions as presented in Figure 2.6. Indeed, we have already presented nonlinear chemical systems in Sections 1.4 and 1.5 for which the corresponding deterministic equations provide a reasonable (approximate) description of the average behaviour of the system. The extra feature of the model (2.19)–(2.21) which makes the ODE model inapplicable is a very low number of the intermediate molecules C in the model. Using (2.30) and the parameter values (2.28)–(2.29), we find that the steady-state value $\overline{C}_s$ is equal to 10^{-3} for $t < 10\,\mathrm{min}$, and 1.98×10^{-3} for $t \geq 10\,\mathrm{min}$. In particular, $\overline{C}_s$ can be no longer successfully interpreted as the number of C molecules in the reactor. The stochastic model predicts that either zero or one molecule of C is in the system (see Figure 2.9(a); the probability that there is more than one molecule of C in the system is very low). It is this low number of C molecules, combined with its participation in a second-order reaction, which is the cause of the large differences in the number of B molecules predicted by the stochastic and deterministic models. To illustrate this point, we consider the model (2.19)–(2.21) with the parameter values

$$k_1 \nu = 10^2 \,\mathrm{sec}^{-1}, \; k_2 = 0.1 \,\mathrm{sec}^{-1}, \; k_3 = 10^{-2} \,\mathrm{sec}^{-1}, \; \frac{k_4}{\nu} = 0.99 \,\mathrm{sec}^{-1} \quad (2.38)$$

and the parameter values (2.29). In Figure 2.8, we compare the time evolution of the system given by the (approximate) system of ODEs (2.25)–(2.27) with the results obtained by the Gillespie SSA (a5)–(d5). We use the same initial condition $[\overline{A}(0), \overline{B}(0), \overline{C}(0)] = [10, 100, 10]$ for both models. Since we use the same values of the rate constants for the chemical reactions (2.21) as before, the time evolution of A is the same as in Figure 2.6(a). However (unlike the case of Figure 2.6), the deterministic description of B matches the stochastic results very well.

(a) (b)

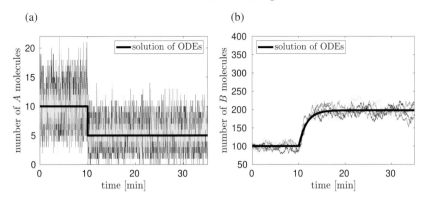

Figure 2.8 *Five realizations of the SSA (a5)–(d5) for the system of chemical reactions (2.19)–(2.21) (coloured lines) and the solution of the system of ODEs (2.25)–(2.27) (black line). The parameters are given by (2.38) and (2.29). (a) Time evolution of the signal A. (b) Time evolution of the product B.*

The values of $\overline{A}_s$ and $\overline{B}_s$ for the second set of parameter values (2.38) are the same as for the rate constants (2.28), i.e. given by (2.31). The difference is in the values of $\overline{C}_s$. Using (2.30) together with the parameter values (2.38) and (2.29), we find that the steady-state value $\overline{C}_s$ is equal to 10 for $t < 10$ min, and 19.8 for $t \geq 10$ min. In particular, $\overline{C}_s$ is already high enough that it can be approximately interpreted as the number of C molecules in the reactor. In Figure 2.9(b), we present the time evolution of C for the second set of parameter values (2.38) and (2.29). We observe that C fluctuates around the values predicted by the ODE model (2.25)–(2.27). This is in contrast with the results obtained by the original set of parameter values (2.28) and (2.29) – see Figure 2.9(a), where we plot the time evolution of C corresponding to the simulation presented in Figure 2.6. Here, the number of molecules of C fluctuates between zero and one and the ODE solution for $\overline{C}$ has values of the order 10^{-3}.

Thus second-order (or higher-order) chemical reactions together with low copy numbers of some chemical species can lead to significant differences between the average behaviours given by the stochastic model and by the corresponding (approximate) deterministic ODEs. Note that in problem (2.19)–(2.21) the signal we observed was the product B, which had molecular abundances of the order of hundreds. However, we still reported significant differences between the stochastic and deterministic description of B because the intermediate chemical C was presented in low copy numbers.

We close this chapter by showing how a good analytical estimate of the average number of molecules of B may be found. Since the production and

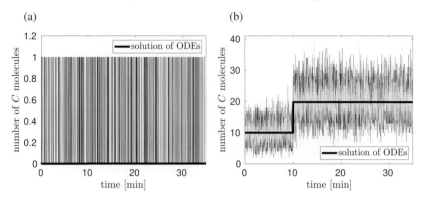

Figure 2.9 *The time evolution of the number of molecules of the intermediate C. Five realizations of the SSA (a5)–(d5) for the system of chemical reactions (2.19)–(2.21) (coloured lines) and the solution of the system of ODEs (2.25)–(2.27) (black line). (a) The parameters are given by (2.28) and (2.29), i.e. we plot the time evolution of C for the simulation presented in Figure 2.6. (b) The parameters are given by (2.38) and (2.29), i.e. we plot the time evolution of C for the simulation presented in Figure 2.8.*

degradation of the signal A do not depend on the number of other molecules in the system, after the initial transient the probability that there are n molecules of A in the system is given by the stationary distribution (1.24), namely

$$\phi_A(n) = \frac{1}{n!} (M_{A,s})^n \exp\left[-M_{A,s}\right].$$ (2.39)

Now, for parameter values (2.28)–(2.29), $k_4/v \gg k_6$, so that the evolution of C is much faster than that of A. Thus, if the number of molecules of A is equal to n, say, C will quickly reach its corresponding stationary distribution (again given by (1.24)), so that the probability that there are m molecules of C is

$$\phi_{C,n}(m) = \frac{\mu_n^m}{m!} \exp\left[-\mu_n\right], \qquad \text{where} \qquad \mu_n = \frac{k_1 v}{k_2 + n k_4/v}.$$ (2.40)

We have already seen that for parameter values (2.28)–(2.29), $\mu_n \ll 1$ and there is either zero or one molecule of C in the system during most of the simulation time (see Figure 2.9(a)). In that case $\phi_{C,n}(1) \approx \mu_n$, i.e. the probability that there is one C molecule in the system when there are n molecules of A is approximately μ_n (and the probability of no molecules of C is approximately $1 - \mu_n$). Combining this with (2.39), we can estimate the average probability of finding a molecule of C in the system as

$$\sum_{n=0}^{\infty} \mu_n \, \phi_A(n) = \sum_{n=0}^{\infty} \frac{k_1 v}{k_2 + n k_4/v} \frac{1}{n!} (M_{A,s})^n \exp\left[-M_{A,s}\right].$$ (2.41)

The stochastic focusing arises because this value is significantly different from that which would arise if the number of molecules of A were fixed at its mean value, namely

$$\mu_{M_{A,s}} = \frac{k_1 v}{k_2 + M_{A,s} k_4 / v}.$$

Since B depends linearly on C and not directly on A, multiplying (2.41) by k_2 / k_3 we estimate the steady-state value of the average number of B molecules as

$$M_{B,s} = \sum_{n=0}^{\infty} \frac{k_1 v k_2}{k_3 (k_2 + n k_4 / v)} \frac{1}{n!} (M_{A,s})^n \exp [-M_{A,s}].$$

Using (2.33), we obtain

$$M_{B,s} \doteq \begin{cases} 113.1, & \text{for } t < 10\,\text{min}, \\ 316.7, & \text{for } t \geq 10\,\text{min}, \end{cases} \tag{2.42}$$

which is a good estimate of $M_{B,s}$ for the rate constants (2.28)–(2.29) – see Figure 2.6(b). The steady-state values $\bar{B}_s$ of the ODEs (2.25)–(2.27) are for the rate constants (2.28)–(2.29) given in (2.31). Comparing (2.31) and (2.42), we conclude that the estimate (2.42) provides a much better approximation of the average behaviour of the system than the results obtained by the deterministic ODEs.

2.4 Designing Stochastic Chemical Systems

Traditionally, chemical reaction networks have been used to study complex intracellular processes. In such applications, one has to first identify important biochemical species, which have been denoted as A, B or C in previous sections, and reactions between them. Then the Gillespie SSA (a5)–(d5) is used to study the stochastic behaviour of the resulting model. It is often an iterative process between biological experiments and modelling, because one also has to find the correct set of parameters that reproduce the observed phenomena.

With the advancement in experimental techniques, computational models are not only useful to explain the observed behaviour, but can be used to design biochemical systems that behave in a desired manner. Engineering reaction networks is the aim of a relatively new interdisciplinary scientific field called synthetic biology. It can be done by altering a pre-existing intracellular network, but one can also engineer a network from scratch. For example, in DNA computing, reaction networks are engineered with chemical species consisting exclusively of DNA molecules, interacting via the so-called toehold-mediated DNA strand-displacement mechanism (Soloveichik et al., 2010). In this way, one can physically represent a very rich set of theoretically designed networks. An abstract network written on a piece of paper may be mapped to a DNA-based physical network provided it consists of up to second-order reactions, with rate coefficients varying over up to six orders of magnitude. A proof-of-concept for DNA computing is a synthetic oscillator called the displacillator, which was implemented in vitro (Srinivas et al., 2017).

There are many interesting mathematical questions connected with the design of chemical systems that are described by deterministic ODEs exhibiting various exotic dynamical phenomena (Plesa et al., 2016): How do we construct a chemical system described by ODEs that undergo specific bifurcations, have multiple limit cycles or multiple fixed points? What types of reactions should we add to a given chemical system so that its behaviour changes in a desired way? Such reaction network constructions are made more challenging by various constraints on their simplicity, where a simple design means that it is easier to engineer the resulting chemical system in experiments.

In this section, we only study the problem of designing and redesigning reaction networks at the (more detailed) stochastic level. We illustrate a general algorithm developed for this purpose by Plesa et al. (2018) on a simple example of redesigning our production–degradation system (1.9). This system is introduced in Section 1.2 in terms of one chemical species A in a container of volume v which is subject to two chemical reactions (1.9). Both of them include symbol $\emptyset$, which denotes chemical species that are of no further interest. However, we can obtain similar dynamics of A if we explicitly consider other chemical species. For example, consider a system of three chemical species A, B and C in a container of volume v which are subject to the following four chemical reactions:

$$A \xrightarrow{k_1} B, \qquad B + C \xrightarrow{k_2} A + C, \qquad (2.43)$$

$$\emptyset \xrightarrow{k_3} C, \qquad B + C \xrightarrow{k_4} B. \qquad (2.44)$$

The first two reactions have rate constants k_1 and k_2 and directly correspond to the two reactions of system (1.9). In the first reaction of both (1.9) and (2.43), molecules of A are removed from the system with rate constant k_1, while they are produced in the second reaction. The production rate is constant in system (1.9), but it is, on the face of it, non-constant in (2.43). It is the third and fourth reactions in (2.44) that make this rate approximately constant for certain choices of parameter values. To see this, we first write the deterministic description of the combined system (2.43)–(2.44) as the system of ODEs

$$\frac{da}{dt} = -k_1\, a + k_2\, bc, \tag{2.45}$$

$$\frac{db}{dt} = k_1\, a - k_2\, bc, \tag{2.46}$$

$$\frac{dc}{dt} = k_3 - k_4\, bc, \tag{2.47}$$

where $a(t)$, $b(t)$ and $c(t)$ are concentrations of A, B and C. Translating this into the (approximate) number of molecules of A, B and C gives

$$\frac{d\overline{A}}{dt} = -k_1\, \overline{A} + \frac{k_2}{v}\, \overline{B}\,\overline{C}, \tag{2.48}$$

$$\frac{d\overline{B}}{dt} = k_1\, \overline{A} - \frac{k_2}{v}\, \overline{B}\,\overline{C}, \tag{2.49}$$

$$\frac{d\overline{C}}{dt} = k_3 v - \frac{k_4}{v}\, \overline{B}\,\overline{C}, \tag{2.50}$$

where $\overline{A}(t) = a(t)v$, $\overline{B}(t) = b(t)v$ and $\overline{C}(t) = c(t)v$. Adding equations (2.48) and (2.49), we observe

$$\frac{d}{dt}(\overline{A} + \overline{B}) = 0,$$

which implies that there exists a constant, denoted N, such that we have a conservation law

$$\overline{A}(t) + \overline{B}(t) = N, \qquad \text{for all } t \geq 0. \tag{2.51}$$

Constant N is determined by the initial conditions, namely $N = \overline{A}(0) + \overline{B}(0)$. Substituting (2.51) into (2.48)–(2.50), we can equivalently rewrite the deterministic ODE model (2.48)–(2.50) as a system of two ODEs for $\overline{A}(t)$ and $\overline{C}(t)$,

$$\frac{d\overline{A}}{dt} = -k_1\, \overline{A} + \frac{k_2}{v}\, (N - \overline{A})\,\overline{C}, \tag{2.52}$$

$$\frac{d\overline{C}}{dt} = k_3 v - \frac{k_4}{v}\, (N - \overline{A})\,\overline{C}, \tag{2.53}$$

together with $\overline{B}(t) = N - \overline{A}(t)$. Let us now assume that $k_3 v = k_4/v$. Then (2.53) can be rewritten as

$$\frac{1}{k_3 v}\frac{d\overline{C}}{dt} = 1 - (N - \overline{A})\,\overline{C}.$$

In particular, considering the limit $k_3 v \to \infty$, we have

$$1 \approx (N - \overline{A})\overline{C}.$$

Substituting this approximation into (2.52), we have

$$\frac{d\overline{A}}{dt} = -k_1 \overline{A} + \frac{k_2}{v}.$$

This ODE looks very similar to equation (1.15) for the time evolution of the average number of molecules of A for system (1.9). There is a small difference in the form of the production term: we have k_2/v above, while we have $k_2 v$ in (1.15). This is caused by the fact that k_2 has different physical units in these two chemical systems, because it is the rate constant of a zeroth-order chemical reaction in (1.9) and of a second-order chemical reaction in (2.43). Put differently, we have shown that, in the case of the deterministic model, the chemical system (2.43)–(2.44) gives the same time evolution of A as the chemical system (1.9), provided that $k_3 v = k_4/v$ and $k_3 v$ is chosen sufficiently large.

Considering the stochastic description of (2.43)–(2.44), we observe that the conservation relation (2.51) still holds, namely

$$A(t) + B(t) = N, \qquad \text{for all } t \geq 0, \tag{2.54}$$

because the sum, $A(t) + B(t)$, does not change if any of the four chemical reactions in (2.43)–(2.44) takes place. The constant N is again given by the initial condition $N = A(0) + B(0)$. In particular, relation (2.54) implies that $A(t) \leq N$. Considering the original model (1.9), we do not have such a restriction on A from above. However, looking at Figure 1.2, we observe that the values of $A(t)$ in model (1.9) "approximately" satisfy $A(t) \leq N$ provided that N is chosen sufficiently large, say $N = 20$ for the parameter values in Figure 1.2. This motivates us to use the set of parameters

$$k_1 = 0.1 \text{ sec}^{-1}, \quad \frac{k_2}{v} = 1 \text{ sec}^{-1}, \quad k_3 v = 100 \text{ sec}^{-1}, \quad \frac{k_4}{v} = 100 \text{ sec}^{-1}, \tag{2.55}$$

and initial conditions

$$A(0) = 0, \qquad B(0) = 20, \qquad C(0) = 0. \tag{2.56}$$

Then $N = A(0) + B(0) = 20$. In Figure 2.10(a), we show the time evolution of A obtained by the Gillespie SSA (a5)–(d5) applied to the system of chemical reactions (2.43)–(2.44) for parameters and initial conditions given by (2.55) and (2.56). We observe that the time evolution of A is comparable to that in Figure 1.2(a).

The chemical system (2.43)–(2.44) extends model (1.9) by introducing additional chemical species B and C without altering the dynamics of chemical species A. We now introduce two additional reactions given as

$$A + B \xrightarrow{k_5} 2A, \qquad\qquad A + B \xrightarrow{k_5} 2B, \qquad\qquad (2.57)$$

and we consider a combined chemical system written, for three chemical species A, B and C, in terms of six chemical reactions (2.43), (2.44) and (2.57). Since we use the same rate constant k_5 in both reactions in (2.57), these additional reactions do not change the deterministic model. It is still given by ODEs (2.45)–(2.47) for any value of parameter k_5, which can be observed by inspecting terms on the right-hand sides of the deterministic ODEs. For example, the first chemical reaction in (2.57) produces molecules of A with rate $k_5 ab$, while the second reaction in (2.57) removes molecules of A with the same rate, $k_5 ab$. Therefore the total contribution of both reactions on the right-hand side of (2.45) is $k_5 ab - k_5 ab = 0$. In particular, all results of the above deterministic ODE-based analysis of chemical system (2.43)–(2.44) also hold for the chemical system consisting of six reactions (2.43), (2.44) and (2.57).

In Figure 2.10, we present results of the Gillespie SSA (a5)–(d5) applied to chemical system (2.43), (2.44) and (2.57) for parameters and initial conditions given by (2.55) and (2.56). While stochastic results in Figure 2.10(a) can be

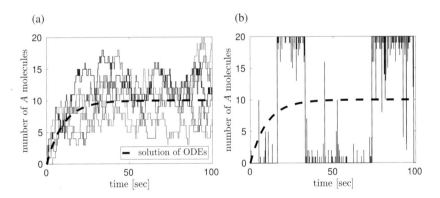

Figure 2.10 (a) *The time evolution of the number of molecules of A obtained by five realizations of the SSA (a5)–(d5) for the system of chemical reactions (2.43)–(2.44) (coloured lines) and the solution of the system of ODEs (2.52)–(2.53) (black line). Parameters and initial conditions are given by (2.55) and (2.56). (b) The time evolution of the number of molecules of A obtained for the extended system of six chemical reactions (2.43), (2.44) and (2.57). One realization (blue line) is plotted for the parameters and initial conditions given by (2.55), (2.56) and $k_5/v = 100 \, sec^{-1}$.*

viewed as simulations for $k_5/\nu = 0$, we use $k_5/\nu = 100 \sec^{-1}$ to get the stochastic result (blue line) in Figure 2.10(b). For clarity, we only plot one realization of the Gillespie SSA (a5)–(d5) in panel (b) and observe that it resembles a system with two favourable states (one for values of A close to 0 and one for values of A close to $N = 20$). Thus we observe that we have a one-parameter family of stochastic models (parametrized by k_5) which have the same deterministic ODE description, but qualitatively different stochastic dynamics. In other words, we have redesigned the stochastic dynamics of the production–degradation example (1.9) without changing its deterministic model. This is further illustrated in Figure 2.11(a), where we plot stationary distributions of the system of chemical reactions (2.43), (2.44) and (2.57) for different values of parameter k_5. For smaller values of k_5 (say, for $k_5 = 0$), we recover the stationary distribution given in Figure 1.2(b). Increasing k_5, we get a system with two favourable states.

To explain this qualitative change in the behaviour of a stochastic model, consider the propensity function of reactions in (2.57). It is given, for both reactions, as $k_5 A(t)B(t)$. This propensity is zero if $A(t) = 0$ or $A(t) = N$. If $A(t) \neq 0$ and $A(t) \neq N$, the propensity of chemical reactions in (2.57) can take arbitrarily large values, if we increase k_5. In particular, the state of the system has a tendency to rapidly change when $A(t) \neq 0$ and $A(t) \neq N$. This biases the observed behaviour towards states when $A(t) = 0$ or $A(t) = N$. If

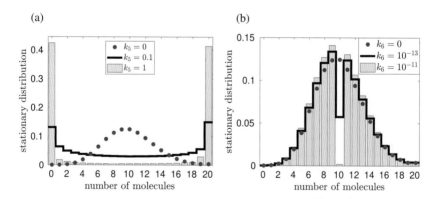

Figure 2.11 (a) *Stationary distributions of molecules of A obtained by the SSA (a5)–(d5) for the system of chemical reactions (2.43), (2.44) and (2.57) for three different values of parameter k_5, namely $k_5/\nu = 0 \sec^{-1}$, $k_5/\nu = 0.1 \sec^{-1}$ and $k_5/\nu = 1 \sec^{-1}$. (b) Stationary distributions of molecules of A for the system of chemical reactions (2.43), (2.44) and (2.58) for three different values of parameter k_6, namely $k_6/\nu^{19} = 0 \sec^{-1}$, $k_6/\nu^{19} = 10^{-13} \sec^{-1}$ and $k_6/\nu^{19} = 10^{-11} \sec^{-1}$. Other parameters and initial conditions are given by (2.55) and (2.56).*

we allow higher-order reactions, then we can generalize this idea in such a way that we can add noise to an arbitrary part of the state space. To illustrate this, consider chemical system (2.43), (2.44) together with the following two chemical reactions:

$$10A + 10B \xrightarrow{k_6} 11A + 9B, \qquad 10A + 10B \xrightarrow{k_6} 9A + 11B. \qquad (2.58)$$

Adding these reactions (with the same rate constant), we again do not change the deterministic model. It is still given by ODEs (2.45)–(2.47) for all values of parameter k_6. To formulate the stochastic model, we have to specify the propensity function of reactions (2.58). It is a higher-order reaction, which we discussed in Section 1.3. The propensity functions of some chemical reactions are given in Tables 1.1 and 1.2. Although chemical reactions (2.58) have not been included in these tables, we can use the same principle as in Section 1.3 (the propensity function is proportional to the number of available combinations of reactants) to deduce that the propensity function of both reactions in (2.58) is

$$\frac{k_6}{v^{19}} A\,(A-1)\,(A-2)\cdots(A-9)\,B\,(B-1)\,(B-2)\cdots(B-9).$$

This propensity is zero, unless $A(t) = 10$. Thus we have introduced a very localized noise term to the stochastic dynamics of A. If k_6 is sufficiently large, chemical reactions (2.58) immediately occur if $A(t) = 10$. In particular, the system does not stay in this state for long. This is illustrated in Figure 2.11(b), where we plot stationary distributions of model (2.43), (2.44) and (2.58) for different values of parameter k_6/v^{19}. For smaller values of k_6 (say, for $k_6 = 0$), we recover the stationary distribution given in Figure 1.2(b). Increasing k_6, we get a system with two favourable states, $\{A(t) \leq 9\}$ and $\{A(t) \geq 11\}$, which are clearly separated by state $\{A(t) = 10\}$. The simulated system spends a very small fraction of time at state $\{A(t) = 10\}$, which is caused by frequent occurrences of reactions (2.58).

Exercises

2.1 Show that $\overline{A}_{s1}$ and $\overline{A}_{s2}$ (given by (2.5)) are stable steady states of the ODE (2.3). Show that $\overline{A}_u$ is the unstable steady state.

2.2 Consider the chemical system (2.19)–(2.20) for the chemical species B and C. Assume that the number of A molecules is given by (2.34) (no fluctuations of the signal). Write the master equation for this reduced system. Show that the equations for the mean values of B and C are exactly given by the system of ODEs (2.35)–(2.36).

2.3 Consider the chemical system (2.1), i.e.

$$3A \underset{k_2}{\overset{k_1}{\rightleftarrows}} 2A, \qquad\qquad A \underset{k_4}{\overset{k_3}{\rightleftarrows}} \emptyset, \qquad\qquad (2.1)$$

in a container of volume v, with the values of rate constants k_1, k_2 and k_3 given as

$$k_1/v^2 = 2.5 \times 10^{-4}\,\text{min}^{-1}, \qquad k_2/v = 0.18\,\text{min}^{-1}, \qquad k_3 = 37.5\,\text{min}^{-1}.$$

In this problem we will vary the value of $k_4 v$ in the range

$$1700\,\text{min}^{-1} \le k_4 v \le 2500\,\text{min}^{-1} \qquad\qquad (2.59)$$

and study the dependence of the behaviour of the chemical system (2.1) on this parameter.

(a) Write the deterministic ODE model describing the system (2.1). Find all steady states of this ODE for (i) $k_4 v = 1750\,\text{min}^{-1}$, (ii) $k_4 v = 2100\,\text{min}^{-1}$, (iii) $k_4 v = 2200\,\text{min}^{-1}$ and (iv) $k_4 v = 2450\,\text{min}^{-1}$. Which of these steady states are stable? Note that the case (iii) has already been considered in Exercise 2.1.

(b) Plot the dependence of the steady states of the deterministic ODE model as a function of $k_4 v$ for $k_4 v$ in range (2.59).

(c) Use the Gillespie SSA (a5)–(d5) to simulate the chemical system (2.1) for (i) $k_4 v = 1750\,\text{min}^{-1}$, (ii) $k_4 v = 2100\,\text{min}^{-1}$, (iii) $k_4 v = 2200\,\text{min}^{-1}$ and (iv) $k_4 v = 2450\,\text{min}^{-1}$. For each parameter value, plot the time evolution of $A(t)$ and estimate the stationary distribution using long-time simulation of the Gillespie SSA (a5)–(d5). Compare with the stationary distribution (2.8) obtained from the chemical master equation. If there are any differences, explain them.

2.4 Consider the chemical system

$$A \xrightarrow{k_1} A + C, \qquad C \xrightarrow{k_2} B \xrightarrow{k_3} \emptyset, \qquad \emptyset \xrightarrow{k_4} A \xrightarrow{k_5} \emptyset,$$

in a container of volume v.

(a) Write down the deterministic ODE model describing the system. Find the steady state.

(b) Use the chemical master equation to estimate the average number of molecules of A, B and C in the system. Can the system exhibit stochastic focusing?

(c) Use the Gillespie SSA (a5)–(d5) to simulate the system when

$$k_1 = 1 \text{ sec}^{-1}, \quad k_2 = 10^3 \text{ sec}^{-1}, \quad k_3 = 10^{-1} \text{ sec}^{-1},$$

$$k_4 v = 10^3 \text{ sec}^{-1}, \quad k_5 = 10^2 \text{ sec}^{-1}$$

to verify your results.

2.5 Consider the chemical system

$$\emptyset \xrightarrow{k_1} C, \qquad A + C \xrightarrow{k_2} A + B,$$

$$B \xrightarrow{k_3} \emptyset, \qquad \emptyset \xrightarrow{k_4} A \xrightarrow{k_5} \emptyset, \qquad C \xrightarrow{k_6} \emptyset,$$

in a container of volume v.

(a) Write down the deterministic ODE model describing the system. Find the steady state.
(b) Suppose that $k_5^2 + k_1 k_5 v \ll k_2 k_4 + k_5 k_6$, so that C evolves more quickly than A, and that the number of molecules of C is small. Find approximately the probability that there is one molecule of C in the system given that there are n molecules of A. Hence find the average probability that there is one molecule of C in the system.
(c) Find the average number of molecules of B in the system given that there are n molecules of A. Deduce the overall average number of molecules of B. Does the system exhibit stochastic focusing?
(d) Use the Gillespie SSA (a5)–(d5) to simulate the system when

$$k_1 v = 300 \text{ sec}^{-1}, \quad \frac{k_2}{v} = 2 \times 10^4 \text{ sec}^{-1}, \quad k_3 = 1 \text{ sec}^{-1},$$

$$k_4 v = 500 \text{ sec}^{-1}, \quad k_5 = 100 \text{ sec}^{-1}, \quad k_6 = 2 \times 10^5 \text{ sec}^{-1}$$

to verify your results.

2.6 Consider the chemical system

$$A \xrightarrow{k_1} A + C \xrightarrow{k_2} A + B,$$

$$B \xrightarrow{k_3} \emptyset, \qquad \emptyset \xrightarrow{k_4} A \xrightarrow{k_5} \emptyset, \qquad C \xrightarrow{k_6} \emptyset,$$

in a container of volume v.

(a) Write down the deterministic ODE model describing the system. Find the steady state.

(b) Suppose that $k_5^2 + k_1 k_4 v \ll k_2 k_4 + k_5 k_6$, so that C evolves more quickly than A, and that the number of molecules of C is small. Find approximately the probability that there is one molecule of C in the system given that there are n molecules of A. Hence find the average probability that there is one molecule of C in the system.

(c) Find the average number of molecules of B in the system given that there are n molecules of A. Deduce the overall average number of molecules of B. Does the system exhibit stochastic focusing?

(d) Use the Gillespie SSA (a5)–(d5) to simulate the system when

$$k_1 = 60 \sec^{-1}, \quad \frac{k_2}{v} = 2 \times 10^4 \sec^{-1}, \quad k_3 = 1 \sec^{-1},$$

$$k_4 v = 500 \sec^{-1}, \quad k_5 = 100 \sec^{-1}, \quad k_6 = 2 \times 10^5 \sec^{-1}$$

to verify your results.

3

Stochastic Differential Equations

The classic (approximate) deterministic description of chemical reactions is based on ordinary differential equations (ODEs). Such equations provide a reasonable description of the system in some cases (as we saw in Chapter 1), while they are inapplicable in some other cases (as shown in Chapter 2).

In this chapter, we introduce stochastic differential equations (SDEs) which are, roughly speaking, ODEs with some additional noise terms. Such equations can be used as approximate models for some of the stochastic chemical systems we saw in Chapter 2. They are easier to solve and analyse (from the mathematical point view) than the (exact) chemical master equation, and they provide a better approximation of a stochastic chemical system than ODEs.

Although we will introduce SDEs with reference to the stochastic approximation of chemical systems, they have many other applications. For example, they can be used for the stochastic modelling of molecular diffusion, which we will discuss in more detail in Chapter 4. Parameters of such SDEs can be derived or estimated from more detailed models, which we will study in the case of molecular diffusion in Chapter 8. In this chapter, we present the derivation of parameters of SDEs from the underlying stochastic models of chemically reacting systems.

We begin by introducing SDEs from the computational point of view, presenting several examples to illustrate the computational definition of the SDE that is used throughout the book. We then introduce the Fokker–Planck and Kolmogorov backward equations, which correspond in some sense to the chemical master equation. We will use these equations to compute the mean transition time between favourable states of SDEs. We then apply the SDE formalism to a chemical system by deriving the chemical Fokker–Planck equation and the corresponding chemical Langevin equation. We use these to further analyse the chemical system from Section 2.1, and then to analyse the self-induced stochastic resonance that was introduced in Section 2.2.

3.1 A Computational Definition of SDE

Consider first a variable $x \equiv x(t) \in \mathbb{R}$ which evolves according to the ODE

$$\frac{dx}{dt} = f(x, t), \tag{3.1}$$

where $f \colon \mathbb{R} \times [0, \infty) \to \mathbb{R}$ is a given function. Given the initial condition $x(0) = x_0$, one can use the ODE (3.1) to find the values of $x(t)$ at times $t > 0$, provided that the function f is "sufficiently nice". Although it is possible to find conditions on f that guarantee existence and uniqueness of the solution of the ODE (3.1), we will not pursue such a rigorous mathematical treatment of ODEs in this book. Instead, we rewrite the ODE (3.1) formally as

$$dx = f(x, t)\, dt. \tag{3.2}$$

Intuitively, the equation (3.2) specifies the infinitesimal change

$$dx \equiv dx(t) = x(t + dt) - x(t)$$

of the x variable in the infinitesimally small time interval $[t, t + dt]$. Thus, the equation (3.2) can also be written as

$$x(t + dt) = x(t) + f(x(t), t)\, dt. \tag{3.3}$$

This equation can be used as a simple way to compute the solution of the original ODE (3.1). Namely, we choose a small time step Δt and we compute the value of $x(t + \Delta t)$ from the value of $x(t)$ by

$$x(t + \Delta t) = x(t) + f(x(t), t)\, \Delta t. \tag{3.4}$$

Given the initial condition $x(0) = x_0$, we can use formula (3.4) to find the values of $x(t)$ at times $t > 0$, iteratively. This iterative approach is called the forward Euler method for solving the ODE (3.1). If f is a "sufficiently nice" function, the forward Euler method will yield an approximate solution of the ODE (3.1). The error of the approximation can be made smaller by choosing the time step Δt smaller. Indeed, the formula (3.4) is exactly equal to (3.3) if we replace the (small) time step Δt by the infinitesimally small time step dt.

Roughly speaking, an SDE is an ODE such as (3.1) with an additional noise term describing stochastic fluctuations. If we consider (3.4) to be a "computational definition" of the ODE (3.1), we may write the computational definition of the corresponding SDE with Gaussian noise as

$$X(t + \Delta t) = X(t) + f(X(t), t)\, \Delta t + g(X(t), t)\, \sqrt{\Delta t}\, \xi, \tag{3.5}$$

where $g \colon \mathbb{R} \times [0, \infty) \to \mathbb{R}$ is a given strength of the noise and ξ is a random number, which is sampled from the normal distribution with zero mean and unit variance. Although other noise terms are possible, we will only be concerned with Gaussian noise. The scaling of the noise term with $\sqrt{\Delta t}$ (and not Δt) may seem strange to those raised on ODEs, but it is in fact the natural scaling, which gives a sensible limit as $\Delta t \to 0$, as we will see in Section 3.2.

Normally distributed random numbers with zero mean and unit variance can be generated by many standard computer languages (e.g. the function randn in Matlab). If a computer language does not include a routine for generating normally distributed random numbers, one can use a generator of uniformly distributed random numbers together with a suitable transformation (see Exercise 3.1) to generate normally distributed random numbers.

Given the initial condition $X(0) = x_0$, we can use formula (3.5) to find the values of $X(t)$ at times $t > 0$ iteratively, by performing two steps at time t:

> **(a6)** Generate a normally distributed (with zero mean and unit variance) random number ξ.
> **(b6)** Compute $X(t + \Delta t)$ from $X(t)$ by (3.5).
> Then continue with step (a6) for time $t + \Delta t$.

The application of the SSA (a6)–(b6) is shown in Section 3.2 where several examples of SDEs are presented. The SSA (a6)–(b6) gives the approximate solution of the SDE which can be formally written in the following form:

$$X(t + dt) = X(t) + f(X(t), t)\, dt + g(X(t), t)\, dW, \qquad (3.6)$$

where dW is so-called white noise (also known as the differential of the Wiener process). If we wanted to provide a rigorous mathematical treatment of SDEs, we would have to rigorously define dW, which would require a formal definition of stochastic integration. However, for the purposes of this book, it is sufficient to assume that the meaning of the SDE (3.6) is given by the corresponding computational definition (3.5). In fact, this is all we need to know to simulate SDEs numerically and to use them for the analysis of reaction–diffusion processes. Consequently, whenever we write SDEs in the form (3.6), we understand them in terms of the computational definition (3.5), i.e. we replace dW by the product of $\sqrt{\Delta t}$ and the random number ξ, which is sampled from a normal distribution with zero mean and unit variance. The SSA (a6)–(b6) is often called the Euler–Maruyama method for solving the SDE (3.6).

If $g(x, t) \equiv 0$, then the SDE (3.6) reduces to the ODE (3.3) which can be equivalently written as (3.1). If $g(x, t) \not\equiv 0$, then the SDE (3.6) and the ODE (3.3) differ by the extra noise term $g(X(t), t)\, dW$. There are many other ways one could add noise to ODEs. However, the noise term $g(X(t), t)\, dW$ is in some sense the most natural choice that appears in applications. We will make this point clear in the following section where we present several examples.

3.2 Examples of SDEs

We consider that both time t and the variable $X(t)$ are dimensionless, i.e. they have no physical units. Let us start by choosing $f(x,t) \equiv 0$ and $g(x,t) \equiv 1$. Then (3.6) reads as follows:

$$X(t + dt) = X(t) + dW, \tag{3.7}$$

which, using the computational definition (3.5), can be interpreted as

$$X(t + \Delta t) = X(t) + \sqrt{\Delta t}\, \xi, \tag{3.8}$$

where Δt is the (small) time step. We choose $\Delta t = 10^{-3}$. Given the initial condition $X(0) = 0$, we compute the time evolution of $X(t)$ by the SSA (a6)–(b6). Six illustrative realizations of the SSA (a6)–(b6) are presented in Figure 3.1(a).

Let $E[\cdot]$ denote the average over (infinitely) many realizations of the SSA (a6)–(b6). Let $M(t)$ be the mean value of $X(t)$ and let $V(t)$ be the variance of $X(t)$, defined by

$$M(t) = E[X(t)], \tag{3.9}$$
$$V(t) = E[(X(t) - M(t))^2] = E[X(t)^2] - M(t)^2. \tag{3.10}$$

Using the formula (3.8), we can compute the average value of $X(t + \Delta t)$ by

$$M(t + \Delta t) = E[X(t + \Delta t)] = E[X(t) + \sqrt{\Delta t}\, \xi]$$
$$= E[X(t)] + \sqrt{\Delta t}\, E[\xi] = E[X(t)] = M(t), \tag{3.11}$$

where we used the fact that ξ is sampled from the normal distribution with zero mean, i.e. $E[\xi] = 0$. The initial condition $X(0) = 0$ implies $M(0) = 0$. Thus, (3.11) implies

$$M(t) = 0. \tag{3.12}$$

Using (3.8), (3.10) and (3.12), we obtain

$$V(t + \Delta t) = E[X(t + \Delta t)^2] - M(t + \Delta t)^2 = E[X(t + \Delta t)^2]$$
$$= E[(X(t) + \sqrt{\Delta t}\, \xi)^2] = E[X(t)^2 + 2X(t)\sqrt{\Delta t}\, \xi + \Delta t\, \xi^2]$$
$$= E[X(t)^2] + 2\, E[X(t)]\sqrt{\Delta t}\, E[\xi] + \Delta t\, E[\xi^2]$$
$$= E[X(t)^2] + \Delta t = E[X(t)^2] - M(t)^2 + \Delta t$$
$$= V(t) + \Delta t, \tag{3.13}$$

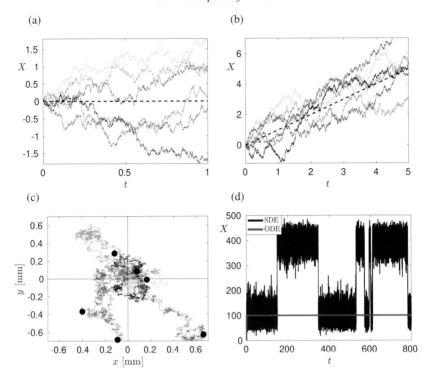

Figure 3.1 (a), (b) *Six realizations computed by the SSA (a6)–(b6) (coloured lines) for the SDE (3.7) (in panel (a)) and the SDE (3.21) (in panel (b)). We use $\Delta t = 10^{-3}$ and the initial condition $X(0) = 0$. The time evolution of the stochastic mean $M(t)$ is plotted as the black dashed line. (c) Solution of (3.18)– (3.19). Six trajectories obtained by the SSA (a6)–(b6) for $D = 10^{-4} mm^2 sec^{-1}$ and $\Delta t = 0.1 sec$. Trajectories were started at the origin and followed for 10 minutes. (d) Solution of the SDE (3.26) obtained by the SSA (a6)–(b6) for $k_1 = 10^{-3}$, $k_2 = 0.75$, $k_3 = 165$, $k_4 = 10^4$, $k_5 = 200$ and the initial condition $X(0) = 0$ (blue line). The solution of the ODE (3.28) for the same parameter values (red line).*

where we have used the fact that $E[\xi] = 0$ and $E[\xi^2] = 1$. Since all realizations start at $X(0) = 0$, we have $V(0) = 0$. Thus, (3.13) implies

$$V(t) = t. \tag{3.14}$$

Consequently, we see that $M(t)$ and $V(t)$ are independent of the time step Δt. Using the fact that ξ is sampled from the normal distribution, one can actually show that any moment $E[X(t)^k], k = 1, 2, 3, \ldots$, is independent of the time step Δt (see Exercise 3.2). This is one of the reasons we chose the noise term in the form $\sqrt{\Delta t}\,\xi$ in the computational definitions (3.5) and (3.8). If the noise term did not scale with the time step as $\sqrt{\Delta t}$, then $V(t)$ would depend on the time step Δt. If ξ was not sampled from the normal distribution, then the third moment, $E[X(t)^3]$, would be dependent on the time step Δt.

The main goal of example (3.7) was to show that solving an SDE of the form (3.6) is relatively straightforward through its computational definition (3.5) and the corresponding SSA (a6)–(b6). In Figure 3.1(a), we had no real application in mind. However, it is also straightforward to add a physical meaning to our computation. All we need to do is to give appropriate physical units to the variable $X(t)$ and time t. For example, assuming that $X(t)$ has units of length we can interpret the SDE (3.7) as a simple evolution equation of the x-coordinate of a diffusing particle. To be more specific, let us consider a protein molecule diffusing in an aqueous environment. Its position can be described by three-dimensional vector $[X(t), Y(t), Z(t)]$. In Chapter 4, we show that the position of the protein molecule at time $t + \Delta t$ can be computed from its position at time t by

$$X(t + \Delta t) = X(t) + \sqrt{2D\Delta t}\, \xi_x, \tag{3.15}$$

$$Y(t + \Delta t) = Y(t) + \sqrt{2D\Delta t}\, \xi_y, \tag{3.16}$$

$$Z(t + \Delta t) = Z(t) + \sqrt{2D\Delta t}\, \xi_z, \tag{3.17}$$

where D is the diffusion constant and ξ_x, ξ_y, ξ_z are random numbers, which are sampled from the normal distribution with zero mean and unit variance. Thus, following our discussion from Section 3.1, we can say that the position of the diffusing protein evolves according to the system of SDEs

$$X(t + dt) = X(t) + \sqrt{2D}\, dW_x, \tag{3.18}$$

$$Y(t + dt) = Y(t) + \sqrt{2D}\, dW_y, \tag{3.19}$$

$$Z(t + dt) = Z(t) + \sqrt{2D}\, dW_z, \tag{3.20}$$

where the subscripts in dW_x, dW_y, dW_z emphasize the fact that the white noises are not correlated, i.e. we need three different random numbers ξ_x, ξ_y, ξ_z to simulate the system of SDEs (3.18)–(3.20) at each time step by (3.15)–(3.17). In particular, the equations (3.15)–(3.17) are not coupled and we can use the SSA (a6)–(b6) to compute their solutions. Choosing $\Delta t = 0.1$ sec, $D = 10^{-4}\,\text{mm}^2\,\text{sec}^{-1}$ (diffusion constant of a typical protein molecule) and $[X(0), Y(0), Z(0)] = [0, 0, 0]$, we plot six realizations of the SSA (a6)–(b6) in Figure 3.1(c). We plot only the x- and y-coordinates. We follow the diffusing molecule for 10 minutes. The position of the molecule at time $t = 10$ min is denoted as a black circle for each trajectory. We clearly see the typical picture of the trajectory of Brownian motion as shown in physics textbooks. We return to this application of SDEs in Chapter 4 where the diffusion process is studied in more detail.

Considering again (3.6), and comparing to (3.3), we see that the function $f(x,t)$ gives the deterministic component of the dynamics. It is often called the *drift coefficient*. Let us now return to dimensionless $X(t)$ and t, and add a non-zero drift to our SDE. We consider the case $f(x,t) \equiv 1$ and $g(x,t) \equiv 1$. Then (3.6) reads as

$$X(t + dt) = X(t) + dt + dW, \tag{3.21}$$

which, using the computational definition (3.5), can be interpreted as

$$X(t + \Delta t) = X(t) + \Delta t + \sqrt{\Delta t}\,\xi, \tag{3.22}$$

where Δt is the (small) time step. We again use $\Delta t = 10^{-3}$ and the initial condition $X(0) = 0$, and compute the time evolution of $X(t)$ by the SSA (a6)–(b6). Six illustrative realizations of the SSA (a6)–(b6) are presented in Figure 3.1(b).

Let $M(t)$ be the mean value of $X(t)$ and let $V(t)$ be the variance of $X(t)$, again defined by (3.9) and (3.10). Using the formula (3.22), we find

$$M(t + \Delta t) = E[X(t + \Delta t)] = E[X(t) + \Delta t + \sqrt{\Delta t}\,\xi] \tag{3.23}$$
$$= E[X(t)] + \Delta t + \sqrt{\Delta t}\,E[\xi] = E[X(t)] + \Delta t = M(t) + \Delta t,$$

where we have again used $E[\xi] = 0$. Using the initial condition $X(0) = 0$, we obtain $M(t) = t$. This line is plotted in Figure 3.1(b) as the black dashed line. We see that the solution of (3.21) fluctuates around the mean value $M(t) = t$, which depends only on the drift coefficient. One can also show that $V(t) = t$ (see Exercise 3.3).

Our final SDE example has two favourable states, and is motivated by the ODE (2.2). We choose

$$f(x,t) \equiv f(x) = -k_1\,x^3 + k_2\,x^2 - k_3\,x + k_4, \tag{3.24}$$
$$g(x,t) \equiv g(x) = k_5, \tag{3.25}$$

where k_1, k_2, k_3, k_4 and k_5 are constants. Then (3.6) reads as follows:

$$X(t + dt) = X(t) + (-k_1\,X(t)^3 + k_2\,X(t)^2 - k_3\,X(t) + k_4)\,dt + k_5\,dW. \tag{3.26}$$

If $k_5 = 0$ then (3.26) is equal to

$$X(t + dt) = X(t) + (-k_1\,X(t)^3 + k_2\,X(t)^2 - k_3\,X(t) + k_4)\,dt. \tag{3.27}$$

There is no noise in this equation, and $X(t)$ is no longer a random variable; it is a deterministic variable, which we will denote as $x(t)$. Equation (3.27) is an ODE which can be equivalently written as

$$\frac{dx}{dt} = -k_1 x^3 + k_2 x^2 - k_3 x + k_4. \tag{3.28}$$

This has the same form as the ODE (2.2) that was studied in Section 2.1. We choose the rate constants as $k_1 = 10^{-3}$, $k_2 = 0.75$, $k_3 = 165$ and $k_4 = 10^4$. Then the ODE (3.28) has three steady states:

$$x_{s1} = 100, \qquad x_u = 250, \qquad x_{s2} = 400. \tag{3.29}$$

The steady states x_{s1} and x_{s2} are stable and the steady state x_u is unstable. Thus we are in the similar situation to the case of the ODE (2.2) from Section 2.1. Starting with $x(0) = 0$, the solution of the ODE (3.28) is plotted in Figure 3.1(d) as the red line. We see that it converges to the stable steady state x_{s1}. Next, we consider the original SDE (3.26). We use the same values of the parameters k_1, k_2, k_3 and k_4 as before, but now we set $k_5 = 200$. Starting with $x(0) = 0$, we compute the solution of the SDE (3.26) by the SSA (a6)–(b6). The result is plotted in Figure 3.1(d) as the blue line.

We observe that the solution of the SDE (3.26) fluctuates around the stable steady states x_{s1} and x_{s2} and occasionally switches between them. The situation is qualitatively similar to the behaviour of the chemical system (2.1) which was described in Section 2.1 (see Figure 2.2(a)). In Section 3.6, we present a calculation of the average switching time between the favourable states of the SDE (3.26). Such a theory will be applicable to chemical systems too because chemical systems can be approximately described by SDEs under some conditions (as we will see in Section 3.7). Before we proceed we need to introduce the Fokker–Planck equation and its consequences.

3.3 Fokker–Planck Equation

Let us suppose that $X(t)$ evolves according to the SDE (3.6). We define its probability distribution function $p(x, t)$ so that $p(x, t)\,dx$ is the probability that $X(t) \in [x, x + dx]$. Since $X(t)$ must be somewhere on the real line, the function $p(x, t)$ satisfies the normalization condition

$$\int_{\mathbb{R}} p(x, t)\,dx = 1 \tag{3.30}$$

for any time t. Roughly speaking, $p(x, t)$ quantifies the chance that the trajectory of the SDE is around the point x at time t. Such information can

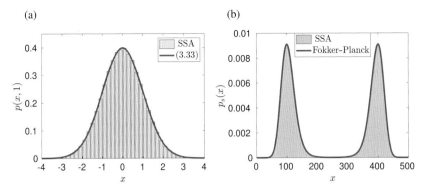

Figure 3.2 (a) *The probability distribution function $p(x, 1)$ for the SDE (3.7) computed by the SSA (a6)–(b6) (grey histogram) and by solving the Fokker–Planck equation (red line) given by (3.33). (b) The stationary distribution $p_s(x)$ of the SDE (3.26) computed by the SSA (a6)–(b6) for $k_1 = 10^{-3}$, $k_2 = 0.75$, $k_3 = 165$, $k_4 = 10^4$, $k_5 = 200$ (grey histogram). The solution (3.39) of the stationary Fokker–Planck equation (red line).*

be computed by many repetitions of the SSA (a6)–(b6). For example, let us consider the SDE example (3.7) and let us compute $p(x, 1)$. To do that we choose the space step Δx and we partition the real line into the bins $[i\Delta x, (i + 1)\Delta x]$ where $i \in \mathbb{Z}$. We run many realizations of the SSA (a6)–(b6), starting at the initial condition $X(0) = 0$ until the time $t = 1$. For each bin, we compute the number of realizations that arrived at this bin at time $t = 1$. We divide the resulting spatial histogram by the number of realizations and the bin size Δx to ensure that the normalization condition (3.30) is satisfied. Using $\Delta x = 0.2$, we plot the result in Figure 3.2(a) as the grey histogram. To compute the histogram, we used 10^5 realizations of the SSA (a6)–(b6) with the time step $\Delta t = 10^{-3}$. If we used fewer realizations, then our results would be more noisy.

Let us now see how the probability distribution function $p(x, t)$ can be computed more efficiently, without performing any stochastic simulation. We will show below that $p(x, t)$ evolves according to the following partial differential equation (PDE):

$$\frac{\partial p}{\partial t}(x, t) = \frac{\partial^2}{\partial x^2}\left(\frac{g^2(x, t)}{2}\, p(x, t)\right) - \frac{\partial}{\partial x}\Big(f(x, t)\, p(x, t)\Big). \tag{3.31}$$

This PDE is often called the Fokker–Planck equation in the literature (Risken, 1989) and we will use this name in this book; other names given to this equation include the Kolmogorov forward equation or the Smoluchowski equation.

We will derive the Fokker–Planck equation later in this section. First, we show some of its applications.

For the SDE (3.7) we have $f(x,t) \equiv 0$ and $g(x,t) \equiv 1$. Thus the corresponding Fokker–Planck equation (3.31) is

$$\frac{\partial p}{\partial t}(x,t) = \frac{1}{2}\frac{\partial^2 p}{\partial x^2}(x,t). \tag{3.32}$$

We imposed previously the initial condition $X(0) = 0$ for the SDE (3.7). This implies that the initial condition of (3.32) is in the form of the Dirac delta function $p(x,0) = \delta(x)$. Solving (3.32) together with the initial condition $p(x,0) = \delta(x)$, we obtain (see Exercise 3.4)

$$p(x,t) = \frac{1}{\sqrt{2\pi t}}\exp\left[-\frac{x^2}{2t}\right]. \tag{3.33}$$

The function $p(x,1)$ is plotted in Figure 3.2(a) for comparison. As expected, it agrees well with the results obtained by many realizations of the SSA (a6)–(b6).

Suppose now that the coefficients f and g of the SDE (3.6) are arbitrary functions of x, but do not depend on time, i.e. $f(x,t) \equiv f(x)$ and $g(x,t) \equiv g(x)$. The long-time behaviour of such SDEs is reasonably characterized by the stationary distribution

$$p_s(x) = \lim_{t\to\infty} p(x,t). \tag{3.34}$$

This function can be obtained as a solution of the stationary problem corresponding to (3.31), which is the ODE

$$\frac{d^2}{dx^2}\left(\frac{g^2(x)}{2}p_s(x)\right) - \frac{d}{dx}\left(f(x)p_s(x)\right) = 0. \tag{3.35}$$

Solving this equation (see Exercise 3.5), we obtain

$$p_s(x) = \frac{C}{g^2(x)}\exp\left[\int_0^x \frac{2f(y)}{g^2(y)}\,dy\right], \tag{3.36}$$

where C is a real constant. To determine the value of C, we note that (3.30) together with (3.34) imply the normalization condition

$$\int_{\mathbb{R}} p_s(x)\,dx = 1. \tag{3.37}$$

Thus

$$C = \left(\int_{\mathbb{R}} \frac{1}{g^2(x)}\exp\left[\int_0^x \frac{2f(y)}{g^2(y)}\,dy\right]dx\right)^{-1}. \tag{3.38}$$

The stationary distribution $p_s(x)$ is a very useful description for SDEs with multiple favourable states, such as SDE (3.26). Substituting (3.24)–(3.25) into (3.36) and integrating over y, we find

$$p_s(x) = \overline{C} \exp\left[\frac{-3k_1\,x^4 + 4k_2\,x^3 - 6k_3\,x^2 + 12k_4\,x}{6k_5^2}\right], \tag{3.39}$$

where

$$\overline{C} = \left(\int_{\mathbb{R}} \exp\left[\frac{-3k_1\,x^4 + 4k_2\,x^3 - 6k_3\,x^2 + 12k_4\,x}{6k_5^2}\right] \mathrm{d}x\right)^{-1}.$$

The function (3.39) is plotted in Figure 3.2(b) as the red line. We compare it with the grey histogram that is obtained by the SSA (a6)–(b6). To compute this histogram, we chose the space step $\Delta x = 1$ and we partitioned the real line into the bins $[i\Delta x, (i+1)\Delta x]$ where $i \in \mathbb{Z}$. Starting at $X(0) = 100$, we ran the SSA (a6)–(b6) for (long) time (until $t = 10^7$). Recording the values of $X(t)$ at equal time intervals (which are time 10^{-1} apart), we computed how many times the simulation visited each bin. To satisfy the normalization condition (3.37), we divided the histogram by Δx and by the number of recordings (which is 10^8 in our case).

Having seen in Figure 3.2 the numerical evidence that the Fokker–Planck equation (3.31) provides a good description of the behaviour of our model SDEs, let us now return to its derivation. We let $p(x, t \mid y, s)\,\mathrm{d}x$ be the probability that $X(t) \in [x, x + \mathrm{d}x)$ under the condition that $X(s) = y$, where s and t are arbitrary time points with $s < t$.

Now consider the value of X at time $t + \Delta t$. The probability that the variable X is in the interval $[z, z + \mathrm{d}z)$ at time $t + \Delta t$ (i.e. $X(t + \Delta t) \in [z, z + \mathrm{d}z)$) is equal to $p(z, t + \Delta t \mid y, s)\,\mathrm{d}z$. We now divide this time interval from s to $t + \Delta t$ into a time interval from s to t, and an interval from t to $t + \Delta t$. At time t the variable X takes on a value in the interval $[x, x + \mathrm{d}x)$ with probability $p(x, t \mid y, s)\,\mathrm{d}x$. It then moves from $[x, x + \mathrm{d}x)$ to the interval $[z, z + \mathrm{d}z)$ over the time interval t to $t + \Delta t$ with probability $p(z, t + \Delta t \mid x, t)\,\mathrm{d}z$. To calculate the probability that X moves from the value y at time s to $[z, z + \mathrm{d}z)$ at time $t + \Delta t$, we need to integrate over all possible intermediate points x. In doing so we obtain the so-called Chapman–Kolmogorov equation

$$p(z, t + \Delta t \mid y, s) = \int_{\mathbb{R}} p(z, t + \Delta t \mid x, t)\, p(x, t \mid y, s)\,\mathrm{d}x, \tag{3.40}$$

where $s < t$. This equation is valid for all Δt, not just small Δt. To derive the Fokker–Planck equation we now take the limit as $\Delta t \to 0$. We first multiply both sides by a smooth test function $\varphi(z)$ and integrate over z, to give

$$\int_{\mathbb{R}} p(z, t + \Delta t \mid y, s)\, \varphi(z)\, dz = \int_{\mathbb{R}} \left[\int_{\mathbb{R}} p(z, t + \Delta t \mid x, t)\, \varphi(z)\, dz \right] p(x, t \mid y, s)\, dx.$$

We rename the integration variable z to be x on the left-hand side to get

$$\int_{\mathbb{R}} p(x, t + \Delta t \mid y, s)\, \varphi(x)\, dx = \int_{\mathbb{R}} \left[\int_{\mathbb{R}} p(z, t + \Delta t \mid x, t)\, \varphi(z)\, dz \right] p(x, t \mid y, s)\, dx.$$

Now using the Taylor expansion of $\varphi(z)$ around the point x on the right-hand side, we get

$$\int_{\mathbb{R}} p(x, t + \Delta t \mid x, s)\, \varphi(x)\, dx$$

$$= \int_{\mathbb{R}} \left[\int_{\mathbb{R}} p(z, t + \Delta t \mid x, t) \left(\varphi(x) + \varphi'(x)(z - x) \right. \right.$$

$$\left. \left. + \varphi''(x)\frac{(z - x)^2}{2} + o((z - x)^2) \right) dz \right] p(x, t \mid y, s)\, dx$$

$$= \int_{\mathbb{R}} \left[\varphi(x) \int_{\mathbb{R}} p(z, t + \Delta t \mid x, t)\, dz + \varphi'(x) \int_{\mathbb{R}} (z - x) p(z, t + \Delta t \mid x, t)\, dz \right.$$

$$+ \frac{1}{2}\varphi''(x) \int_{\mathbb{R}} (z - x)^2 p(z, t + \Delta t \mid x, t)\, dz$$

$$\left. + \int_{\mathbb{R}} o((z - x)^2) p(z, t + \Delta t \mid x, t)\, dz \right] p(x, t \mid y, s)\, dx. \tag{3.41}$$

Next, we simplify the integrals on the right-hand side. Starting at $X(t) = x$, the probability that the trajectory arrives somewhere at time $t + \Delta t$ is equal to 1 (we do not lose the trajectory). Thus

$$\int_{\mathbb{R}} p(z, t + \Delta t \mid x, t)\, dz = 1. \tag{3.42}$$

The average value of $X(t + \Delta t) - x$ provided that $X(t) = x$ is given by the integral

$$\mathrm{E}[X(t + \Delta t) - x \mid X(t) = x] = \int_{\mathbb{R}} (z - x) p(z, t + \Delta t \mid x, t)\, dz. \tag{3.43}$$

Using the computational definition (3.5), we obtain

$$\mathrm{E}[X(t + \Delta t) - x \mid X(t) = x] = \mathrm{E}[f(x, t)\, \Delta t + g(x, t)\, \sqrt{\Delta t}\, \xi]$$
$$= f(x, t)\, \Delta t + g(x, t)\, \sqrt{\Delta t}\, \mathrm{E}[\xi]$$
$$= f(x, t)\, \Delta t,$$

where we have used the fact that $\mathrm{E}[\xi] = 0$. Thus (3.43) implies

$$\int_{\mathbb{R}} (z - x) p(z, t + \Delta t \mid x, t)\, dz = f(x, t)\, \Delta t. \tag{3.44}$$

The average value of $(X(t + \Delta t) - x)^2$ provided that $X(t) = x$ is given by the integral

$$E[(X(t + \Delta t) - x)^2 \mid X(t) = x] = \int_{\mathbb{R}} (z - x)^2 p(z, t + \Delta t \mid x, t) \, dz. \qquad (3.45)$$

Using the computational definition (3.5), we obtain

$$\begin{aligned}
E[(X(t + \Delta t) - x)^2 \mid X(t) = x] &= E[(f(x, t) \Delta t + g(x, t) \sqrt{\Delta t}\, \xi)^2] \\
&= f(x, t)^2 \Delta t^2 + 2f(x, t)g(x, t) \Delta t^{3/2} E[\xi] + g(x, t)^2 \Delta t \, E[\xi^2] \\
&= g(x, t)^2 \Delta t + O(\Delta t^2),
\end{aligned}$$

where we have used $E[\xi] = 0$ and $E[\xi^2] = 1$. Consequently, (3.45) yields

$$\int_{\mathbb{R}} (z - x)^2 p(z, t + \Delta t \mid x, t) \, dz = g(x, t)^2 \Delta t + O(\Delta t^2). \qquad (3.46)$$

Substituting (3.42), (3.44) and (3.46) into (3.41) we obtain

$$\begin{aligned}
&\int_{\mathbb{R}} p(x, t + \Delta t \mid y, s) \varphi(x) \, dx \\
&= \int_{\mathbb{R}} \left[\varphi(x) + \varphi'(x) f(x, t) \Delta t + \varphi''(x) \frac{g(x, t)^2}{2} \Delta t \right] p(x, t \mid y, s) \, dx + O(\Delta t^2).
\end{aligned}$$

A simple algebraic manipulation yields

$$\begin{aligned}
&\int_{\mathbb{R}} \frac{p(x, t + \Delta t \mid y, s) - p(x, t \mid y, s)}{\Delta t} \varphi(x) \, dx \\
&= \int_{\mathbb{R}} \varphi'(x) f(x, t) p(x, t \mid y, s) \, dx + \int_{\mathbb{R}} \varphi''(x) \frac{g(x, t)^2}{2} p(x, t \mid y, s) \, dx + O(\Delta t).
\end{aligned}$$

Using integration by parts on the right-hand side, we get

$$\begin{aligned}
&\int_{\mathbb{R}} \frac{p(x, t + \Delta t \mid y, s) - p(x, t \mid y, s)}{\Delta t} \varphi(x) \, dx \\
&= -\int_{\mathbb{R}} \varphi(x) \frac{\partial}{\partial x} \Big(f(x, t) p(x, t \mid y, s) \Big) \, dx \\
&\quad + \int_{\mathbb{R}} \varphi(x) \frac{\partial^2}{\partial x^2} \left(\frac{g(x, t)^2}{2} p(x, t \mid y, s) \right) dx + O(\Delta t),
\end{aligned}$$

which can be rewritten as one integral

$$\begin{aligned}
0 = \int_{\mathbb{R}} \varphi(x) \times \Bigg\{ &- \frac{p(x, t + \Delta t \mid y, s) - p(x, t \mid y, s)}{\Delta t} \\
&- \frac{\partial}{\partial x} \Big(f(x, t) p(x, t \mid y, s) \Big) + \frac{\partial^2}{\partial x^2} \left(\frac{g(x, t)^2}{2} p(x, t \mid y, s) \right) \Bigg\} dx + O(\Delta t).
\end{aligned}$$

Since the test function $\varphi(x)$ is arbitrary, we conclude that the term inside curly brackets must be zero. Thus

$$\frac{p(x, t + \Delta t \mid y, s) - p(x, t \mid y, s)}{\Delta t}$$

$$= \frac{\partial^2}{\partial x^2} \left(\frac{g(x, t)^2}{2} p(x, t \mid y, s) \right) - \frac{\partial}{\partial x} \Big(f(x, t) \, p(x, t \mid y, s) \Big) + O(\Delta t).$$

Passing to the limit $\Delta t \to 0$, we obtain the Fokker–Planck equation in the form

$$\frac{\partial}{\partial t} p(x, t \mid y, s) = \frac{\partial^2}{\partial x^2} \left(\frac{g(x, t)^2}{2} p(x, t \mid y, s) \right) - \frac{\partial}{\partial x} \Big(f(x, t) \, p(x, t \mid y, s) \Big). \quad (3.47)$$

Although the derivation above is formal, we note that the Fokker–Planck equation (3.47) can be justified rigorously.

At the beginning of this section, we defined the function $p(x, t)$ so that $p(x, t) \, dx$ is the probability that $X(t) \in [x, x + dx)$. In this definition, we implicitly assumed that the trajectory of the SDE (3.6) started at a given initial condition $X(0) = x_0$. Thus what we wrote as $p(x, t)$ should now be more properly written as $p(x, t \mid x_0, 0)$. Choosing $y = x_0$ and $s = 0$ in (3.47), we arrive at the Fokker–Planck equation in the form (3.31).

3.4 Boundary Conditions on the Fokker–Planck Equation

The Fokker–Planck equation (3.47) was derived on an infinite domain, i.e. one in which the variable X can take any real value. Often we are interested in SDEs on finite domains, typically with either adsorbing or reflective boundary conditions at the edge of the domain (we will consider more complicated *partially adsorbing* boundaries in Chapter 4).

Let us consider first a reflecting boundary at $x = 0$, say. In that case the simple SSA (a6)–(b6) needs to be updated to include a third step to account for the reflection:

(a7) Generate a normally distributed (with zero mean and unit variance) random number ξ.

(b7) Compute possible position $X(t + \Delta t)$ at time $t + \Delta t$ by (3.5).

(c7) If $X(t + \Delta t)$ computed by (3.5) is less than 0, then set instead
$$X(t + \Delta t) = -X(t) - f(X(t), t) \, \Delta t - g(X(t), t) \sqrt{\Delta t} \, \xi.$$
Then continue with step (a7) for time $t + \Delta t$.

Since the variable X can now only take positive values, the Chapman–Kolmogorov equation is (compare with equation (3.40))

$$p(z, t + \Delta t \mid y, s) = \int_0^\infty p(z, t + \Delta t \mid x, t)\, p(x, t \mid y, s)\, dx, \qquad (3.48)$$

where, as before, $s < t$. We multiply both sides by a smooth test function $\varphi(z)$ and integrate over z, to give (after again relabelling z by x on the left-hand side)

$$\int_0^\infty p(x, t + \Delta t \mid y, s)\, \varphi(x)\, dx = \int_0^\infty \left[\int_0^\infty p(z, t + \Delta t \mid x, t)\, \varphi(z)\, dz \right] p(x, t \mid y, s)\, dx.$$

The key step is now to observe that we may write

$$\int_0^\infty p(z, t + \Delta t \mid x, t)\, \varphi(z)\, dz$$

$$= \int_0^\infty \bar{p}(z, t + \Delta t \mid x, t)\, \varphi(z)\, dz + \int_{-\infty}^0 \bar{p}(z, t + \Delta t \mid x, t)\, \varphi(-z)\, dz, \qquad (3.49)$$

where $\bar{p}(z, t+\Delta t \mid x, t)$ is the transition probability of the *unconstrained* process, i.e. that given by steps (a7)–(b7), without the reflective step (c7). The first integral on the right-hand side of (3.49) corresponds to those values of $X(t+\Delta t)$ that remained positive, while the second corresponds to those that need to be reflected (so that φ is evaluated at $-z$ rather than z). Thus, if we impose that our test function has a smooth even extension, so that $\varphi(z) = \varphi(-z)$, then we may write

$$\int_0^\infty p(x, t + \Delta t \mid y, s)\, \varphi(x)\, dx = \int_0^\infty \left[\int_{-\infty}^\infty \bar{p}(z, t + \Delta t \mid x, t)\, \varphi(z)\, dz \right] p(x, t \mid y, s)\, dx.$$

Taylor expanding $\varphi(z)$ around the point x we can evaluate the resulting integrals exactly as was done in equation (3.41) before (since they involve $\bar{p}(z, t+\Delta t \mid x, t)$ and not $p(z, t+\Delta t \mid x, t)$), to give, after taking the limit $\Delta t \to 0$,

$$\int_0^\infty \frac{\partial}{\partial t} p(x, t \mid y, s)\, \varphi(x)\, dx$$

$$= \int_0^\infty \varphi'(x)\, f(x, t)\, p(x, t \mid y, s)\, dx + \int_0^\infty \varphi''(x)\, \frac{g(x, t)^2}{2}\, p(x, t \mid y, s)\, dx.$$

Now using integration by parts, noting that $\varphi(x) = \varphi(-x)$ implies $\varphi'(0) = 0$, we find

$$\int_0^\infty \frac{\partial}{\partial t} p(x,t \mid y,s) \varphi(x) \, dx$$

$$= -\int_0^\infty \varphi(x) \frac{\partial}{\partial x} \Big(f(x,t) \, p(x,t \mid y,s) \Big) \, dx$$

$$+ \int_0^\infty \varphi(x) \frac{\partial^2}{\partial x^2} \left(\frac{g(x,t)^2}{2} \, p(x,t \mid y,s) \right) dx$$

$$+ \left[\phi(x) f(x,t) \, p(x,t \mid y,s) - \phi(x) \frac{\partial}{\partial x} \left(\frac{g(x,t)^2}{2} \, p(x,t \mid y,s) \right) \right]_{x=0}^{x=\infty}.$$

Since the test function $\varphi(x)$ is arbitrary, the integral terms imply that $p(x,t \mid y,s)$ must again satisfy the Fokker–Planck equation (3.47). However, since $\varphi(0)$ is arbitrary, the remaining terms imply that $p(x,t \mid y,s)$ must also satisfy the boundary condition

$$f(x,t) \, p(x,t \mid y,s) - \frac{\partial}{\partial x} \left(\frac{g(x,t)^2}{2} \, p(x,t \mid y,s) \right) = 0 \quad \text{on } x = 0. \tag{3.50}$$

The Fokker–Planck equation (3.47) may be thought of as a conservation equation for probability density: it may be written as

$$\frac{\partial}{\partial t} p(x,t \mid y,s) + \frac{\partial Q}{\partial x} = 0,$$

where Q is the probability flux given by

$$Q = f(x,t) \, p(x,t \mid y,s) - \frac{\partial}{\partial x} \left(\frac{g(x,t)^2}{2} \, p(x,t \mid y,s) \right).$$

We see that the boundary condition (3.50) corresponds to $Q(0) = 0$ as we would expect: there is no flux of probability out of the domain since trajectories are reflected at the boundary.

Reflective boundary conditions will be further discussed in Section 4.1, where we study molecular diffusion described by SDEs in a finite domain. Other boundary conditions considered in this book include adsorbing boundaries in Section 4.4, partially adsorbing boundaries in Section 4.5 or periodic boundary conditions in Section 8.6. They can be specified using their algorithmic implementations (as was done in the SSA (a7)–(c7)) or as boundary conditions for the corresponding probability distribution functions (as was done in equation (3.50)). For example, if $\{x = 0\}$ is an adsorbing boundary, then the boundary condition will simply be $p(0,t \mid y,s) = 0$: there is no chance to find a trajectory with $X(t) = 0$ because it is immediately terminated. In this case the integral of $p(x,t)$ over the domain is not conserved but decays in time: there is an increasing probability that the trajectory has left the domain.

3.5 Kolmogorov Backward Equation

Sometimes we want to know how the likelihood of ending up in a given state depends on the starting state. In such cases the end position is known, and the starting position is undetermined: in a sense we are thinking about the problem backwards. It is then very useful to be able to see how the conditional probability distribution function $p(x, t \mid y, s)$ depends on y and s. Using a slightly modified approach to that which led to the Fokker–Planck equation, we can derive an evolution equation for the distribution function $p(x, t \mid y, s)$ in terms of the initial time s and initial value y. To do that, we rename variables in the Chapman–Kolmogorov equation (3.40) to obtain

$$p(x, t \mid y, s - \Delta s) = \int_{\mathbb{R}} p(x, t \mid z, s)\, p(z, s \mid y, s - \Delta s)\, dz. \tag{3.51}$$

As before, this equation is valid for any Δs, but we will now take the limit in which $\Delta s \to 0$. Using the Taylor expansion about the point $z = y$ we rewrite the term $p(x, t \mid z, s)$ as

$$p(x, t \mid z, s) = p(x, t \mid y, s) + (z - y)\frac{\partial p}{\partial y}(x, t \mid y, s)$$
$$+ \frac{(z - y)^2}{2}\frac{\partial^2 p}{\partial y^2}(x, t \mid y, s) + o((z - y)^2).$$

Substituting into the right-hand side of (3.51) gives

$$p(x, t \mid y, s - \Delta s) = p(x, t \mid y, s) \times \int_{\mathbb{R}} p(z, s \mid y, s - \Delta s)\, dz$$
$$+ \frac{\partial p}{\partial y}(x, t \mid y, s) \times \int_{\mathbb{R}} (z - y)p(z, s \mid y, s - \Delta s)\, dz$$
$$+ \frac{\partial^2 p}{\partial y^2}(x, t \mid y, s) \times \int_{\mathbb{R}} \frac{(z - y)^2}{2} p(z, s \mid y, s - \Delta s)\, dz + O(\Delta s^2).$$

Using (3.42), (3.44) and (3.46), we obtain

$$\frac{p(x, t \mid y, s - \Delta s) - p(x, t \mid y, s)}{\Delta s}$$
$$= f(y, s)\frac{\partial p}{\partial y}(x, t \mid y, s) + \frac{g(y, s)^2}{2}\frac{\partial^2 p}{\partial y^2}(x, t \mid y, s) + O(\Delta s).$$

Passing to the limit $\Delta s \to 0$, we derive the so-called Kolmogorov backward equation

$$-\frac{\partial p}{\partial s}(x, t \mid y, s) = f(y, s)\frac{\partial p}{\partial y}(x, t \mid y, s) + \frac{g(y, s)^2}{2}\frac{\partial^2 p}{\partial y^2}(x, t \mid y, s). \tag{3.52}$$

Both the Fokker–Planck equation (3.47) and the Kolmogorov backward equation (3.52) provide an exact description of $p(x, t \mid y, s)$ corresponding to the SDE (3.6). To simplify the notation, we define the *diffusion coefficient* by

$$d(y, s) = \frac{g(y, s)^2}{2}.\qquad(3.53)$$

Then the Fokker–Planck equation (3.31) is

$$\frac{\partial p}{\partial t}(x, t) = \frac{\partial^2}{\partial x^2}\big(d(x, t)\, p(x, t)\big) - \frac{\partial}{\partial x}\big(f(x, t)\, p(x, t)\big),\qquad(3.54)$$

the Kolmogorov backward equation (3.52) can be rewritten as

$$-\frac{\partial p}{\partial s}(x, t \mid y, s) = f(y, s)\frac{\partial p}{\partial y}(x, t \mid y, s) + d(y, s)\frac{\partial^2 p}{\partial y^2}(x, t \mid y, s)\qquad(3.55)$$

and the stationary distribution (3.36) is given by

$$p_s(x) = \frac{C}{d(x)}\exp\left[\int_0^x \frac{f(y)}{d(y)}\,dy\right],\qquad(3.56)$$

where the normalization constant is

$$C = \left(\int_{\mathbb{R}} \frac{1}{d(x)}\exp\left[\int_0^x \frac{f(y)}{d(y)}\,dy\right]dx\right)^{-1}.$$

3.6 SDEs with Multiple Favourable States

The SDE (3.26) is a simple example of an SDE with two favourable states. As seen in Figure 3.1(d), the value of $X(t)$ stays either close to x_{s1} or close to x_{s2}, where x_{s1} and x_{s2} (together with x_u) are given by (3.29). One of the most important characteristics of a system with multiple favourable states is the average time it takes to switch between two of those states. Having the SDE example (3.26) in mind, we focus on the switch from x_{s1} to x_{s2}. While it is clear to the eye from looking at Figure 3.1(d) which state the system is in at any given time, it is not immediately clear how to define *mathematically* when the system is in a given state. In the deterministic version of the equation the system will sit exactly at $x = x_{s1}$ or $x = x_{s2}$, but in the stochastic model the system will jump around near one state or the other, and will rarely be *exactly* at x_{s1} or x_{s2}. We could define a neighbourhood of each state, and define the system to be in state 1 when it lies in the neighbourhood of x_{s1}. We would hope that the transition time was not too sensitive to how large we chose the neighbourhood, which would be the case if the noise is not too large. Here we adopt a slightly different view. We denote by $\bar{\tau}$ the average time for simulation to reach the point x_u provided that it started at $X(0) = x_{s1}$. We have seen that this is the point that separates the deterministic trajectories from those which will settle in x_{s1} and those which will settle in x_{s2}. If the simulation reaches x_u, there is a 50% chance that it will return back to x_{s1} and a 50% chance to continue on to x_{s2}. Thus the mean switching time from x_{s1} to x_{s2} is simply $2\bar{\tau}$.

Using many realizations of the SSA (a6)–(b6), we can estimate $\bar{\tau}$ as follows. We start each realization of the SSA (a6)–(b6) at $X(0) = x_{s1}$ and we wait until the first time the simulation leaves the interval $(-\infty, x_u)$. We compute $\bar{\tau}$ as an average of the simulation time over all realizations of the SSA (a6)–(b6). Using the parameter values from Figure 3.1(d), $\Delta t = 10^{-3}$ and averaging over 10^5 realizations, we obtain that $\bar{\tau}$ is approximately equal to 64.7. We denote this estimate of $\bar{\tau}$ as $\bar{\tau}_{sim} = 64.7$.

Next, we derive an analytical formula for $\bar{\tau}$. This formula will give us the value of $\bar{\tau}$ without the necessity of performing (computationally intensive) stochastic simulation. We will derive a general formula for any SDE of the form (3.6) where $f(x, t) \equiv f(x)$ and $g(x, t) \equiv g(x)$. Let $h(y, t)$ be the probability that $X(t') \in (-\infty, x_u)$ for all times $0 < t' < t$ given that it started at $X(0) = y \in (-\infty, x_u)$. Then

$$h(y, t) = \int_{-\infty}^{x_u} p(x, t \mid y, 0)\, dx, \qquad (3.57)$$

where $p(x, t \mid y, s)\, dx$ represents the probability that the trajectory remains in $(-\infty, x_u)$ and lies in the interval $[x, x + dx)$ at time t given that it started at position y at time $s < t$. Distribution p satisfies the Fokker–Planck and Kolmogorov backward equations with the boundary conditions $p(x_u, t \mid y, s) = p(x, t \mid x_u, s) = 0$, so that $p(x, t \mid y, s) = 0$ if $y \geq x_u$ or $x \geq x_u$. Since we assume that the coefficients $f(x)$ and $g(x)$ of the SDE (3.6) do not depend explicitly on time, we can shift time in p in (3.57) to give

$$h(y, t) = \int_{-\infty}^{x_u} p(x, 0 \mid y, -t)\, dx. \qquad (3.58)$$

Using the transformation s to $-t$ in the Kolmogorov backward equation (3.55), we obtain

$$\frac{\partial p}{\partial t}(x, 0 \mid y, -t) = f(y)\frac{\partial p}{\partial y}(x, 0 \mid y, -t) + d(y)\frac{\partial^2 p}{\partial y^2}(x, 0 \mid y, -t),$$

where the diffusion coefficient $d(y)$ is defined by (3.53). Integrating over x and using (3.58), we obtain

$$\frac{\partial h}{\partial t}(y, t) = f(y)\frac{\partial h}{\partial y}(y, t) + d(y)\frac{\partial^2 h}{\partial y^2}(y, t). \qquad (3.59)$$

Let $\tau(y)$ be the average time to first leave the interval $(-\infty, x_u)$ given that initially $X(0) = y$. The probability that X first leaves the interval $(-\infty, x_u)$ during the time interval $[t, t + dt)$ (under the condition that $X(0) = y$) is given by

$$h(y, t) - h(y, t + dt) \approx -\frac{\partial h}{\partial t}(y, t) \, dt.$$

Thus $\tau(y)$ can be computed as follows:

$$\tau(y) = -\int_0^\infty t \frac{\partial h}{\partial t}(y, t) \, dt = \int_0^\infty h(y, t) \, dt,$$

where we have used the integration by parts on the right-hand side. Integrating equation (3.59) over t and using the fact that $h(y, 0) = 1$ (since $y \in (-\infty, x_u)$) and $h(y, \infty) = 0$ (since the trajectory must cross x_u eventually), we obtain

$$-1 = f(y)\frac{d\tau}{dy}(y) + d(y)\frac{d^2\tau}{dy^2}(y) \qquad \text{for } y \in (-\infty, x_u). \tag{3.60}$$

To solve this equation we need to impose some boundary conditions. If we start the trajectory further and further to the left we expect that the exit time does not depend on the starting position: the trajectories in this region are far from equilibrium and are dominated by the deterministic potential. Thus we impose

$$\frac{d\tau}{dy}(-\infty) = 0. \tag{3.61}$$

The second boundary condition comes from the fact that $p(x, t \mid x_u, s) = 0$, which implies $h(x_u, t) = 0$, which gives

$$\tau(x_u) = 0. \tag{3.62}$$

This is to be expected: if the trajectory starts at $x = x_u$ it leaves the region $(-\infty, x_u)$ in zero time. Using an integrating factor to integrate (3.60) subject to the condition (3.61) we obtain

$$\frac{d\tau}{dy}(y) = -\exp\left[-\int_0^y \frac{f(z)}{d(z)} \, dz\right]\int_{-\infty}^y \frac{1}{d(x)} \exp\left[\int_0^x \frac{f(z)}{d(z)} \, dz\right] dx$$

$$= -\frac{1}{d(y)p_s(y)} \int_{-\infty}^y p_s(x) \, dx,$$

where the stationary distribution p_s is given by (3.56). Now integrating over y and using (3.62) we obtain

$$\tau(y) = \int_y^{x_u} \frac{1}{d(z)p_s(z)} \int_{-\infty}^z p_s(x) \, dx \, dz. \tag{3.63}$$

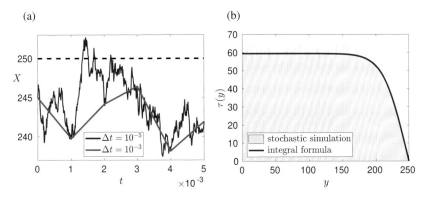

Figure 3.3 (a) *The trajectory of (3.26) computed by the SSA (a6)–(b6) for* $\Delta t =$ 10^{-5} *(blue line). The coarser trajectory with* $\Delta t = 10^{-3}$ *(red line). (b) The exit time* $\tau(y)$ *obtained by the integral formula (3.63) (blue line) and by the SSA (a6)–(b6) (yellow histogram). The parameter values are* $k_1 = 10^{-3}$, $k_2 = 0.75$, $k_3 = 165$, $k_4 = 10^4$ *and* $k_5 = 200$.

Thus $\bar{\tau}$ can be computed as

$$\bar{\tau} = \tau(x_{s1}) = \int_{x_{s1}}^{x_u} \frac{1}{d(z)p_s(z)} \int_{-\infty}^{z} p_s(x)\,\mathrm{d}x\,\mathrm{d}z. \tag{3.64}$$

The stationary distribution p_s for the SDE example (3.26) was computed as (3.39). Substituting p_s into (3.64) and evaluating the resulting integrals numerically, we obtain $\bar{\tau} = 59.45$. We see that there is an error (approximately 9%) between the theoretical value of $\bar{\tau}$ and the value $\bar{\tau}_{sim} = 64.7$ estimated previously using the SSA (a6)–(b6). The main contribution to the error is due to the fact that we simulated the SDE (3.26) using the SSA (a6)–(b6) with the finite time step $\Delta t = 10^{-3}$. Decreasing the time step Δt improves the accuracy of the SSA (a6)–(b6). In Figure 3.3(a), we plot an example of the stochastic trajectory with $\Delta t = 10^{-5}$ (blue line). This trajectory crosses the boundary $x_u = 250$. However, if we consider the same trajectory using the coarser time step $\Delta t = 10^{-3}$ (red line), then the coarser trajectory never leaves the domain $(-\infty, x_u)$. This explains why the SSA (a6)–(b6) with $\Delta t = 10^{-3}$ overestimates the value of $\bar{\tau}$. The stochastic trajectories (which should be terminated) stay in the domain $(-\infty, x_u)$ for a longer time if the time step is larger.

Decreasing the time step Δt is one possible way to improve the accuracy of the estimation of $\bar{\tau}$ by the SSA (a6)–(b6). However, it makes the simulation computationally intensive. Another option is to consider that there is a nonzero probability that the simulation left the domain $(-\infty, x_u)$ during the time interval $(t, t + \Delta t)$ even if $X(t + \Delta t) < x_u$ and $X(t) < x_u$. To estimate this probability we suppose that during this time interval the particle is diffusing only, with a diffusion constant $D = k_5^2/2$. This is a good approximation close to the point x_u, where the drift coefficient is zero. We can then estimate the desired probability as $\exp[-(X(t)-x_u)(X(t+\Delta t)-x_u)/(D\Delta t)]$ (see Exercise 3.6). Thus we can estimate $\bar{\tau}$ by executing many realizations of the SSA (a6)–(b6) as follows. We start each realization of the SSA (a6)–(b6) at $X(0) = x_{s1}$. If $X(t + \Delta t) \geq x_u$, we terminate the trajectory. If $X(t + \Delta t) < x_u$, then we generate a random number uniformly distributed in $(0, 1)$. If this number is less than $\exp[-(X(t)-x_u)(X(t+\Delta t)-x_u)/(D\Delta t)]$, we terminate the trajectory. We compute $\bar{\tau}$ as an average of the simulation time over all realizations. Using the parameter values from Figure 3.1(d), $\Delta t = 10^{-3}$ and averaging over 10^5 realizations, we obtain that $\bar{\tau}$ is approximately equal to 59.2, which compares well with the theoretical value $\bar{\tau}$ obtained by formula (3.64).

In Figure 3.3(b), we plot the function $\tau(y)$ given by the formula (3.63). It compares well with the results obtained by stochastic simulation (yellow histogram). To get the yellow histogram, we estimated $\tau(y)$ for each integer value y as an average of 1000 realizations of the SSA (a6)–(b6). We used $\Delta t = 10^{-3}$ and we computed the simulation results with the help of the correction probability $\exp[-(X(t) - x_u)(X(t + \Delta t) - x_u)/(D\Delta t)]$ just described.

3.7 Chemical Fokker–Planck Equation

In this section, we show how SDEs such as (3.6) can be used to give us an approximate description of chemical systems. The methods developed in this chapter can then be used to determine, for example, the mean switching time between equilibrium states of a chemical system.

Let us reconsider example (1.9) from Section 1.2, that is, we consider a chemical species A in a container of volume v which is subject to the following two chemical reactions:

$$A \xrightarrow{k_1} \emptyset, \qquad \emptyset \xrightarrow{k_2} A. \qquad (3.65)$$

The chemical master equation for this problem is given by (1.11), and is an infinite system of ODEs for $p_n(t)$, $n = 0, 1, 2, 3, \ldots$, where $p_n(t)$ is the probability that $A(t) = n$. We can rewrite (1.11) as

$$\frac{dp_n}{dt}(t) = h_1(n + 1, t) - h_1(n, t) + h_2(n - 1, t) - h_2(n, t), \qquad (3.66)$$

where

$$h_1(n, t) = k_1 n\, p_n(t), \qquad h_2(n, t) = k_2 v\, p_n(t). \qquad (3.67)$$

Now let us suppose that most of the action occurs for n large. We quantify this by introducing a fixed large number ω, so that we are interested in the values of p_n for n around the same size as ω. We now write $n = \eta\omega$ and $p_n(t) = p(\eta, t)$, treating η as a continuous variable. The same change of variables in (3.67) gives the real-valued functions

$$h_1(\eta, t) = k_1 \eta \omega \, p(\eta, t), \qquad h_2(\eta, t) = k_2 v \, p(\eta, t) \tag{3.68}$$

and (3.66) reads as follows:

$$\frac{\partial p}{\partial t}(\eta, t) = h_1\left(\eta + \frac{1}{\omega}, t\right) - h_1(\eta, t) + h_2\left(\eta - \frac{1}{\omega}, t\right) - h_2(\eta, t). \tag{3.69}$$

Since ω is large we can approximate the right-hand side by a Taylor series about the point (η, t):

$$h_1\left(\eta + \frac{1}{\omega}, t\right) = h_1(\eta, t) + \frac{1}{\omega}\frac{\partial h_1}{\partial \eta}(\eta, t) + \frac{1}{2\omega^2}\frac{\partial^2 h_1}{\partial \eta^2}(\eta, t) + O\left(\frac{h_1}{\omega^3}\right),$$

$$h_2\left(\eta - \frac{1}{\omega}, t\right) = h_2(\eta, t) - \frac{1}{\omega}\frac{\partial h_2}{\partial \eta}(\eta, t) + \frac{1}{2\omega^2}\frac{\partial^2 h_2}{\partial \eta^2}(\eta, t) + O\left(\frac{h_2}{\omega^3}\right).$$

Substituting into (3.69), and truncating at $O(1/\omega^2)$, we get

$$\frac{\partial p}{\partial t}(\eta, t) = \frac{1}{\omega}\frac{\partial h_1}{\partial \eta}(\eta, t) + \frac{1}{2\omega^2}\frac{\partial^2 h_1}{\partial \eta^2}(\eta, t) - \frac{1}{\omega}\frac{\partial h_2}{\partial \eta}(\eta, t) + \frac{1}{2\omega^2}\frac{\partial^2 h_2}{\partial \eta^2}(\eta, t).$$

The rescaling of n with ω when defining η allowed us to asymptotically expand in inverse powers of the large number ω. However, now that we have truncated the expansion it is often more convenient to write the equations in terms of $x = \eta\omega$, which can be thought of as an extension of n to non-integer values. This gives

$$\frac{\partial p}{\partial t}(x, t) = \frac{\partial h_1}{\partial x}(x, t) + \frac{1}{2}\frac{\partial^2 h_1}{\partial x^2}(x, t) - \frac{\partial h_2}{\partial x}(x, t) + \frac{1}{2}\frac{\partial^2 h_2}{\partial x^2}(x, t); \tag{3.70}$$

we need to be careful to remember that this equation is only valid for large values of x. Using (3.68) this can be written as the Fokker–Planck equation

$$\frac{\partial p}{\partial t}(x, t) = \frac{\partial^2}{\partial x^2}\big(d(x)\, p(x, t)\big) - \frac{\partial}{\partial x}\big(f(x)\, p(x, t)\big), \tag{3.71}$$

where the drift and diffusion coefficients are given by

$$f(x) = -k_1 x + k_2 v, \tag{3.72}$$

$$d(x) = (k_1 x + k_2 v)/2. \tag{3.73}$$

Now that we have obtained the Fokker–Planck equation (3.71) for the chemical system (3.65), we can apply the theory we developed earlier in this chapter to it. We found previously that the stationary distribution is given by (3.56). Substituting (3.72)–(3.73) into (3.56), we obtain

$$
\begin{aligned}
p_s(x) &= \frac{2C}{k_1 x + k_2 v} \exp\left[2 \int_0^x \frac{-k_1 y + k_2 v}{k_1 y + k_2 v}\, dy\right] \\
&= \frac{2C}{k_1 x + k_2 v} \exp\left[-2x + 4k_2 v \int_0^x \frac{1}{k_1 y + k_2 v}\, dy\right] \\
&= \frac{2C}{k_1 x + k_2 v} \exp\left[-2x + \frac{4k_2 v}{k_1} \log(k_1 x + k_2 v)\right] \\
&= 2C \exp\left[-2x + \left(\frac{4k_2 v}{k_1} - 1\right) \log(k_1 x + k_2 v)\right].
\end{aligned}
\tag{3.74}
$$

Since x approximates the number of molecules, equation (3.74) is only meaningful for positive values of x, i.e. the normalization constant $2C$ can be obtained by integrating $p_s(x)$ over interval $(0, \infty)$ to get

$$
2C = \left(\int_0^\infty \exp\left[-2x + \left(\frac{4k_2 v}{k_1} - 1\right) \log(k_1 x + k_2 v)\right] dx\right)^{-1}.
$$

The function $p_s(x)$ given by (3.74) is plotted as the red line in Figure 3.4(a). It compares well with the results obtained by the Gillespie SSA (a3)–(d3) (the grey histogram), which we computed earlier in Section 1.2 (see Figure 1.2(b)).

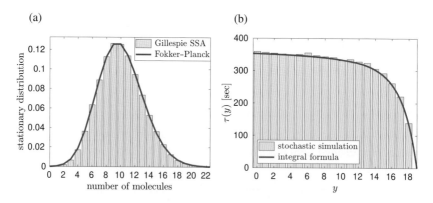

(a)

(b)

Figure 3.4 *System of chemical reactions (3.65) for $k_1 = 0.1\,sec^{-1}$ and $k_2 v = 1\,sec^{-1}$. (a) Stationary distribution obtained by the formula (3.74) (red solid line) and by long-time simulation of the SSA (a3)–(d3) (grey histogram). (b) The mean exit time $\tau(y)$ obtained by the integral formula (3.75) (red solid line) and by the Gillespie SSA (a3)–(d3) (grey histogram).*

Let us see how well the chemical Fokker–Planck equation does at estimating mean transition times. Define $\tau(n)$, $n = 1, 2, \ldots, 18$, to be the average time the Gillespie SSA (a3)–(d3) requires to leave the interval $[0, 18]$ when the initial condition is $A(0) = n$. The values of $\tau(n)$ estimated as the averages of 10^4 realizations (which were started at $A(0) = n$ and continued until the first time t satisfying $A(t) = 19$) are plotted in Figure 3.4(b) as the grey histogram. The number $\tau(n)$ can also be approximated by the formula (3.63) which was derived from the Fokker–Planck equation (3.71). Using $x_u = 19$ in (3.63) (i.e. we are looking for the average time to leave the interval $(-\infty, 19)$ where we postulate $p_s(x) \equiv 0$ on $(-\infty, 0)$), we get

$$\tau(y) = \int_y^{19} \frac{1}{d(z)p_s(z)} \int_0^z p_s(x) \, dx \, dz, \qquad (3.75)$$

where $p_s(z)$ is given by (3.74) and $d(z)$ is given by (3.73). Evaluating the integrals in (3.75) numerically, we present the function $\tau(y)$ in Figure 3.4(b) as the red solid line for comparison. We see that the formula (3.75) is a good approximation of $\tau(n)$.

We derived the chemical Fokker–Planck equation (3.71) by asymptotically expanding the chemical master equation when the number of molecules was large. An alternative approach is to approximate the stochastic process which leads to the chemical master equation directly. Such an approach was carried out by Gillespie (2000), and leads to the so-called chemical Langevin equation, which, for the model problem (3.65), can be written as the SDE

$$X(t + dt) = X(t) + (-k_1 x + k_2 v) \, dt - \sqrt{k_1 x} \, dW_1 + \sqrt{k_2 v} \, dW_2. \qquad (3.76)$$

The idea behind this approximation is similar to that of tau-leaping introduced by Gillespie (2001), which is a numerical method for speeding up the Gillespie algorithm (see Chapter 5). We suppose that there are enough molecules of each species that during the small but finite time step Δt the propensity functions do not change appreciably, even though there are many reactions that take place. Since the reaction rates are then constant for this time interval, the number of times that each reaction occurs is given by a Poisson distribution. If we now assume that each reaction occurs many times then this Poisson distribution is well approximated by a normal distribution. This leads to the equation

$$X(t + \Delta t) = X(t) + (-k_1 x + k_2 v) \Delta t - \sqrt{k_1 x} \sqrt{\Delta t} \, \xi_1 + \sqrt{k_2 v} \sqrt{\Delta t} \, \xi_2, \qquad (3.77)$$

where ξ_1 and ξ_2 are two random numbers which are sampled from the normal distribution with zero mean and unit variance. Equation (3.76) is recovered in the limit $\Delta t \to 0$. In fact, we can consider (3.77) as the computational definition of (3.76). This equation is slightly different from the SDE (3.6) studied previously, in that it has two independent white noises, dW_1 and dW_2. Fortunately, having two independent noise terms in the SDE does not complicate the derivation of the corresponding Fokker–Planck equation. In fact, one can prove (see Exercise 3.7) that the Fokker–Planck equation corresponding to the SDE (3.76) is given by (3.71).

So far we have seen the chemical Fokker–Planck equation only for the specific chemical system (3.65). Let us now consider a general well-stirred mixture of $N \geq 1$ molecular species that chemically interact through $q \geq 1$ chemical reactions R_j, $j = 1, 2, \ldots, q$. The state of this system is described by $\mathbf{X} = [X_1, X_2, \ldots, X_N]$, where $X_i \equiv X_i(t)$ is the number of molecules of the ith chemical species, $i = 1, 2, \ldots, N$. Let $\alpha_j(\mathbf{x})$ be the propensity function of the chemical reaction R_j, $j = 1, 2, \ldots, q$, i.e. $\alpha_j(\mathbf{x}) \, dt$ is the probability that, given $\mathbf{X}(t) = \mathbf{x}$, one R_j reaction will occur in the next infinitesimal time interval $[t, t + dt)$. Let ν_{ji} be the change in the number of X_i by one occurrence of the reaction R_j, $i = 1, 2, \ldots, N$, $j = 1, 2, \ldots, q$. Then the chemical system can be described by the chemical master equation

$$\frac{\partial}{\partial t} p(\mathbf{x}, t) = \sum_{j=1}^{q} \left[\alpha_j(\mathbf{x} - \boldsymbol{\nu}_j) \, p(\mathbf{x} - \boldsymbol{\nu}_j, t) - \alpha_j(\mathbf{x}) \, p(\mathbf{x}, t) \right], \qquad (3.78)$$

where $p(\mathbf{x}, t)$ is the probability that $\mathbf{X}(t) = \mathbf{x}$. Then the chemical Langevin equation can be written in the following form (Gillespie, 2000):

$$dX_i = \left(\sum_{j=1}^{q} \nu_{ji} \alpha_j(\mathbf{X}(t)) \right) dt + \sum_{j=1}^{q} \nu_{ji} \sqrt{\alpha_j(\mathbf{X}(t))} \, dW_j, \qquad (3.79)$$

which corresponds to the chemical Fokker–Planck equation

$$\frac{\partial}{\partial t} p(\mathbf{x}, t) = \sum_{i=1}^{N} \frac{\partial}{\partial x_i} \left[-\left(\sum_{j=1}^{q} \nu_{ji} \alpha_j(\mathbf{x}) \right) p(\mathbf{x}, t) \right]$$

$$+ \frac{1}{2} \sum_{i=1}^{N} \frac{\partial^2}{\partial x_i^2} \left[\left(\sum_{j=1}^{q} \nu_{ji}^2 \alpha_j(\mathbf{x}) \right) p(\mathbf{x}, t) \right] \qquad (3.80)$$

$$+ \sum_{i=1}^{N} \sum_{k=1}^{i-1} \frac{\partial^2}{\partial x_i \partial x_k} \left[\left(\sum_{j=1}^{q} \nu_{ji} \nu_{jk} \alpha_j(\mathbf{x}) \right) p(\mathbf{x}, t) \right].$$

3.8 Analysis of Problem from Section 2.1

In Section 2.1 we considered the chemical A in the container of volume v which was subject to the following four chemical reactions (see also (2.1)):

$$3A \underset{k_2}{\overset{k_1}{\rightleftarrows}} 2A, \qquad\qquad A \underset{k_4}{\overset{k_3}{\rightleftarrows}} \emptyset. \qquad (3.81)$$

We simulated (3.81) using the Gillespie SSA and we found out that the system randomly switches between its two favourable states. The SDE theory presented in the previous sections will help us to further analyse this model. In particular we will use the chemical Fokker–Planck equation to find an approximate formula for the mean switching time between these states.

The propensity functions of chemical reactions (3.81) are given by (see Table 1.1 and Table 1.2)

$$\alpha_1(x) = x(x-1)(x-2)k_1/v^2, \qquad \alpha_2(x) = x(x-1)k_2/v,$$
$$\alpha_3(x) = xk_3, \qquad\qquad\qquad \alpha_4(x) = k_4v.$$

Substituting the propensities into (3.80), we obtain the chemical Fokker–Planck equation

$$\frac{\partial p}{\partial t}(x, t) = \frac{\partial^2}{\partial x^2}\Big(d(x)\,p(x, t)\Big) - \frac{\partial}{\partial x}\Big(f(x)\,p(x, t)\Big), \qquad (3.82)$$

where the drift and diffusion coefficients are given by

$$f(x) = -\alpha_1(x) + \alpha_2(x) - \alpha_3(x) + \alpha_4(x) \qquad (3.83)$$
$$= -x(x-1)(x-2)k_1/v^2 + x(x-1)k_2/v - xk_3 + k_4v,$$
$$d(x) = [\alpha_1(x) + \alpha_2(x) + \alpha_3(x) + \alpha_4(x)]/2 \qquad (3.84)$$
$$= [x(x-1)(x-2)k_1/v^2 + x(x-1)k_2/v + xk_3 + k_4v]/2.$$

As in Section 2.1, we choose the rate constants as follows: $k_1/v^2 = 2.5 \times 10^{-4}\,\text{min}^{-1}$, $k_2/v = 0.18\,\text{min}^{-1}$, $k_3 = 37.5\,\text{min}^{-1}$ and $k_4v = 2200\,\text{min}^{-1}$. The stationary distribution is given by (3.56). Substituting (3.83)–(3.84) into (3.56) and evaluating the resulting integral numerically, we obtain an approximation of the stationary distribution, which is plotted in Figure 3.5(a) as the blue solid line. We also plot the stationary distribution (yellow histogram) obtained by long-time simulation of the Gillespie SSA (a5)–(d5), which we have already seen in Figure 2.2(b).

The stationary distribution $p_s(x)$ has two local maxima $x_{s1} = 95$ and $x_{s2} = 392$. The local minimum between them is attained at the point $x_u = 235$. Let $\tau(n)$ be the average time to reach the point x_u given that the simulation

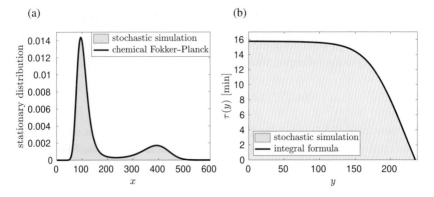

Figure 3.5 (a) *Stationary distribution of the chemical system* (3.81) *obtained by the Gillespie SSA (yellow histogram) for parameters* $k_1/v^2 = 2.5 \times 10^{-4}\,min^{-1}$, $k_2/v = 0.18\,min^{-1}$, $k_3 = 37.5\,min^{-1}$ *and* $k_4v = 2200\,min^{-1}$. *The blue solid line shows* $p_s(x)$ *computed by* (3.56) *where* $f(x)$ *and* $d(x)$ *are given by* (3.83)– (3.84). (b) *Mean exit time from the interval* $[0, 235]$ *computed by the Gillespie SSA (yellow histogram) and by the formula* (3.85) *(blue solid line).*

started at $X(0) = n$. The values of $\tau(n)$, $n = 0, 1, 2, \ldots, 234$, estimated as averages over 10^4 exits are given in Figure 3.5(b) (yellow histogram). Using (3.63), we can estimate $\tau(n)$ by the integral

$$\tau(y) = \int_y^{x_u} \frac{1}{d(z)p_s(z)} \int_0^z p_s(x)\,dx\,dz, \tag{3.85}$$

where $p_s(z)$ is given by (3.56) and $d(z)$ by (3.84). Since we have already computed $p_s(z)$ (given in Figure 3.5(a)), we can evaluate the integral (3.85) numerically. The resulting function $\tau(y)$ is plotted in Figure 3.5(b) as the blue solid line. It compares well with the results of stochastic simulation.

We are mostly interested in the mean transition time from x_{s1} to x_{s2}. If the simulation reaches x_u, there is a 50% chance to return back to x_{s1} and a 50% chance to continue to x_{s2}. Thus the mean switching time from x_{s1} to x_{s2} is simply $2\tau(x_{s1})$ where $\tau(x_{s1})$ is given by (3.64). Using our parameter values we find $\tau(x_{s1}) = \tau(95) = 15.6\,min$.

Before we close this section, we introduce one more approximation which is occasionally useful for systems with two favourable states such as (3.81). The formula (3.64) is an exact expression for the mean exit time (given the SDE (3.26), which, as we have seen, may itself be an approximation). We can approximate (3.64) if necessary as follows. First we define the potential

$$\Phi(x) = -\int_0^x \frac{f(y)}{d(y)}\,dy + \log(d(x)), \tag{3.86}$$

so that (3.56) implies $p_s(x) = C \exp[-\Phi(x)]$. Then x_{s1} is a local minimum of Φ, and x_u is a local maximum of Φ. Rewritten in terms of Φ and restricting to positive values of x, formula (3.64) is

$$\tau(x_{s1}) = \int_{x_{s1}}^{x_u} \frac{\exp[\Phi(z)]}{d(z)} \int_0^z \exp[-\Phi(x)]\, dx\, dz. \tag{3.87}$$

Now, if Φ is large, then the exponential $\exp[-\Phi(x)]$ is dominated by the region near x_{s1} where Φ is minimum, while $\exp[\Phi(x)]$ is dominated by the region near x_u where Φ is maximum. In this case we introduce only a small error when we replace the integration interval $(0, z)$ of the inner integral by $(0, x_u)$, since the main contribution to the inner integral comes from the region close to x_{s1}. Thus we can approximate

$$\tau(x_{s1}) \approx \left(\int_{x_{s1}}^{x_u} \frac{\exp[\Phi(z)]}{d(z)}\, dz \right) \left(\int_0^{x_u} \exp[-\Phi(x)]\, dx \right). \tag{3.88}$$

The main contribution to the first integral comes from the region close to x_u, while the main contribution to the second integral comes from the region close to x_{s1}. We can now use Laplace's method to evaluate these integrals. This involves Taylor expanding the exponents about x_u and x_{s1} as

$$\Phi(z) \approx \Phi(x_u) - (z - x_u)^2\ |\Phi''(x_u)|\ /2,$$
$$\Phi(x) \approx \Phi(x_{s1}) + (x - x_{s1})^2\ \Phi''(x_{s1})/2,$$

remembering that $\Phi'(x_{s1}) = \Phi'(x_u) = 0$, while $\Phi''(x_{s1}) > 0$, $\Phi''(x_u) < 0$. Substituting these into (3.88) we introduce only small errors when we replace the lower limit of the first integral by $-\infty$ and the domain of integration of the second integral to $\mathbb{R}$, to give

$$\tau(x_{s1}) \approx \frac{1}{d(x_u)} \int_{-\infty}^{x_u} \exp\left[\Phi(x_u) - \frac{1}{2}(z - x_u)^2\ |\Phi''(x_u)| \right] dz$$
$$\times \int_{\mathbb{R}} \exp\left[-\Phi(x_{s1}) - \frac{1}{2}(x - x_{s1})^2\ \Phi''(x_{s1}) \right] dx.$$

Using $\int_{-\infty}^{\infty} \exp[-x^2]\, dx = \sqrt{\pi}$, the integrals above can be evaluated explicitly to obtain the so-called Kramers approximation of the mean exit time (Hänggi et al., 1990)

$$\tau(x_{s1}) \approx \frac{\pi\ \exp[\Phi(x_u) - \Phi(x_{s1})]}{d(x_u)\ \sqrt{|\Phi''(x_u)|\Phi''(x_{s1})}}.$$

For the parameter values of this section, the Kramers approximation gives $\tau(x_{s1}) \approx 11.8\,\mathrm{min}$, which is of the same order of magnitude as the exact value $\tau(x_{s1}) = 15.6\,\mathrm{min}$.

3.9 Analysis of Problem from Section 2.2

In Section 2.2 we studied a chemical system consisting of two chemical species A and B in the volume v subject to the following four chemical reactions (see also (2.9)):

$$2A + B \xrightarrow{k_1} 3A, \qquad \emptyset \underset{k_3}{\overset{k_2}{\rightleftharpoons}} A, \qquad \emptyset \xrightarrow{k_4} B. \qquad (3.89)$$

Using the parameter values (2.14), we observed that the solution of the deterministic ODE model (2.12)–(2.13) converged to a steady state while the stochastic model had oscillatory solutions (see Figure 2.5(a)). The mean period of oscillations can be estimated (using the Gillespie SSA (a5)–(d5)) as $\tau_{per} \doteq 7.33 \min \doteq 7 \min 20 \sec$, which is an average over 10^5 periods. Actually, in giving this estimate we have to define what we mean by an oscillation, since in any given realization there are many small excursions due to the noise. For the estimate above, we start the computation of the period when A first becomes greater than 10^3; then we wait until $A < 20$ before we test the condition $A > 10^3$ again. The condition $A > 10^3$ ensures that the trajectory is far from the steady state of the deterministic model, while the condition $A < 20$ guarantees that the random fluctuations around the point $A = 10^3$ are not counted as two or more periods. The main goal of this section is to show that we can get a reasonable estimate of the period of oscillations by analysing the chemical Fokker–Planck equation.

Let $p(x, y, t)$ be the probability that $A(t) = x$ and $B(t) = y$. As before, we smoothly extend the values of $p(x, y, t)$ to non-integer values of x and y. The propensity functions of the chemical reactions (3.89) are

$$\alpha_1(x, y) = x(x - 1)y\, k_1 / v^2, \qquad\qquad \alpha_2(x, y) = k_2 v,$$
$$\alpha_3(x, y) = x k_3, \qquad\qquad\qquad\qquad \alpha_4(x, y) = k_4 v,$$

where x corresponds to number of A molecules and y to number of B molecules. Substituting the propensities into (3.80), we obtain a two-dimensional chemical Fokker–Planck equation

$$\frac{\partial}{\partial t} p(x, y, t) = \frac{\partial}{\partial x} \Big[\big(-x(x - 1)y\, k_1 / v^2 - k_2 v + x k_3 \big)\, p(x, y, t) \Big]$$
$$+ \frac{\partial}{\partial y} \Big[\big(x(x - 1)y\, k_1 / v^2 - k_4 v \big)\, p(x, y, t) \Big]$$
$$+ \frac{1}{2} \frac{\partial^2}{\partial x^2} \Big[\big(x(x - 1)y\, k_1 / v^2 + k_2 v + x k_3 \big)\, p(x, y, t) \Big]$$

$$+ \frac{1}{2}\frac{\partial^2}{\partial y^2}\left[(x(x-1)y\,k_1/v^2 + k_4v)\,p(x,y,t)\right]$$

$$- \frac{\partial^2}{\partial x\partial y}\left[x(x-1)y\,k_1/v^2\,p(x,y,t)\right]. \tag{3.90}$$

This equation is more complicated to analyse and solve than the one-dimensional Fokker–Planck equations we have previously studied. Our main goal is to estimate the mean period of oscillations. We could get this estimate by systematically analysing (3.90), as it has been shown by Erban et al. (2009). However, we will adopt a more intuitive approach. We see from Figure 2.5(a) that the number of B molecules is slowly growing while the number of A molecules fluctuates rapidly. We will use this difference in time scales to further approximate the system. Let us suppose that the period of the oscillation is controlled by the fluctuations in A.

We begin by fixing the number of B molecules as $B = b$ where b is a constant. This simplifies the system of equations (3.89) to

$$2A \xrightarrow{\;k_1 b/v\;} 3A, \qquad \emptyset \underset{k_3}{\overset{k_2}{\rightleftarrows}} A. \tag{3.91}$$

Using (3.80), the chemical Fokker–Planck equation corresponding to (3.91) can be written as

$$\frac{\partial p}{\partial t}(x,t;b) = \frac{\partial^2}{\partial x^2}\Big(d(x;b)\,p(x,t;b)\Big) - \frac{\partial}{\partial x}\Big(f(x;b)\,p(x,t;b)\Big), \tag{3.92}$$

where we explicitly include the dependence of $p(x,t)$ on b in the notation $p(x,t;b)$. The drift and diffusion coefficients are given by

$$f(x;b) = (k_1/v^2)\,x(x-1)b + k_2v - k_3x, \tag{3.93}$$

$$d(x;b) = [(k_1/v^2)\,x(x-1)b + k_2v + k_3x]/2. \tag{3.94}$$

Let us suppose that the number of molecules $A(t)$ evolves according to the model (3.91) with fixed b. Let $\tau(x;b)$ be the average time for $A(t)$ to leave the interval $[0,10^3)$ provided that initially $A(0) = x$. Using (3.63), the mean exit time $\tau(x;b)$ can be computed by

$$\tau(x;b) = \int_x^{1000} \frac{1}{d(z;b)} \int_0^z \exp[\Phi(z;b) - \Phi(y;b)]\,dy\,dz, \tag{3.95}$$

where the potential $\Phi(x;b)$ is given by

$$\Phi(x;b) = -\int_0^x \frac{f(y;b)}{d(y;b)}\,dy + \log(d(x;b)). \tag{3.96}$$

The local extrema of the potential $\Phi(x; b)$ can be found as the solutions of $\Phi'(x; b) = 0$, where the prime denotes differentiation with respect to x. Differentiating (3.96), we obtain

$$-f(x; b) + d'(x; b) = 0,$$

which is the quadratic equation

$$x^2 - \left(2 + \frac{k_3 v^2}{k_1 b}\right)x + \frac{k_1 b + 2k_2 v^3 - k_3 v^2}{2k_1 b} = 0.$$

We label the two solutions

$$x_f(b) = 1 + \frac{k_3\, v^2}{2k_1 b} - \frac{1}{2}\sqrt{\left(2 + \frac{k_3 v^2}{k_1 b}\right)^2 - \frac{2k_1 b + 4k_2 v^3 - 2k_3 v^2}{k_1 b}},$$

$$x_u(b) = 1 + \frac{k_3\, v^2}{2k_1 b} + \frac{1}{2}\sqrt{\left(2 + \frac{k_3 v^2}{k_1 b}\right)^2 - \frac{2k_1 b + 4k_2 v^3 - 2k_3 v^2}{k_1 b}}. \tag{3.97}$$

Roughly speaking, the values of A in the interval $(0, x_u(b))$ correspond to fluctuations around the locally favourable state $x_f(b)$ and the values of A larger than $x_u(b)$ correspond to the excited state. Let us consider only values of A in the interval $(0, x_u(b))$. The stationary distribution (3.56) conditioned on the fact that A is in $(0, x_u(b))$ can be written as

$$p_s(x; b) = \exp[-\Phi(x; b)]\left(\int_0^{x_u(b)} \exp[-\Phi(z; b)]\, dz\right)^{-1}. \tag{3.98}$$

Since A evolves much more rapidly than B, from now on we will assume that when $B(t) = b$ and A is in $(0, x_u(b))$ the number of A molecules is distributed according to $p_s(x; b)$. Then the average time to leave the interval $(0, 10^3)$, given that $B(t) = b$, is equal to the average of $\tau(x; b)$ over all possible starting positions x, namely

$$\tau(b) = \int_0^{x_u(b)} p_s(x; b)\, \tau(x; b)\, dx. \tag{3.99}$$

So far we have been treating b as a constant; we now allow it to vary slowly. Since the fluctuations in B are much smaller than those in A, we model the change in b deterministically, averaging the rate of the first reaction in (3.89) over all possible values of A. Then b evolves according to the ODE

$$\frac{db}{dt} = w(b), \tag{3.100}$$

where

$$w(b) = -\frac{k_1\, b}{v^2}\int_0^{x_u(b)} x(x - 1)p_s(x; b)\, dx + k_4\, v. \tag{3.101}$$

In particular, the average time to reach $B(t) = b$ provided that $B(0) = 0$ can be computed (under the condition that the system did not fire) as

$$\int_0^b \frac{1}{w(z)} \, dz. \tag{3.102}$$

Let us now put all this together. For a given b, the probability to leave the A interval $(0, 10^3)$ (i.e. the probability to fire) during the time interval $[t, t + dt)$ is equal to $dt/\tau(b)$. Using (3.100), we conclude that the probability to leave the A interval $(0, 10^3)$ during the infinitesimally small B interval $[b, b + db)$ is equal to

$$\frac{db}{\tau(b)w(b)}$$

and the probability to arrive at b without previously firing is therefore equal to

$$\exp\left[-\int_0^b \frac{dz}{\tau(z)w(z)} \right]$$

(compare to the analysis of degradation in Section 1.1). Thus the probability of *first* firing during the interval $[b, b + db)$ is

$$\exp\left[-\int_0^b \frac{dz}{\tau(z)w(z)} \right] \frac{db}{\tau(b)w(b)}. \tag{3.103}$$

Combining (3.103) with (3.102), we can estimate the period of oscillations as

$$\tau_{per} \approx \int_{1000}^{10\,000} \frac{1}{\tau(b)w(b)} \left(\int_0^b \frac{1}{w(z)} \, dz \right) \exp\left[-\int_0^b \frac{dz}{\tau(z)w(z)} \right] db, \tag{3.104}$$

where $w(b)$ is given by (3.101) and $\tau(b)$ is given by (3.99), and the limits of the b integration come from observing the relevant range of B in Figure 2.5(a). Evaluating this integral numerically, we obtain $\tau_{per} \approx 7.3$ min, which compares well with the mean period of oscillations estimated by the Gillespie SSA.

In this section, we have analysed a system of two chemical species (3.89) which could be described by the two-dimensional chemical Fokker–Planck equation (3.90). We have estimated the mean period of oscillations by using a separation of time scales, which simplifies the problem to solving one-dimensional PDEs. If this simplification was not possible, we could apply numerical methods for solving higher-dimensional PDEs, as has been done in the literature for the chemical Fokker–Planck equation by Sjöberg et al. (2009), Cotter et al. (2013) and Liao et al. (2015). Considering numerical approaches, one could also apply them directly to the original chemical master equation (Munsky and Khammash, 2006; Kazeev et al., 2014). The advantage of PDEs is that we can choose different discretizations in different parts of the domain, while solving the chemical master equation is necessary for reaction systems that are not well approximated by the corresponding chemical Fokker–Planck equation; see works of Grima et al. (2011), Grima (2011), Schnoerr et al. (2014) and Duncan et al. (2015) for further discussion.

Exercises

3.1   Suppose that r_1 and r_2 are two random numbers uniformly distributed in $(0, 1)$. Define

$$\xi_1 = \sqrt{2|\log r_1|}\,\cos(2\pi r_2), \qquad \xi_2 = \sqrt{2|\log r_1|}\,\sin(2\pi r_2).$$

Show that ξ_1 and ξ_2 are random numbers sampled from normal distribution with zero mean and unit variance.

Note: This method for sampling normally distributed random numbers is called the Box–Muller transform or the Box–Muller algorithm.

3.2   Consider the SDE (3.7). Compute $E[X(t)^k]$ for $k \in \mathbb{N}$.

Hint: Formula (3.8) implies that $X(t)$ is a sum of normally distributed random variables. Conclude that $X(t)$ is itself normally distributed with zero mean and variance $V(t)$ given by (3.14). Thus the odd moments are zero and the even moments $E[X(t)^{2k}]$ are equal to $(V(t)/2)^k (2k)!/k!$ for $k \in \mathbb{N}$.

3.3   Consider the SDE (3.21). Show that the variance $V(t)$ of $X(t)$ is given by $V(t) = t$.

Hint: Use (3.22) together with (3.23), $E[\xi] = 0$ and $E[\xi^2] = 1$.

3.4   Verify that (3.33) is a solution of (3.32).

3.5   Verify that (3.36) is a solution of (3.35).

3.6   Consider the SDE (3.18). Let us suppose that $X(t) < x_u$, and $X(t + \Delta t) < x_u$, where $x_u \in \mathbb{R}$. What is the probability that the trajectory left the interval $(-\infty, x_u)$ during the time step $(t, t + \Delta t)$?

Hint: Use similar arguments as in Exercise 4.8.

3.7 Show that the Fokker–Planck equation corresponding to the SDE (3.76) is given by (3.71).

Hint: Follow the method from Section 3.3.

3.8 Consider again the chemical system (2.1), i.e.

$$3A \underset{k_2}{\overset{k_1}{\rightleftarrows}} 2A, \qquad A \underset{k_4}{\overset{k_3}{\rightleftarrows}} \emptyset, \qquad (2.1)$$

with the values of (dimensionless) rate constants k_1, k_2 and k_3 as

$$k_1/v^2 = 2.5 \times 10^{-4}, \qquad k_2/v = 0.18, \qquad k_3 = 37.5.$$

(a) Write the chemical Fokker–Planck equation corresponding to the system (2.1). Plot the stationary distribution p_s given by the chemical Fokker–Planck equation for (i) $k_4 v = 1750$, (ii) $k_4 v = 2100$, (iii) $k_4 v = 2200$ and (iv) $k_4 v = 2450$. Compare with the results computed in Exercise 2.3(c) (using the Gillespie SSA and the chemical master equation) for the same numerical values of parameters. Note that we use the units of rate constants as $[\text{min}^{-1}]$ in Exercise 2.3, but we have dropped physical units in this exercise for simplicity.

(b) Plot the graph of local maxima and minima of p_s as a function of $k_4 v$ for $k_4 v \in (1700, 2500)$. Compare with the graph of the steady states of the deterministic ODE model obtained in Exercise 2.3(a). If there are any differences, explain them.

(c) Plot the average switching time between two favourable states of the system as a function of $k_4 v \in (1700, 2500)$. To obtain this plot, use formula (3.85). Compare the results with the Gillespie SSA for (i) $k_4 v = 2100$ and (ii) $k_4 v = 2200$.

(d) Use $k_4 v = 2100$. Plot the stationary distribution. Now, change v to $10v$. That means that we will use the following parameters:

$$k_1/v^2 = 2.5 \times 10^{-4}/100 = 2.5 \times 10^{-6}, \qquad k_2/v = 0.18/10 = 0.018,$$

$$k_3 = 37.5, \qquad k_4 v = 2100 \times 10 = 21\,000.$$

Plot the stationary distribution. What is the effect of changing the volume v on the shape of the stationary distribution?

3.9   A particle moves in a one-dimensional interval $[0, L]$, where $L > 0$, according to the discretized stochastic differential equation

$$X(t + \Delta t) = X(t) + f(X(t)) \Delta t + \sqrt{2\Delta t}\, \xi,$$

where $X(t) \in [0, L]$ is the particle position at time $t \geq 0$, $f: [0, L] \to \mathbb{R}$, $f(0) = 0$, $\Delta t > 0$ is the time step and ξ is a random number, which is sampled from the normal distribution with zero mean and unit variance. The initial condition is $X(0) = 0$. The boundary $x = 0$ is reflective, i.e. whenever the particle crosses this boundary, it is reflected back to the domain. In parts (a) and (b), we will consider $L = \infty$, which means that $X(t) \in [0, \infty)$.

(a) Let $L = \infty$. Let $p(x, t)$ be the distribution of the probability that $X(t) = x$. Write the *Fokker–Planck equation* satisfied by $p(x, t)$ in the limit $\Delta t \to 0$. Write the boundary condition for the Fokker–Planck equation at $x = 0$.

(b) Let $L = \infty$ and $f(x) = -x$. Find the stationary distribution.

(c) Let $L = 10$ and $f(x) \equiv 0$ for $x \in [0, L]$. Let $X(0) = 0$. What is the average time to reach the right boundary, i.e. the point $x = L$?

(d) Let $L = 1$ and $f(x) = x(1 - x)$, with a reflective boundary condition at $x = L$. Find the stationary distribution. Express it in the form $p_s(x) = Ch(x)$, where C is the normalization constant and $h(x)$ is a function of x. You need not calculate the normalization constant C explicitly.

4

Diffusion

In Section 3.2, we discussed that diffusion (Brownian motion) of small particles can be modelled using SDEs. In this chapter, we present several alternative stochastic models of molecular diffusion together with some important properties of the diffusion process. We start by revisiting the SDE model in Section 4.1. Then we present a compartment-based approach in Section 4.2, which will be analysed using the diffusion master equation. In Section 4.3, we present diffusion models based on velocity-jump processes rather than position-jump processes.

Having introduced some of the main approaches to modelling diffusion, we move on to study some of the properties of these models. We begin by considering systems in which molecules are adsorbed by surfaces. We then move on to study molecular diffusion in a container with a chemically reactive boundary (for example, a living cell surrounded by its membrane), showing how to impose the reactive boundary condition for each model of diffusion. We conclude with the derivation of the Einstein–Smoluchowski relation, which is a special case of the so-called fluctuation–dissipation theorem relating the diffusion constant to the size of the diffusing molecule, absolute temperature and viscosity of the solution. The modelling approaches to diffusion that are introduced in this chapter will be coupled later in Chapter 6 with models of chemical reactions to obtain methods for stochastic modelling of reaction–diffusion processes.

In Chapter 8, we will return back to the diffusion process by considering its more detailed microscopic models, which explicitly describe both the diffusing particle and surrounding solvent molecules. The most detailed models discussed in this book are given in terms of molecular dynamics, describing all solvent atoms in Section 8.6. In Chapter 8, we will derive stochastic diffusion models from the underlying microscopic molecular dynamics description of the system and present the generalized fluctuation–dissipation theorem.

4.1 Diffusion Modelled by SDEs

Let us consider a typical protein molecule immersed in the aqueous medium of a living cell. As with any small particle, it has a non-zero kinetic energy that is proportional to the absolute temperature. In particular, the protein molecule has a non-zero instantaneous speed. However, it cannot travel too far before it bumps into other molecules (e.g. water molecules) in the solution. As a result, the trajectory of the molecule is not straight but it executes a random walk – the well-known Brownian motion. In Figure 3.1(c), we plotted six possible trajectories of the Brownian motion of the protein molecule using six different colours. All trajectories started at the origin and were followed for 10 minutes. We obtained these trajectories as a solution of the system of SDEs (3.18)–(3.20), namely

$$X(t + dt) = X(t) + \sqrt{2D}\, dW_x, \tag{4.1}$$

$$Y(t + dt) = Y(t) + \sqrt{2D}\, dW_y, \tag{4.2}$$

$$Z(t + dt) = Z(t) + \sqrt{2D}\, dW_z, \tag{4.3}$$

where $[X(t), Y(t), Z(t)] \in \mathbb{R}^3$ is the position of the diffusing molecule at time t and D is the diffusion constant. To solve the SDEs (4.1)–(4.3), we chose the time step Δt and we used the computational definition (3.5) of SDEs, i.e. we computed the solution iteratively by (3.15)–(3.17).

Two natural questions arise: Why is the system of SDEs (4.1)–(4.3) a good model of Brownian motion? And, how can we compute the diffusion constant D, the single parameter in the model? The first question will be answered in Section 4.3, where we outline the derivation of the SDEs (4.1)–(4.3) from Newton's second law of motion. The answer to the second question will be given by the Einstein–Smoluchowski relation in Section 4.6. In the rest of this section, we relate the model (4.1)–(4.3) to the classic deterministic description of diffusion.

Let $p_x(x,t)\, dx$ be the probability that $X(t) \in [x, x + dx)$. Since $X(t)$ evolves according to the SDE (4.1), the probability density $p_x(x,t)$ satisfies the corresponding Fokker–Planck equation (compare with (3.31))

$$\frac{\partial p_x}{\partial t}(x,t) = D\frac{\partial^2 p_x}{\partial x^2}(x,t). \tag{4.4}$$

Similarly, if $p_y(y,t)\, dy$ and $p_z(z,t)\, dz$ are the probability that $Y(t) \in [y, y + dy)$ and $Z(t) \in [z, z+dz)$, respectively, then the Fokker–Planck equations for $p_y(y,t)$ and $p_z(z,t)$ are

$$\frac{\partial p_y}{\partial t}(y,t) = D\frac{\partial^2 p_y}{\partial y^2}(y,t), \tag{4.5}$$

$$\frac{\partial p_z}{\partial t}(z, t) = D \frac{\partial^2 p_z}{\partial z^2}(z, t). \tag{4.6}$$

Now let $p(x, y, z, t) \, dx \, dy \, dz$ be the probability that $X(t) \in [x, x + dx)$, $Y(t) \in [y, y + dy)$ and $Z(t) \in [z, z + dz)$ at time t. Since the SDEs (4.1)–(4.3) are independent (i.e. the behaviour of coordinates is independent) of each other, we have $p(x, y, z, t) \, dx \, dy \, dz = p_x(x, t) \, dx \times p_y(y, t) \, dy \times p_z(z, t) \, dz$. Thus

$$p(x, y, z, t) = p_x(x, t) \, p_y(y, t) \, p_z(z, t). \tag{4.7}$$

Multiplying equation (4.4) by $p_y p_z$, equation (4.5) by $p_x p_z$, equation (4.6) by $p_x p_y$ and adding we obtain

$$\frac{\partial}{\partial t}\Big(p_x(x, t) \, p_y(y, t) \, p_z(z, t)\Big)$$

$$= D \frac{\partial^2 p_x}{\partial x^2}(x, t) \, p_y(y, t) \, p_z(z, t) + D \, p_x(x, t) \frac{\partial^2 p_y}{\partial y^2}(y, t) \, p_z(z, t)$$

$$+ D \, p_x(x, t) \, p_y(y, t) \frac{\partial^2 p_z}{\partial z^2}(z, t).$$

Using (4.7) we see that this is just

$$\frac{\partial p}{\partial t} = D \left(\frac{\partial^2 p}{\partial x^2} + \frac{\partial^2 p}{\partial y^2} + \frac{\partial^2 p}{\partial z^2} \right), \tag{4.8}$$

which is the diffusion equation in three-dimensional space.

In Figure 3.1(c), we plotted six possible trajectories of the protein molecule. Since the trajectories all started at the origin, we can get the corresponding probability distribution function $p(x, y, z, t)$ by solving (4.8) with the initial condition $p(x, y, z, 0) = \delta(x, y, z)$, where δ is the Dirac distribution at the origin. We obtain (see Exercise 4.1)

$$p(x, y, z, t) = \frac{1}{(4D\pi t)^{3/2}} \exp\left[-\frac{x^2 + y^2 + z^2}{4Dt}\right]. \tag{4.9}$$

In order to visualize this probability distribution function we reduce the number of independent coordinates by integrating it over z to get a so-called marginal probability distribution function,

$$\psi(x, y, t) = \int_{\mathbb{R}} p(x, y, z, t) \, dz = \frac{1}{4D\pi t} \exp\left[-\frac{x^2 + y^2}{4Dt}\right]. \tag{4.10}$$

By integrating over z we have lost the information about the value of $Z(t)$; $\psi(x, y, t) \, dx \, dy$ is the probability that $X(t) \in [x, x + dx)$ and $Y(t) \in [y, y + dy)$ at time t with any value of $Z(t)$. The function $\psi(x, y, t)$ at time $t = 10$ min is plotted in Figure 4.1(a) for $D = 10^{-4} \, \text{mm}^2 \, \text{sec}^{-1}$. It can be also obtained by computing many trajectories $[X(t), Y(t), Z(t)]$ of the SDEs (4.1)–(4.2) (i.e.

(a)   (b)

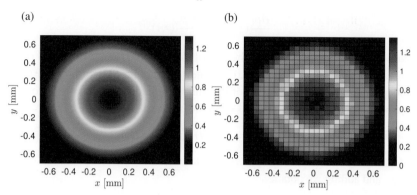

Figure 4.1 (a) *Probability distribution function* $\psi(x,y,t)$ *given by* (4.10) *at time*
$t = 10\,min$ *for* $D = 10^{-4}\,mm^2\,sec^{-1}$. (b) *The histogram of positions at time* $t =$
$10\,min$ *computed by one million realizations of the SSA* (a6)–(b6) *applied to the*
SDEs (4.1)–(4.2).

using many realizations of the SSA (a6)–(b6) applied to the SDEs (4.1)–(4.2))
and plotting the histogram of possible positions of a molecule at time 10 min;
such positions were denoted as black circles for the six illustrative trajectories in Figure 3.1(c). To compute the histogram, we choose the mesh step
$h = 50\,\mu m$ and we divide the plane $\mathbb{R}^2$ into squares $[(i-1)h, ih) \times [(j-1)h, jh)$,
where i and j are integers. Considering one million trajectories, we calcu-
late the number of trajectories that end up (at time $t = 10\,min$) in each
square $[(i-1)h, ih) \times [(j-1)h, jh)$, $i \in \mathbb{Z}$, $j \in \mathbb{Z}$. This number is divided
by the square area (h^2) and by the number of realizations (10^6). The resulting histogram is plotted in Figure 4.1(b), which compares well with the
formula (4.10).

   In Figure 3.1(c), we observed that different realizations of the SSA (a6)–(b6)
for the SDEs (4.1)–(4.3) yield different results. To get reproducible quantities, we reformulated the problem in terms of probability densities, such
as $p(x,y,z,t)$ given by (4.7). The probability densities evolve according to
deterministic PDEs such as the diffusion equation (4.8). As stated, the function $p(x,y,z,t)$ is the probability (density) of finding the diffusing molecule
at the point (x,y,z,t) at time t. If we consider the behaviour of many (non-
interacting) molecules whose trajectories follow the SDEs (4.1)–(4.3), we are
more interested in the densities of molecules (concentration profiles) than in
the trajectories of individual molecules. However, such concentration profiles
are proportional to the probability densities studied above, simply because
the trajectories of many non-interacting molecules (which are all released at
the origin) can be also interpreted as many realizations of the SSA for the
one-particle system.

The SDEs (4.1)–(4.3) are not coupled. To compute the time evolution of $X(t)$, we do not need to know the time evolution of $Y(t)$ or $Z(t)$. Thus, in the next few sections, we focus on the time evolution of the xth coordinate only, effectively studying one-dimensional problems. Two-dimensional or three-dimensional problems can be treated similarly. One important issue which was not addressed previously is that molecules diffuse in bounded volumes (e.g. in a cell bounded by its membrane). Considering one-dimensional problems, we study the motion of molecules in the one-dimensional interval $[0, L]$. Since the domain of interest has boundaries (namely $x = 0$ and $x = L$), suitable boundary conditions must be implemented. The simplest possibility is to assume that molecules are reflected at the boundaries $x = 0$ and $x = L$. Then the SSA can be formulated as follows. Given the initial condition $X(0) = x_0$, we compute the values of $X(t)$ at times $t > 0$ iteratively, by performing three steps at time t:

(a8) Generate a normally distributed (with zero mean and unit variance) random number ξ.

(b8) Compute the position of the molecule at time $t + \Delta t$ by

$$X(t + \Delta t) = X(t) + \sqrt{2D \Delta t} \, \xi. \qquad (4.11)$$

(c8) If $X(t + \Delta t)$ computed by (4.11) is less than 0, then

$$X(t + \Delta t) = -X(t) - \sqrt{2D \Delta t} \, \xi.$$

If $X(t + \Delta t)$ computed by (4.11) is greater than L, then

$$X(t + \Delta t) = 2L - X(t) - \sqrt{2D \Delta t} \, \xi.$$

Then continue with step (a8) for time $t + \Delta t$.

The first two steps (a8)–(b8) are the same as the SSA (a6)–(b6). The boundary condition implemented in the step (c8) is the reflective boundary condition or zero-flux boundary condition previously seen in (c7), imposed at both $x = 0$ and $x = L$. It can be used when there is no chemical interaction between the boundary and diffusing molecules. More complicated boundary conditions are discussed in Sections 4.4 and 4.5.

Choosing $D = 10^{-4}$ mm^2 sec^{-1}, $L = 1$ mm, $X(0) = 0.4$ mm and $\Delta t = 0.1$ sec, we plot ten realizations of the SSA (a8)–(c8) in Figure 4.2(a). Let us assume that we have a system of 1000 molecules which are released at position $x = 0.4$ mm at time $t = 0$. Then Figure 4.2(a) can be viewed as a plot of the trajectories of ten representative molecules. Considering 1000 molecules, the trajectories of individual molecules are of no special interest. We are rather interested in spatial histograms (density of molecules, or concentration profiles). An example of such a plot is given in Figure 4.2(b) where we simulated 1000 molecules,

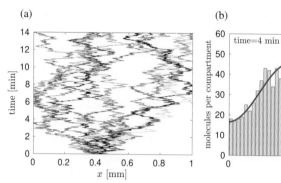

Figure 4.2 (a) *Ten trajectories computed by the SSA (a8)–(c8) for* $D = 10^{-4}\,mm^2\,sec^{-1}$, $L = 1\,mm$, $X(0) = 0.4\,mm$ *and* $\Delta t = 0.1\,sec$. (b) *Numbers of molecules in bins of length* $h = 25\,\mu m$ *at time* $t = 4\,min$. *The solution of* (4.4) *with the boundary conditions* (4.12) *is plotted as the red solid line.*

each following the SSA (a8)–(c8). At time $t = 4$ min, we divided the domain of interest $[0, L]$ into 40 bins of length $h = L/40 = 25\,\mu$m. We calculated the number of molecules in each bin $[(i-1)h, ih)$, $i = 1, 2, \ldots, 40$, at time $t = 4$ min and plotted them as a grey histogram.

Recall that the deterministic counterpart to the SSA (a8)–(c8) is a solution of the diffusion equation (4.4) with zero-flux boundary conditions (compare with (3.50))

$$\frac{\partial p_x}{\partial x}(0, t) = \frac{\partial p_x}{\partial x}(L, t) = 0. \tag{4.12}$$

The solution of (4.4) with the boundary conditions (4.12) and the Dirac-like initial condition at $x = 0.4$ mm is plotted as the red solid line in Figure 4.2(b) for comparison.

4.2 Compartment-Based Approach to Diffusion

In Section 4.1, we simulated the behaviour of 1000 molecules by computing the individual trajectories of every molecule (using the SSA (a8)–(c8)). At the end of the simulation, we divided the computational domain $[0, L]$ into $K = 40$ compartments and we plotted numbers of molecules in each compartment in Figure 4.2(b). In particular, most of the computed information (1000 trajectories) was not used for the final result – the spatial histogram. We visualized only 40 numbers (the number of molecules in each compartment) instead of 1000 computed positions of molecules. In this section, we present a different SSA for the simulation of molecular diffusion. We redo the example from Section 4.1, but instead of simulating the 1000 positions of the individual molecules, we are going to simulate directly the time evolution of 40 compartments.

To do that, we divide the computational domain $[0, L]$ into $K = 40$ compartments of length $h = L/K = 25\,\mu$m. We denote the number of molecules of chemical species A in the ith compartment $[(i-1)h, ih)$ by A_i, $i = 1, 2, \ldots, K$. As a result of Brownian motion, molecules jump between neighbouring compartments. Thus we model diffusion as the following chain of "chemical reactions":

$$A_1 \overset{d}{\underset{d}{\rightleftharpoons}} A_2 \overset{d}{\underset{d}{\rightleftharpoons}} A_3 \overset{d}{\underset{d}{\rightleftharpoons}} \cdots \overset{d}{\underset{d}{\rightleftharpoons}} A_K, \tag{4.13}$$

where, as usual,

$$A_i \overset{d}{\underset{d}{\rightleftharpoons}} A_{i+1} \qquad \text{means that} \qquad A_i \overset{d}{\longrightarrow} A_{i+1} \text{ and } A_{i+1} \overset{d}{\longrightarrow} A_i.$$

The system of chemical reactions (4.13) can be simulated by the Gillespie SSA (a5)–(d5). We will shortly show that the Gillespie SSA of (4.13) is a suitable model for diffusion provided that the rate constant d in (4.13) is chosen as

$$d = \frac{D}{h^2}, \tag{4.14}$$

where D is the diffusion constant and h is the compartment length. In order to implement the Gillespie SSA (a5)–(d5) more efficiently, let us observe that (4.13) is a system of $(2K - 2)$ chemical reactions which have only K different propensities. Denoting

$$\alpha_i(t) \equiv A_i(t)\, d = A_i(t)\, \frac{D}{h^2}, \qquad \text{for } i = 1, 2, \ldots, K, \tag{4.15}$$

the function $\alpha_i(t)$, $i = 2, 3, \ldots, K-1$, is the propensity function of two different reactions:

$$A_i \overset{d}{\longrightarrow} A_{i+1} \text{ (jump to the right)} \quad \text{and} \quad A_i \overset{d}{\longrightarrow} A_{i-1} \text{ (jump to the left)}.$$

The total propensity $\alpha_0 \equiv \alpha_0(t)$ is given by (compare with (1.53))

$$\alpha_0 = \sum_{i=1}^{K-1} \alpha_i(t) + \sum_{i=2}^{K} \alpha_i(t), \tag{4.16}$$

where the first term on the right-hand side corresponds to reactions $A_i \to A_{i+1}$ (jumps to the right) and the second term corresponds to reactions $A_i \to A_{i-1}$ (jumps to the left). Using α_0, we compute the time $t+\tau$ when the next "chemical reaction" occurs in the step (c5) of the Gillespie SSA (a5)–(d5). If the next reaction is the jump to the right,

$$A_j \xrightarrow{d} A_{j+1},$$

then the numbers of molecules in the compartments change as follows:

$$A_j(t + \tau) = A_j(t) - 1,$$
$$A_{j+1}(t + \tau) = A_{j+1}(t) + 1, \qquad (4.17)$$
$$A_i(t + \tau) = A_i(t), \quad \text{for } i \neq j, \ i \neq j + 1.$$

If the next reaction is the jump to the left,

$$A_j \xrightarrow{d} A_{j-1},$$

then the numbers of molecules in the compartments change as follows:

$$A_j(t + \tau) = A_j(t) - 1,$$
$$A_{j-1}(t + \tau) = A_{j-1}(t) + 1, \qquad (4.18)$$
$$A_i(t + \tau) = A_i(t), \quad \text{for } i \neq j, \ i \neq j - 1.$$

In both cases, there are only two compartments for which the number of molecules changes. Thus only two of the propensity functions (4.15) change and need to be updated in the step (b5) of the Gillespie SSA (a5)–(d5). Moreover, the formula (4.16) can be simplified as

$$\alpha_0 = \sum_{i=1}^{K-1} \alpha_i(t) + \sum_{i=2}^{K} \alpha_i(t) = 2\left(\sum_{i=1}^{K} \alpha_i(t) \right) - \alpha_1(t) - \alpha_K(t)$$

$$= 2d\left(\sum_{i=1}^{K} A_i(t) \right) - \alpha_1(t) - \alpha_K(t) = 2dN - \alpha_1(t) - \alpha_K(t), \qquad (4.19)$$

where N is the total number of molecules in the simulation (this number is conserved because there is no creation or degradation of the molecules in the system). In particular, we need to recompute α_0 only when there is a change in $\alpha_1(t)$ or $\alpha_K(t)$, i.e. whenever the boundary compartments were involved in the previous reaction (diffusive jump). Taking the above remarks into account, the Gillespie SSA (a5)–(d5) for the system of chemical reactions (4.13) can be formulated as the following compartment-based SSA for molecular diffusion. Starting with a given initial condition for $A_i(0)$, $i = 1, 2, \ldots, K$, we compute the system evolution iteratively by performing six steps at time t:

(a9) Generate two random numbers r_1, r_2 uniformly distributed in $(0, 1)$.

(b9) Compute propensity functions of reactions $\alpha_i(t)$ by (4.15). Compute α_0 by (4.19).

(c9) Compute the time at which the next chemical reaction takes place as $t + \tau$ where τ is given by (1.54).

(d9) If $r_2 < \sum_{i=1}^{K-1} \alpha_i/\alpha_0$, then find $j \in \{1, 2, \ldots, K-1\}$ such that

$$r_2 \geq \frac{1}{\alpha_0} \sum_{i=1}^{j-1} \alpha_i \quad \text{and} \quad r_2 < \frac{1}{\alpha_0} \sum_{i=1}^{j} \alpha_i.$$

Then compute the number of molecules at time $t + \tau$ by (4.17).

(e9) If $r_2 \geq \sum_{i=1}^{K-1} \alpha_i/\alpha_0$, then find $j \in \{2, 3, \ldots, K\}$ such that

$$r_2 \geq \frac{1}{\alpha_0} \left(\sum_{i=1}^{K-1} \alpha_i + \sum_{i=2}^{j-1} \alpha_i \right) \quad \text{and} \quad r_2 < \frac{1}{\alpha_0} \left(\sum_{i=1}^{K-1} \alpha_i + \sum_{i=2}^{j} \alpha_i \right).$$

Then compute the number of molecules at time $t + \tau$ by (4.18).

(f9) Continue with step (a9) for time $t + \tau$.

To implement the algorithm efficiently, the functions $\alpha_i(t)$ should be recomputed in the step (b9) only for those two compartments which were involved in the previous reaction. The total propensity function α_0 should be updated by (4.19) only when the previous reaction changes the numbers of molecules in the boundary compartments. The time of the next chemical reaction is computed in the step (c9) using formula (1.54) derived previously. The decision about which reaction takes place is done in the steps (d9)–(e9) with the help of random number r_2. Jumps to the right are implemented in the step (d9) and jumps to the left in the step (e9).

We now revisit the example from Section 4.1 using this compartment-based model, with $K = 40$ compartments. We need to simulate 1000 molecules starting from position 0.4 mm in the interval $[0, L]$ for $L = 1$ mm. Since 0.4 mm is exactly a boundary between the 16th and 17th compartments, the initial condition is given by $A_{16}(0) = 500$, $A_{17}(0) = 500$ and $A_i(0) = 0$ for $i \neq 16$, $i \neq 17$. As $D = 10^{-4}$ mm^2 sec^{-1}, we have $d = D/h^2 = 0.16$ sec^{-1}, using the formula (4.14). The numbers $A_i(t)$, $i = 1, 2, \ldots, K$, at time $t = 4$ min are plotted in Figure 4.3(b) as a histogram. This panel can be directly compared with Figure 4.2(b). The red solid line is the solution of (4.4) with the boundary conditions (4.12) and the Dirac-like initial condition at $x = 0.4$ mm.

The SSA (a9)–(f9) does not compute the trajectories of individual molecules. However, we can still compute a plot comparable with Figure 4.2(a). To

(a)   (b)

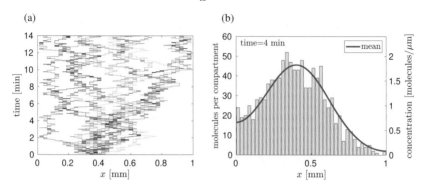

Figure 4.3 *Compartment-based SSA model of diffusion.* (a) *Ten realizations of the simulation of an individual molecule by the SSA (a9)–(f9).* (b) *Numbers $A_i(t)$, $i = 1, 2, \ldots, K$, at time $t = 4\,min$ obtained by the SSA (a9)–(f9). We use $d = D/h^2 = 0.16\,sec^{-1}$, $K = 40$ and initial condition $A_{16}(0) = 500$, $A_{17}(0) = 500$ and $A_i(0) = 0$ for $i \neq 16$, $i \neq 17$. The solution of (4.4) with the boundary conditions (4.12) and the Dirac-like initial condition at $x = 0.4$ mm is plotted as the red solid line.*

do that, we repeat the simulation with one molecule instead of 1000. Then, at given time t, exactly one of numbers $A_i(t)$, $i = 1, 2, \ldots, K$, is non-zero and equal to 1. This is a position of the molecule at time t. Ten realizations of the SSA (a9)–(f9) with one molecule released at 0.4 mm at $t = 0$ are plotted in Figure 4.3(a). This panel can be directly compared with Figure 4.2(a).

We will now show how to relate the hopping rate d to the diffusion coefficient D. We denote by $p(\mathbf{n}, t)$ the joint probability that $A_i(t) = n_i$, $i = 1, 2, \ldots, K$, where $\mathbf{n} = [n_1, n_2, \ldots, n_K]$. Let us define the operators $R_i, L_i : \mathbb{N}^K \to \mathbb{N}^K$ (where $\mathbb{N}$ is the set of non-negative integers) by

$$R_i : [n_1, \ldots, n_i, n_{i+1}, \ldots, n_K] \to [n_1, \ldots, n_i + 1, n_{i+1} - 1, \ldots, n_K], \quad (4.20)$$

for $i = 1, 2, \ldots, K - 1$, and

$$L_i : [n_1, \ldots, n_{i-1}, n_i, \ldots, n_K] \to [n_1, \ldots, n_{i-1} - 1, n_i + 1, \ldots, n_K], \quad (4.21)$$

for $i = 2, 3, \ldots, K$. Then the chemical master equation, which corresponds to the system of chemical reactions given by (4.13), can be written as follows:

$$\frac{\partial p(\mathbf{n})}{\partial t} = d \sum_{j=1}^{K-1} \left\{ (n_j + 1)\, p(R_j \mathbf{n}) - n_j\, p(\mathbf{n}) \right\}$$

$$+ d \sum_{j=2}^{K} \left\{ (n_j + 1)\, p(L_j \mathbf{n}) - n_j\, p(\mathbf{n}) \right\}. \quad (4.22)$$

The mean is defined as the vector $\mathbf{M}(t) \equiv [M_1, M_2, \ldots, M_K]$ where

$$M_i(t) = \sum_{\mathbf{n}} n_i \, p(\mathbf{n}, t) \equiv \sum_{n_1=0}^{\infty} \sum_{n_2=0}^{\infty} \cdots \sum_{n_K=0}^{\infty} n_i \, p(\mathbf{n}, t) \qquad (4.23)$$

gives the mean number of molecules in the ith compartment, $i = 1, 2, \ldots, K$. To derive an evolution equation for the mean vector $\mathbf{M}(t)$, we can follow the method from Section 1.2 – see the derivation of (1.15) from the chemical master equation (1.11). Multiplying (4.22) by n_i and summing over $\mathbf{n}$, we obtain (leaving the details as Exercise 4.2) a system of equations for M_i of the form

$$\frac{\partial M_i}{\partial t} = d\,(M_{i+1} + M_{i-1} - 2M_i), \qquad i = 2, 3, \ldots, K - 1, \qquad (4.24)$$

$$\frac{\partial M_1}{\partial t} = d(M_2 - M_1), \qquad \frac{\partial M_K}{\partial t} = d(M_{K-1} - M_K). \qquad (4.25)$$

The classic deterministic description of diffusion is written in terms of concentration $c(x, t)$, which can be approximated as $c(x_i, t) \approx M_i(t)/h$, where x_i is the centre of the ith compartment, $i = 1, 2, \ldots, K$. Dividing (4.24) by h, we obtain

$$\frac{\partial c}{\partial t}(x_i, t) \approx d\left(c(x_i + h, t) + c(x_i - h, t) - 2c(x_i, t)\right).$$

By Taylor expanding the right-hand side we get

$$\frac{\partial c}{\partial t}(x_i, t) \approx dh^2 \frac{\partial^2 c}{\partial x^2}(x_i, t),$$

which is equivalent to equation (4.4) provided that $d = D/h^2$. Hence we have derived the relation (4.14) between the rate constant d in (4.13), the diffusion constant D and the compartment length h.

The analysis above is exactly analogous to that for first-order chemical reactions in Chapter 1. There we found that the stochastic mean satisfied the deterministic ODE version of the chemical system; here we find that the mean satisfies the (deterministic) diffusion equation. In Chapter 1 we were able to extract further information from the chemical master equation by considering in addition the variance of the stochastic solution. A similar analysis is also possible in the present case, and we sketch it briefly.

The noise is described by the variance vector $\mathbf{V}(t) \equiv [V_1, V_2, \ldots, V_K]$ where

$$V_i(t) = \sum_{\mathbf{n}} (n_i - M_i(t))^2 \, p(\mathbf{n}, t) \equiv \sum_{n_1=0}^{\infty} \sum_{n_2=0}^{\infty} \cdots \sum_{n_K=0}^{\infty} (n_i - M_i(t))^2 \, p(\mathbf{n}, t) \quad (4.26)$$

gives the variance of the number of molecules in the ith compartment, $i = 1, 2, \ldots, K$. To derive the evolution equation for the vector $\mathbf{V}(t)$, we define more generally the covariance matrix $\{V_{i,j}\}$ by

$$V_{i,j} = \sum_{\mathbf{n}} n_i n_j \, p(\mathbf{n}, t) - M_i M_j, \qquad \text{for } i, j = 1, 2, \ldots, K. \tag{4.27}$$

From (4.26) we see that the variance vector comprises the diagonal entries of this matrix: $V_i = V_{i,i}$ for $i = 1, 2, \ldots, K$. Multiplying (4.22) by n_i^2 and summing over $\mathbf{n}$, we obtain

$$\frac{\partial}{\partial t} \sum_{\mathbf{n}} n_i^2 p(\mathbf{n}) = d \sum_{j=1}^{K-1} \left\{ \sum_{\mathbf{n}} n_i^2 (n_j + 1) \, p(R_j \mathbf{n}) - \sum_{\mathbf{n}} n_i^2 n_j \, p(\mathbf{n}) \right\}$$

$$+ d \sum_{j=2}^{K} \left\{ \sum_{\mathbf{n}} n_i^2 (n_j + 1) \, p(L_j \mathbf{n}) - \sum_{\mathbf{n}} n_i^2 n_j \, p(\mathbf{n}) \right\}. \tag{4.28}$$

Let us consider the case that $i \in \{2, 3, \ldots, K - 1\}$. We evaluate first the term corresponding to $j = i$ in the first sum on the right-hand side. We get

$$\sum_{\mathbf{n}} n_i^2 (n_i + 1) \, p(R_i \mathbf{n}) - \sum_{\mathbf{n}} n_i^2 n_i \, p(\mathbf{n}) = \sum_{\mathbf{n}} (n_i - 1)^2 n_i \, p(\mathbf{n}) - \sum_{\mathbf{n}} n_i^2 n_i \, p(\mathbf{n})$$

$$= \sum_{\mathbf{n}} (-2n_i^2 + n_i) \, p(\mathbf{n}) = -2V_i - 2M_i^2 + M_i.$$

Here we first changed indices in the first sum $R_i \mathbf{n} \to \mathbf{n}$ and then used definitions (4.23) and (4.26). Similarly, the term corresponding to $j = i - 1$ in the first sum on the right-hand side of (4.28) can be rewritten as

$$\sum_{\mathbf{n}} n_i^2 (n_{i-1} + 1) \, p(R_{i-1}\mathbf{n}) - \sum_{\mathbf{n}} n_i^2 n_{i-1} \, p(\mathbf{n}) = \sum_{\mathbf{n}} (2n_i n_{i-1} + n_{i-1}) \, p(\mathbf{n})$$

$$= 2V_{i,i-1} + 2M_i M_{i-1} + M_{i-1}.$$

Other terms (corresponding to $j \neq i, i-1$) in the first sum on the right-hand side of (4.28) are equal to zero. The second sum on the right-hand side of (4.28) can be handled analogously, to give, finally

$$\frac{\partial}{\partial t} \sum_{\mathbf{n}} n_i^2 p(\mathbf{n}) = d \{ 2V_{i,i-1} + 2M_i M_{i-1} + M_{i-1} - 2V_i - 2M_i^2 + M_i \}$$

$$+ d \{ 2V_{i,i+1} + 2M_i M_{i+1} + M_{i+1} - 2V_i - 2M_i^2 + M_i \}. \tag{4.29}$$

Now using (4.26) and (4.24) on the left-hand side of (4.29), we obtain

$$\frac{\partial}{\partial t} \sum_{\mathbf{n}} n_i^2 p(\mathbf{n}) = \frac{\partial V_i}{\partial t} + 2M_i \frac{\partial M_i}{\partial t} = \frac{\partial V_i}{\partial t} + d(2M_i M_{i+1} + 2M_i M_{i-1} - 4M_i^2).$$

Substituting this into (4.29), we get

$$\frac{\partial V_i}{\partial t} = 2d\left\{V_{i,i+1} + V_{i,i-1} - 2V_i\right\} + d\left\{M_{i+1} + M_{i-1} + 2M_i\right\} \qquad (4.30)$$

for $i = 2, 3, \ldots, K - 1$. A similar analysis gives

$$\frac{\partial V_1}{\partial t} = 2d\left\{V_{1,2} - V_1\right\} + d\left\{M_2 + M_1\right\}, \qquad (4.31)$$

$$\frac{\partial V_K}{\partial t} = 2d\left\{V_{K,K-1} - V_K\right\} + d\left\{M_{K-1} + M_K\right\}. \qquad (4.32)$$

We see that the evolution equation for the variance vector $\mathbf{V}(t)$ depends on the mean $\mathbf{M}$, on the variance $\mathbf{V}$ and on non-diagonal terms of the covariance matrix $\{V_{i,j}\}$. To get a closed system of equations, we have to derive evolution equations for $V_{i,j}$ too. This can be done by multiplying (4.22) by $n_i n_j$, summing over $\mathbf{n}$ and following the same arguments as before.

We note that, although the system of equations we have found for the variance is not so simple as that we found for a single first-order reaction in Chapter 1, it is still much simpler than the chemical master equation (4.22). The covariance matrix $\{V_{i,j}\}$ has two indices i and j, which range from 1 to K, the number of compartments. On the other hand, the probability distribution function $p(\mathbf{n})$ has K indices, each of which ranges from 1 to N, the total number of molecules. Thus for our example problem with 40 compartments and 1000 molecules, $\{V_{i,j}\}$ involves 1600 numbers, while $p(\mathbf{n})$ involves 10^{120} numbers!

We conclude this section with consequences of equations (4.24)–(4.25) and (4.30)–(4.32). Looking at the steady states of equations (4.24)–(4.25), we obtain $M_i = N/K$, $i = 1, 2, \ldots, K$, where N is the total number of diffusing molecules. Moreover, the variance equations imply that $V_i = (N/K) - (N/K^2)$, for $i = 1, 2, \ldots, K$, and $V_{i,j} = -(N/K^2)$ for $i \neq j$, at the steady state. In particular, M_i is approximately equal to V_i for larger values of K. The same steady-state results can be confirmed by solving the corresponding stationary master equation – see Exercise 4.3, where the stationary probability mass function is obtained as a multinomial distribution; see equation (4.98) and the Appendix.

4.3 Diffusion and Velocity-Jump Processes

The diffusion models from Section 4.1 and Section 4.2 were simple examples of position-jump processes. Such processes can be characterized by abrupt changes (jumps) in the position of diffusing particles at random instants of time. In particular, it does not make sense to talk about particle velocities for position-jump processes. We could try to formally define the velocity as the derivative of the position, namely[*]

$$V(t) = \frac{dX}{dt}(t), \qquad (4.33)$$

but such a derivative is infinite whenever a position jump occurs instantaneously, and zero when a jump does not occur. In this section, we present some models for diffusion based on so-called velocity-jump processes (Othmer et al., 1988). For such processes, the particle position $X(t)$ evolves according to equation (4.33), with the particle velocity $V(t)$ subject to a stochastic process. As before, we focus on the time evolution of the x-coordinate only, studying effectively one-dimensional models. A simple example of a diffusion model based on a velocity-jump process can be formulated as follows. We assume that a particle moves along the x-axis at a constant speed s. It can go either to the left or to the right. The diffusing particle reverses its direction at random instants of time. The probability of turning in the infinitesimally small time interval $[t, t + dt)$ is equal to $\lambda\, dt$, where the turning frequency λ is given by

$$\lambda = \frac{s^2}{2D} \qquad (4.34)$$

and D is the diffusion constant. We will shortly justify that this velocity-jump process is a model of molecular diffusion provided that the turning frequency λ is chosen according to the formula (4.34). Before doing that, we reformulate the model as an SSA.

Since the speed s of the particle is constant, the particle velocity can have only two values, $V(t) = \pm s$. The particle position evolves according to equation (4.33). Choosing a small time step Δt, a discretized version of equation (4.33) can be written as

$$X(t + \Delta t) = X(t) + V(t)\,\Delta t. \qquad (4.35)$$

If the time step Δt is small enough (satisfying $\lambda\,\Delta t \ll 1$), we can reformulate the above definition of the velocity-jump process as the following SSA. Starting with the initial conditions $X(0) = x_0$ and $V(0) = v_0$, we perform three steps at time t:

(a10) Generate a random number r uniformly distributed in $(0, 1)$.
(b10) Compute the position of the molecule at time $t + \Delta t$ by (4.35).
(c10) If $r < \lambda\,\Delta t$, then put $V(t + \Delta t) = -V(t)$. Otherwise, $V(t + \Delta t) = V(t)$.
Then continue with step (a10) for time $t + \Delta t$.

---

[*] Note that $V(t)$ denotes the velocity in this section (and later in Section 4.6 and Chapter 8). In particular, we do not introduce special symbols to denote variances of stochastic variables (as it has been done in parts of the book where velocities of particles do not appear).

Since r is a random number uniformly distributed in the interval $(0, 1)$, the probability that $r < \lambda \Delta t$ is equal to $\lambda \Delta t$. Consequently, step (c10) says that the probability that the particle turns in the time interval $[t, t + \Delta t)$ is equal to $\lambda \Delta t$. Thus the step (c10) correctly implements the definition of the velocity-jump process provided that Δt is small, so that $\lambda \Delta t \ll 1$.

The velocity-jump process described by the SSA (a10)–(c10) is characterized by two parameters, the diffusion constant D and the particle speed s. The diffusion constant of a typical protein molecule is $D \approx 10^{-4}$ mm^2 sec^{-1} (this is the value we used in Section 4.1). To estimate a realistic value of the speed s, let us note that the kinetic energy associated with the movement of a small particle along the x-axis is equal to $k_B T / 2$, where T is the absolute temperature and $k_B = 1.38 \times 10^{-14}$ g mm^2 sec^{-2} K^{-1} is the Boltzmann constant (Einstein, 1905; Berg, 1983). Thus $m_p s^2 / 2 \approx k_B T / 2$ where m_p is the particle mass. Solving for s, we obtain

$$s \approx \sqrt{\frac{k_B T}{m_p}}. \tag{4.36}$$

Consequently, the average speed of the protein molecule of the molecular mass 25 kDa at the temperature of the human body $T = 310$ K can be estimated as $s \approx 10^4$ mm sec^{-1}. Choosing $D = 10^{-4}$ mm^2 sec^{-1} and $s = 10^1$ mm sec^{-1}, the formula (4.34) implies that $\lambda = 5 \times 10^{11}$ sec^{-1}. The SSA (a10)–(c10) was derived under the condition $\lambda \Delta t \ll 1$. Using $\Delta t = 2 \times 10^{-14}$ sec, we obtain $\lambda \Delta t = 10^{-2}$, i.e. the assumption $\lambda \Delta t \ll 1$ of the SSA (a10)–(c10) is satisfied. The time evolution of $X(t)$ obtained by the SSA (a10)–(c10) is given in Figure 4.4(a). Starting at $X(0) = 0$, we plot six realizations of the SSA (a10)–(c10). The initial direction of the molecule (i.e. its initial velocity) is chosen randomly. Its time evolution is calculated with relatively small time step $\Delta t = 2 \times 10^{-14}$ sec, which is equal to 20 femtoseconds. Later, in Section 8.6, we will discuss molecular dynamics simulations that use an even smaller time step, 1 femtosecond. This is the price one pays for considering more detailed models and we need to think carefully about designing efficient algorithms.

The derivation of the SSA (a10)–(c10) is analogous to the derivation of the "naive" SSA (a1)–(b1). We substituted the infinitesimally small time step dt by the (small) finite time step Δt to get these algorithms. In Section 1.1, we were able to replace the "naive" SSA (a1)–(b1) by the Gillespie SSA (a2)–(c2) which was less computationally intensive. The key point was the derivation of equation (1.5), which specifies the time when the next chemical reaction occurs. The SSA (a10)–(c10) can be improved in a similar way. Instead of using many random numbers between successive turns of the diffusing particle

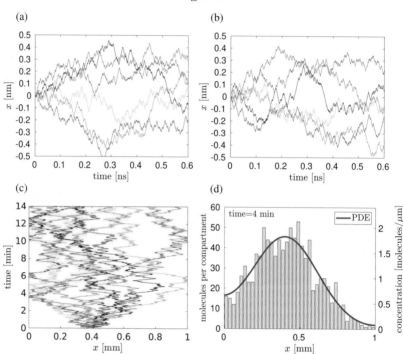

Figure 4.4 (a) *Six trajectories computed by the SSA (a10)–(c10) for* $D = 10^{-4} mm^2 sec^{-1}$, $s = 10^4 mm sec^{-1}$, $\Delta t = 2 \times 10^{-14} sec$ *and* $X(0) = 0$. *The initial velocity is chosen randomly.* (b) *Six trajectories computed by the SSA (a11)–(c11) using the same parameter values.* (c) *Ten trajectories computed by the SSA (a12)–(d12) for* $D = 10^{-4} mm^2 sec^{-1}$, $s = 10^{-2} mm sec^{-1}$, $L = 1 mm$, $X(0) = 0.4 mm$ *and* $\Delta t = 0.1 sec$. (d) *Numbers of molecules in bins of length* $h = 25 \mu m$ *at time* $t = 4 min$. *The solution of* (4.4) *with the boundary conditions* (4.12) *and the Dirac-like initial condition at* $x = 0.4 mm$ *is plotted as the red solid line.*

(in the step (c10)), we can calculate the time of the next turn directly using only one random number. Assuming that the diffusing particle turned at time t, we can compute the time of its next turn as $t + \tau$, where

$$\tau = \frac{1}{\lambda} \ln \left[\frac{1}{r} \right] \qquad (4.37)$$

and r is a random number uniformly distributed in $(0, 1)$. The derivation of this formula is similar to the derivation of (1.5). We leave it to the reader as Exercise 4.4. Since the particle velocity is equal to $V(t)$ in the interval $[t, t+\tau)$, we compute its position at time $t + \tau$ by

$$X(t + \tau) = X(t) + V(t) \tau. \qquad (4.38)$$

Thus, we derived the following SSA. Starting with the initial conditions $X(0) = x_0$ and $V(0) = v_0$, we perform three steps at time t:

(a11) Generate a random number r uniformly distributed in $(0, 1)$.

(b11) Compute the time of the next turn of the molecule as $t + \tau$, where τ is given by (4.37).

(c11) Compute the position of the molecule at time $t + \tau$ by (4.38). Put
$V(t + \tau) = -V(t)$.
Then continue with step (a11) for time $t + \tau$.

The time evolution of $X(t)$ obtained by the SSA (a11)–(c11) is given in Figure 4.4(b) for $D = 10^{-4}\,\text{mm}^2\,\text{sec}^{-1}$ and $s = 10^4\,\text{mm}\,\text{sec}^{-1}$. Starting at $X(0) = 0$, we plot six realizations of the SSA (a11)–(c11). The initial direction of the molecule is chosen randomly.

To plot the time evolution of one realization in Figure 4.4(a) we had to repeat the steps (a10)–(c10) 30 000 times. In Figure 4.4(b) we computed the time evolution using the SSA (a11)–(c11), and had to repeat the steps (a11)–(c11) approximately 300 times for each realization. Thus the computational efficiency was improved by a factor of 100 by the modified algorithm.

In Figure 4.4(b) we are studying the time evolution of the system over a time scale of nanoseconds. Since the average time to the next turn is $1/\lambda$, and we estimated $\lambda = 5 \times 10^{11}\,\text{sec}^{-1}$, there are in fact about 500 turns every nanosecond. Every turn corresponds to one repeat of the steps (a11)–(c11). Thus the computation of the trajectory over a few seconds or minutes (as we did in Section 4.1) would be significantly more computationally intensive than the SSA (a8)–(c8). In Chapter 8, we will study more detailed molecular dynamics models of the diffusion process that are even more computationally intensive. However, we will also see later that SSAs that are based on velocity-jump processes are suitable for modelling the behaviour of larger objects than molecules, when the computation of their trajectories over a few seconds will not be an issue. For example, in Section 7.3 we will present an application of velocity-jump processes to the modelling of bacteria, where we will be able to compute the time evolution of the system over several hours.

We can use velocity-jump processes as a starting point of the derivation of the position-jump diffusion models. We will outline such a derivation of the SSA (a8)–(c8) at the end of this section, but first we justify the formula (4.34).

Let $p^+(x, t)\,\mathrm{d}x$ be the probability that $X(t) \in [x, x + \mathrm{d}x)$ and $V(t) = s$ and let $p^-(x, t)\,\mathrm{d}x$ be the probability that $X(t) \in [x, x + \mathrm{d}x)$ and $V(t) = -s$. Looking at the SSA (a10)–(c10), there are two possible ways for $X(t + \Delta t)$ to be in the interval $[x, x + \mathrm{d}x)$ and moving to the right: either $X(t) \in [x - s\Delta t, x - s\Delta t + \mathrm{d}x)$,

$V(t) = s$ and the particle did not turn in the step (c10), or $X(t) \in [x + s\Delta t, x + s\Delta t + dx)$, $V(t) = -s$ and the particle did turn in the step (c10). Thus we have

$$p^+(x, t + \Delta t)\, dx$$
$$= p^+(x - s\,\Delta t, t)\, dx \times (1 - \lambda\,\Delta t) + p^-(x + s\,\Delta t, t)\, dx \times \lambda\,\Delta t. \quad (4.39)$$

By Taylor expanding, we approximate

$$p^+(x, t + \Delta t) = p^+(x, t) + \Delta t\,\frac{\partial p^+}{\partial t}(x, t) + o(\Delta t^2),$$

$$p^+(x - s\,\Delta t, t) = p^+(x, t) - s\,\Delta t\,\frac{\partial p^+}{\partial x}(x, t) + o(\Delta t^2),$$

$$p^-(x + s\,\Delta t, t) = p^-(x, t) + o(\Delta t).$$

Substituting into (4.39) gives

$$\Delta t\,\frac{\partial p^+}{\partial t}(x, t) + s\,\Delta t\,\frac{\partial p^+}{\partial x}(x, t) = -\lambda\, p^+(x, t)\,\Delta t + \lambda\, p^-(x, t)\,\Delta t + o(\Delta t^2).$$

Dividing by Δt and passing to the limit $\Delta t \to 0$, we obtain

$$\frac{\partial p^+}{\partial t} + s\,\frac{\partial p^+}{\partial x} = -\lambda p^+ + \lambda p^-. \quad (4.40)$$

Using similar arguments (see Exercise 4.5), we derive the evolution equation for $p^-(x, t)$ as

$$\frac{\partial p^-}{\partial t} - s\,\frac{\partial p^-}{\partial x} = \lambda p^+ - \lambda p^-. \quad (4.41)$$

If we are given the initial probability densities

$$p^+(x, 0) = p_0^+(x), \qquad p^-(x, 0) = p_0^-(x), \quad (4.42)$$

then the coupled system of PDEs (4.40)–(4.41) uniquely describes the time evolution of the probability densities p^+ and p^-.

  Now suppose we are interested only in the position of the molecules, not in their instantaneous velocities. We let $p(x, t)\, dx$ be the probability that $X(t) \in [x, x + dx)$. Since this includes molecules travelling both to the left and to the right, $p(x, t)$ is given by

$$p(x, t) = p^+(x, t) + p^-(x, t). \quad (4.43)$$

To write the evolution equation for $p(x, t)$, we first define the auxiliary variable $q = p^+ - p^-$. Then, by adding and subtracting equations (4.40)–(4.41), we find

$$\frac{\partial p}{\partial t} + s\,\frac{\partial q}{\partial x} = 0, \quad (4.44)$$

$$\frac{\partial q}{\partial t} + s\,\frac{\partial p}{\partial x} = -2\lambda q. \quad (4.45)$$

Eliminating q by differentiating (4.45) with respect to x and using (4.44), we obtain the so-called telegraph equation (Kac, 1974) for p:

$$\frac{1}{2\lambda}\frac{\partial^2 p}{\partial t^2} + \frac{\partial p}{\partial t} = \frac{s^2}{2\lambda}\frac{\partial^2 p}{\partial x^2}. \tag{4.46}$$

Equation (4.46) is a hyperbolic PDE, and is different from the diffusion equation we were hoping for. However, it can be shown (Zauderer, 1983; Karch, 2000) that the long-time behaviour of solutions of this equation satisfies the parabolic equation

$$\frac{\partial p}{\partial t} = \frac{s^2}{2\lambda}\frac{\partial^2 p}{\partial x^2}, \tag{4.47}$$

which is obtained by simply dropping the first term. This approximation is good for times $t \gg 1/2\lambda$ (which is what we mean by "long-time" above). In fact, in our example $\lambda = 5 \times 10^{11}\ \text{sec}^{-1}$, and this "long time" starts on a time scale of nanoseconds!

Equation (4.47) is equal to the diffusion equation (4.4) provided that

$$D = \frac{s^2}{2\lambda}.$$

Thus we have derived the formula (4.34) relating the diffusion constant D with the particle speed s and the turning frequency λ of the velocity-jump process.

The initial condition (4.42) is equivalent to the following initial condition for the telegraph equation (4.46):

$$p(x,0) = p_0^+(x) + p_0^-(x), \qquad \frac{\partial p}{\partial t}(x,0) = -s\frac{dp_0^+}{dx}(x) + s\frac{dp_0^-}{dx}(x). \tag{4.48}$$

To determine uniquely the time evolution of $p(x,t)$ given by (4.46), we have to specify not only the initial density $p(\cdot,0)$ but also the initial rate of change $\partial p/\partial t(\cdot,0)$. This is analogous to the initialization of the SSA (a10)–(c10) or the SSA (a11)–(c11), where we have to prescribe not only the initial positions of the diffusing particles but also their initial velocities. However, the additional initial condition (initial velocity) is not important if we study the behaviour of the system for sufficiently large times (in fact, we chose the initial velocity randomly in our computations and it did not significantly influence our results). Roughly speaking, the information about the initial velocity is quickly forgotten after a few turns of the particle. This is another manifestation of the fact that the behaviour of (4.46) is described for times $t \gg 1/2\lambda$ by the diffusion equation (4.47), which requires only the initial density $p(\cdot,0)$ to be known.

Let us now revisit the example from Figure 4.2 in Section 4.1 using this velocity-jump process as the model for diffusion. We study the time evolution of the x-coordinates of molecules diffusing in the interval $[0, L]$, and reflecting at the boundary points $x = 0$ and $x = L$. We can implement this boundary condition by modifying the SSA (a10)–(c10) as follows. Given the initial condition $X(0) = x_0$ and $V(0) = v_0$, we compute the values of $X(t)$ and $V(t)$ at times $t > 0$ iteratively, by performing four steps at time t:

(a12) Generate a random number r uniformly distributed in $(0, 1)$.
(b12) Compute the position of the molecule at time $t + \Delta t$ by (4.35).
(c12) If $X(t + \Delta t)$ computed by (4.35) is less than 0, then
$$X(t + \Delta t) = -X(t) - V(t)\,\Delta t \text{ and } V(t + \Delta t) = -V(t).$$
If $X(t + \Delta t)$ computed by (4.35) is greater than L, then
$$X(t + \Delta t) = 2L - X(t) - V(t)\,\Delta t \text{ and } V(t + \Delta t) = -V(t).$$
(d12) If $r < \lambda \Delta t$, then put $V(t + \Delta t) = -V(t)$.
Otherwise, put $V(t + \Delta t) = V(t)$.
Then continue with step (a12) for time $t + \Delta t$.

The boundary condition is implemented in the step (c12). The position of the molecule is reflected around boundary points as in the step (c8). We also reverse the direction of the particle after the reflection, putting $V(t + \Delta t) = -V(t)$. We choose $D = 10^{-4}\,\text{mm}^2\,\text{sec}^{-1}$ as in Figure 4.2. In addition we have to specify the speed s. As discussed before, the speed s calculated by (4.36) could be of the order of $10\,\text{m}\,\text{sec}^{-1}$ for a typical protein molecule, which would make the SSA (a12)–(d12) computationally intensive. The diffusion model will be less computationally intensive if we consider the behaviour of slower (heavier) particles. Thus we choose $s = 10^{-2}\,\text{mm}\,\text{sec}^{-1}$ and $D = 10^{-4}\,\text{mm}^2\,\text{sec}^{-1}$ for our computations. Choosing $L = 1$ mm, $X(0) = 0.4$ mm and $\Delta t = 0.1$ sec, we plot ten realizations of the SSA (a12)–(d12) in Figure 4.4(c). This panel can be directly compared with results in Figures 4.2(a) and 4.3(a). In Figure 4.4(d), we present results of simulation of 1000 molecules, each following the SSA (a12)–(d12) for the same parameters as before. At time $t = 4$ min, we divide the domain of interest $[0, L]$ into 40 bins of length $h = L/40 = 25\ \mu m$. We calculate the number of molecules in each bin $[(i-1)h, ih)$, $i = 1, 2, \ldots, 40$, at time $t = 4$ min and plot them as a grey histogram. This panel can be directly compared with Figures 4.2(b) and 4.3(b). The red solid line is the solution of the diffusion equation (4.4) with the boundary conditions (4.12).

The SSAs (a10)–(c10), (a11)–(c11) and (a12)–(d12) present velocity-jump processes that have a very small set of possible velocity values, containing only two elements $\{s, -s\}$. In the rest of this section, we will study another model for diffusion, based on a velocity-jump process that allows any real values of velocities. We choose a small time step Δt. Let $X(t)$ and $V(t)$ be the position and velocity of the diffusing particle, respectively. To evolve them, we compute the position $X(t + \Delta t)$ and the velocity $V(t + \Delta t)$ from the position $X(t)$ and the velocity $V(t)$ by the formulae

$$X(t + \Delta t) = X(t) + V(t) \Delta t, \tag{4.49}$$

$$V(t + \Delta t) = V(t) - \beta V(t) \Delta t + \beta \sqrt{2D\Delta t} \, \xi, \tag{4.50}$$

where D is the diffusion constant, β is the so-called friction coefficient and ξ is a normally distributed random variable with zero mean and unit variance. Equation (4.49) is the same as equation (4.35), that is, it is the discretized version of the ODE (4.33). Comparing (4.50) with (3.5), we conclude that (4.50) is the discretized version of the SDE

$$V(t + dt) = V(t) - \beta V(t) \, dt + \beta \sqrt{2D} \, dW. \tag{4.51}$$

This equation can be interpreted as Newton's second law of motion (see Section 4.6) in the following sense. The rate of change of momentum is proportional to the net force on the diffusing particle. This force has two components: a deterministic friction and a random forcing corresponding to collisions by surrounding (solvent) molecules. In the limit of many small collisions the resulting impulse will be normally distributed. We will present a derivation of equation (4.51) from a more detailed model of collisions in Section 8.4.

In the limit $\beta \rightarrow \infty$, equation (4.51) formally gives $V(t) \, dt \approx \sqrt{2D} \, dW$. Substituting this value for $V(t) \, dt$ into (4.49), we obtain the SDE (4.1). Thus it is possible to derive formally the SDE (4.1) from the more complicated velocity-jump process (4.49)–(4.50).

We can also consider the limit $\beta \rightarrow \infty$ in the probability distribution function. To that end we let $f(x, v, t)$ be the probability density that the diffusing molecule is at the position x with the velocity v at time t, i.e. $f(x, v, t) \, dx \, dv$ is the probability that $X(t) \in [x, x + dx]$ with velocity between v and $v + dv$ at time t. It can be shown (Exercise 4.6) that f satisfies the PDE

$$\frac{\partial f}{\partial t} + v \frac{\partial f}{\partial x} = \beta \frac{\partial}{\partial v} \left(vf + \beta D \frac{\partial f}{\partial v} \right). \tag{4.52}$$

This PDE is often also called the Fokker–Planck equation. We will call it the position–velocity Fokker–Planck equation to distinguish it from the Fokker–Planck equation (3.31). Now suppose we are interested only in the distribution of the position of molecules, and not in their velocity distribution. The probability density for finding a diffusing molecule at the point x at time t (with any velocity) is given by integrating f over all possible velocities:

$$p(x, t) = \int_{\mathbb{R}} f(x, v, t) \, dv. \tag{4.53}$$

We hope to derive the diffusion equation for p in the limit that β is large. To this end we rescale the velocity variable by setting

$$v = \eta \sqrt{\beta}, \qquad \overline{f}(x, \eta, t) = f(x, v, t),$$

to give

$$\frac{1}{\beta} \frac{\partial \overline{f}}{\partial t} + \frac{1}{\sqrt{\beta}} \eta \frac{\partial \overline{f}}{\partial x} = \frac{\partial}{\partial \eta} \left(\eta \overline{f} + D \frac{\partial \overline{f}}{\partial \eta} \right). \tag{4.54}$$

We now expand $\overline{f}$ in inverse powers of $\sqrt{\beta}$ as

$$\overline{f}(x, \eta, t) = f_0(x, \eta, t) + \frac{1}{\sqrt{\beta}} f_1(x, \eta, t) + \frac{1}{\beta} f_2(x, \eta, t) + \cdots . \tag{4.55}$$

Substituting (4.55) into (4.54) and equating coefficients of powers of β we obtain the hierarchy of equations

$$\frac{\partial}{\partial \eta} \left(\eta f_0 + D \frac{\partial f_0}{\partial \eta} \right) = 0, \tag{4.56}$$

$$\frac{\partial}{\partial \eta} \left(\eta f_1 + D \frac{\partial f_1}{\partial \eta} \right) = \eta \frac{\partial f_0}{\partial x}, \tag{4.57}$$

$$\frac{\partial}{\partial \eta} \left(\eta f_2 + D \frac{\partial f_2}{\partial \eta} \right) = \eta \frac{\partial f_1}{\partial x} + \frac{\partial f_0}{\partial t}. \tag{4.58}$$

Solving equation (4.56) gives

$$f_0(x, \eta, t) = \varrho(x, t) \exp \left[-\frac{\eta^2}{2D} \right], \tag{4.59}$$

where the function $\varrho(x, t)$ is independent of η. Since f_0 has separated into a function of position multiplied by a function of velocity, we see that in the limit $\beta \to \infty$ the position and velocity of a particle are independent. Introducing the normalization constant we see that velocities are distributed according to the distribution function

$$\frac{1}{\sqrt{2D\beta\pi}} \exp \left[-\frac{v^2}{2D\beta} \right], \tag{4.60}$$

which is known as Maxwell's velocity distribution. However, as yet we have no information on the function $\varrho(x, t)$, which (using (4.53)) is proportional to the positional probability density $p(x, t)$. Proceeding with our expansion we solve equation (4.57) to give

$$f_1(x, \eta, t) = -\frac{\partial \varrho}{\partial x}(x, t) \eta \exp \left[-\frac{\eta^2}{2D} \right]. \tag{4.61}$$

Substituting (4.59) and (4.61) into (4.58) gives

$$\frac{\partial}{\partial\eta}\left(\eta f_2 + D\frac{\partial f_2}{\partial\eta}\right) = -\frac{\partial^2\varrho}{\partial x^2}\,\eta^2\exp\left[-\frac{\eta^2}{2D}\right] + \frac{\partial\varrho}{\partial t}\exp\left[-\frac{\eta^2}{2D}\right].$$

Now integrating over η and using the fact that $f_2 \to 0$ as $\eta \to \pm\infty$ gives the solvability condition

$$\frac{\partial\varrho}{\partial t} = D\frac{\partial^2\varrho}{\partial x^2}.$$

Since p is proportional to ϱ we see that $p(x,t)$ satisfies the diffusion equation (4.4) for large β.

We could now revisit the example of Figures 4.2, 4.3 and 4.4 using the model (4.49)–(4.50). However, we leave the design and implementation of the corresponding SSA to the reader (Exercise 4.7).

4.4 Diffusion to Adsorbing Surfaces

In Sections 4.1, 4.2 and 4.3, we studied diffusion in containers with reflective boundary conditions. In this section, we consider that the boundary (or part of it) is fully adsorbing, i.e. whenever a molecule hits the boundary it is removed from the system. Examples of the sort of questions that we might want to answer are: What is the probability that the molecule is captured by the adsorbing boundary? What is the average time for a molecule to be adsorbed? If we keep the density of molecules constant far from the boundary, what is the average rate of adsorption (i.e. how many molecules are adsorbed per unit of time)? The answers to these questions depend on the geometry of the domain. We will study both one-dimensional and three-dimensional problems in this section. Some of our results will be useful later, in Chapter 6, for development of stochastic approaches to reaction–diffusion modelling. Chemically reactive boundaries (that is, boundaries that are partially adsorbing and partially reflecting) will be studied in Section 4.5.

The SDEs (4.1)–(4.3) give the time evolution of trajectories of Brownian motion in the whole space $\mathbb{R}^3$. Let us consider that the movement of the diffusing particles is restricted to the half space with a positive x-coordinate, i.e. $[X(t), Y(t), Z(t)] \in \{x > 0\}$, where we use $\{x > 0\}$ to denote the three-dimensional space $(0, \infty) \times \mathbb{R} \times \mathbb{R}$. The boundary $\{x = 0\}$ is assumed to be fully adsorbing. Whenever the molecule hits the boundary $\{x = 0\}$ it is removed from the solution. Since the SDEs (4.1)–(4.3) are not coupled and there are no obstacles in the y- and z-directions, we can simulate the time evolution of

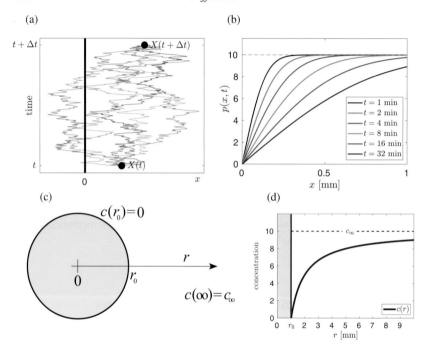

Figure 4.5   (a) *Schematic of five possible trajectories between $X(t)$ and $X(t + \Delta t)$.*
(b) *The concentration profile $p_x(x, t)$ computed by (4.65) at several instants of
time. We use $D = 10^{-4}\ mm^2\ sec^{-1}$ and $p_{in} = 10$.* (c) *Schematic of the adsorbing
sphere.* (d) *The concentration $c(x, y, z) \equiv c(r)$ given by (4.70) for $c_\infty = 10$ and
$r_0 = 1\ mm$.*

the y-coordinate and z-coordinate by formulae (3.16)–(3.17) which we previously studied. Thus we can focus on the evolution of the x-coordinate only.
As before, we choose a small time step Δt and we compute $X(t)$ iteratively by
equation (4.11), that is,

$$X(t + \Delta t) = X(t) + \sqrt{2D\,\Delta t}\ \xi, \tag{4.62}$$

where ξ is a normally distributed random number with zero mean and unit
variance. If $X(t + \Delta t)$ is negative or zero, then the particle has hit the boundary and should be removed from the system. If $X(t + \Delta t)$ is positive, then the
computed position of the particle is at time $t + \Delta t$ in the computational domain
$\{x > 0\}$. However, there is still a non-zero probability that the particle has hit
the boundary $\{x = 0\}$ sometimes during the time interval $(t, t+\Delta t)$. The situation
is schematically shown in Figure 4.5(a). The positions of the molecule at times
t and $t + \Delta t$ are denoted as black circles. They were computed by (4.62). We

also plot five possible trajectories given by the SDE (4.1) that start at $X(t)$ and end at $X(t + \Delta t)$. Some of these trajectories (red and green) cross the boundary $\{x = 0\}$ and should be removed from the system. Other trajectories stay in the half space $\{x > 0\}$ during the time interval $(t, t + \Delta t)$. Thus there is a non-zero probability that the molecule has hit the boundary $\{x = 0\}$ even if $X(t + \Delta t)$, computed by the discretization (4.62), is positive. We leave it to the reader to compute (Exercise 4.8) that the probability that the particle hit the boundary during the time interval $(t, t + \Delta t)$ is equal to

$$\exp\left[\frac{-X(t)X(t + \Delta t)}{D\Delta t}\right]. \tag{4.63}$$

Thus, we can implement the fully adsorbing boundary condition at $\{x = 0\}$ for the SDE model (4.1)–(4.3) as follows. We compute the position $[X(t + \Delta t), Y(t + \Delta t), Z(t + \Delta t)]$ from the position $[X(t), Y(t), Z(t)]$ by the formulae (3.15)–(3.17). If $X(t + \Delta t)$ computed by (3.15) is negative or zero, then we remove the molecule from the solution. If $X(t + \Delta t)$ computed by (3.15) is positive, then we generate a random number r uniformly distributed in $(0, 1)$ and we remove the molecule from the solution if $r < \exp[-X(t)X(t + \Delta t)/(D\Delta t)]$.

Let us now consider a system of many molecules that diffuse in the domain $\{x > 0\}$ with the adsorbing boundary condition at $x = 0$. Let $p_x(x, t)\, dx$ be the probability of finding a molecule in $[x, x + dx)$, which, as we have seen, satisfies the diffusion equation (4.4). The boundary condition for p_x at $x = 0$ is $p_x(0, t) = 0$ for $t > 0$. If we suppose that the initial density $p_x(\cdot, 0)$ is equal to the constant p_{in}, then the time evolution of $p_x(x, t)$ can be computed as the solution of the diffusion equation (4.4) subject to the boundary conditions

$$p_x(0, t) = 0, \qquad \lim_{x \to \infty} p_x(x, t) = p_{in} \tag{4.64}$$

for $t \in [0, \infty)$, and the initial condition $p_x(x, 0) \equiv p_{in}$ for $x \in (0, \infty)$. The solution is given by (see Exercise 4.9)

$$p_x(x, t) = p_{in}\, \mathrm{erf}\left(\frac{x}{2\sqrt{Dt}}\right), \tag{4.65}$$

where the error function is defined by the integral

$$\mathrm{erf}(z) = \frac{2}{\sqrt{\pi}} \int_0^z \exp[-\zeta^2]\, d\zeta.$$

Choosing $D = 10^{-4}\,\mathrm{mm}^2\,\mathrm{sec}^{-1}$ and $p_{in} = 10$, we plot the profile p_x given by equation (4.65) at several instants of time in Figure 4.5(b). We see that the adsorbing boundary $x = 0$ is sucking molecules from the solution. At any fixed point in space $[x, y, z]$ with $x > 0$ we have

$$\lim_{t \to \infty} p_x(x, t) = 0. \tag{4.66}$$

Even though the constant initial condition $p(\cdot, 0) \equiv p_{in} > 0$ means that we have an infinite number of molecules available in the half space $\{x > 0\}$, we also have very large adsorbing boundary – the infinite plane $\{x = 0\}$. Thus the adsorbing boundary has the potential to remove molecules from any point of the plane. In particular, the formula (4.66) implies that the solution does not converge to a non-trivial steady-state profile.

The situation is more interesting when we consider adsorption of molecules to finite size objects. In Figure 4.5(c), there is a schematic of the adsorption of molecules by a sphere of radius r_0. We study the diffusion of molecules in the domain exterior to the sphere. Such an adsorbing sphere could be a caricature of a living cell surrounded by its membrane with receptors. Our goal is to find how the fully adsorbing sphere influences the concentration of molecules in its neighbourhood. In this case we will see that there is a well-defined steady-state concentration profile around the adsorbing sphere. To that end let $c(x, y, z)$ be the (average) concentration of molecules in the solution at the steady state. Then c satisfies the steady-state version of the diffusion equation (4.8), which is the Laplace equation

$$D \left(\frac{\partial^2 c}{\partial x^2} + \frac{\partial^2 c}{\partial y^2} + \frac{\partial^2 c}{\partial z^2} \right) = 0. \tag{4.67}$$

We assume that the concentration far from the sphere is held constant, so that the boundary conditions on (4.67) are

$$c(x, y, z) \equiv c(r) = 0, \qquad \text{for } r = r_0, \tag{4.68}$$

$$c(x, y, z) \equiv c(r) = c_\infty, \qquad \text{for } r \to \infty, \tag{4.69}$$

where

$$r = \sqrt{x^2 + y^2 + z^2}.$$

The solution of the problem (4.67)–(4.69) is given by (see Exercise 4.10)

$$c(x, y, z) = c_\infty \left(1 - \frac{r_0}{\sqrt{x^2 + y^2 + z^2}} \right) = c_\infty \left(1 - \frac{r_0}{r} \right) \qquad \text{for } r > r_0. \tag{4.70}$$

Choosing $r_0 = 1$ mm and $c_\infty = 10$, we plot the steady-state concentration profile as a function of r in Figure 4.5(d). An important characteristic of the adsorption process is the average number of molecules that are removed by the adsorbing surface per unit of time. Since the solution (4.70) is radially symmetric, the flux (that is, the average number of molecules that are transported per unit area per unit time) is constant along the boundary $r = r_0$. We can compute the boundary flux density as

$$J(r_0) = D \frac{\partial c}{\partial r}(r_0, 0, 0) = \frac{D c_\infty}{r_0}.$$

Consequently, the average number of molecules that are removed by the adsorbing sphere per unit time is given as a product of $J(r_0)$ and the area of the sphere $4\pi r_0^2$, so that

$$J_{\text{sphere}} = 4\pi D c_\infty r_0. \tag{4.71}$$

We mentioned that the adsorbing sphere can be considered as a useful caricature of a cell covered with receptors. In reality, only part of the cell is covered with receptors. Thus we can get an improved understanding of the process by considering a (non-adsorbing) sphere that is covered by small adsorbing discs – see Figure 4.6(a). We suppose as before that the radius of the sphere is r_0, and that now there are k fully adsorbing discs of diameter s on its surface. If $s \ll r_0$ and the area occupied by receptors is a small fraction of the total surface area then the average number of molecules that are removed by the sphere per unit of time is

$$J_{\text{sphere with receptors}} = \frac{4\pi D c_\infty r_0 ks}{ks + 2\pi r_0} = \frac{J_{\text{sphere}}}{1 + 2\pi r_0 / ks}. \tag{4.72}$$

The derivation of this formula can be done in several ways (Berg and Purcell, 1977; Shoup and Szabo, 1982; Berg, 1983). One possibility is to use the theory of partially adsorbing boundary conditions which is introduced in Section 4.5: a molecule is either adsorbed (if it hits an adsorbing disc) or reflected. This, roughly speaking, resembles the definition of partially adsorbing boundaries.

Following Berg (1983), let us investigate some of the consequences of (4.72). We consider a cell with radius $r_0 = 5 \ \mu$m covered by $k = 25\,000$ receptors (adsorbing discs) each of diameter $s = 4$ nm. The surface of the cell has the area $4\pi r_0^2$, the area of an adsorbing disc is $\pi s^2 / 4$ and the fraction of the surface covered by receptors is

$$\frac{k\pi s^2 / 4}{4\pi r_0^2} = \frac{ks^2}{16 r_0^2} = 10^{-3}.$$

Thus only 0.1% of the surface is covered by the receptors. Using our parameter values, formula (4.72) implies

$$J_{\text{sphere with receptors}} = \frac{J_{\text{sphere}}}{1 + \pi / 10} \approx 0.76 \times J_{\text{sphere}}.$$

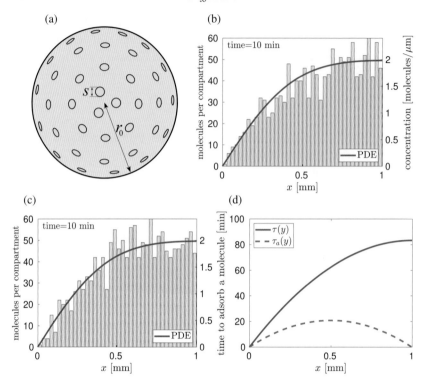

Figure 4.6 (a) *Sphere with adsorbing discs.* (b) *Numbers of molecules in bins of length* $h = 25\ \mu m$ *at time* $t = 10\,min$*. We have initially 2000 molecules in the solution, each following the SSA (a13)–(e13). The solution of (4.4) with the boundary conditions (4.73) is plotted as the red solid line. (c) The results obtained by the SSA (a5)–(d5) applied to the system of chemical reactions (4.13) and (4.74). (d) Average time to adsorb a molecule. The functions* $\tau(y)$ *and* $\tau_a(y)$ *are given by (4.77) and (4.78). We use* $D = 10^{-4}\ mm^2\ sec^{-1}$ *and* $L = 1\,mm$*.*

Consequently, a small fraction of the surface (0.1%) has 76% of the "adsorption power" of a fully adsorbing sphere. This is a remarkable fact explaining why the cellular membrane can accommodate many specialized receptors, each adsorbing molecules of a specific kind with an efficiency of the same order as the efficiency of a surface fully covered by these receptors.

We will further build on our discussion of diffusion in a three-dimensional space exterior to an adsorbing sphere in Section 6.6, when we discuss the reaction radius of bimolecular diffusion-limited reactions. In that case, it is also important to distinguish between one-dimensional and three-dimensional diffusion processes.

In the rest of this section, we return to SSAs for diffusion of molecules in a one-dimensional interval $[0, L]$, and impose a fully adsorbing boundary at $x = 0$. We assume that the boundary $x = L$ is reflecting. Then, using (4.63), we can reformulate the SSA (a8)–(c8) as follows. Given the initial condition $X(0) = x_0$, we compute the value of $X(t)$ at times $t > 0$ iteratively, by performing five steps at time t:

(a13) Generate a normally distributed (with zero mean and unit variance) random number ξ.

(b13) Compute the position of the molecule at time $t + \Delta t$ by (4.62).

(c13) If $X(t + \Delta t) < 0$, then terminate the trajectory.

(d13) If $X(t + \Delta t) \in [0, L]$, then generate a random number r uniformly distributed in the interval $(0, 1)$ and terminate the trajectory if $r < \exp[-X(t)X(t + \Delta t)/(D\Delta t)]$.

(e13) If $X(t + \Delta t)$ computed by (4.62) is greater than L, then
$$X(t + \Delta t) = 2L - X(t) - \sqrt{2D \Delta t}\, \xi.$$
Then continue with step (a13) for time $t + \Delta t$.

We choose $D = 10^{-4} \, \text{mm}^2 \, \text{sec}^{-1}$, $L = 1 \, \text{mm}$ and $\Delta t = 0.1 \, \text{sec}$. We consider 2000 molecules following the SSA (a13)–(e13). The initial position of each molecule is chosen randomly in $[0, L]$, i.e. we consider that the concentration of molecules is initially uniform. At time $t = 10 \, \text{min}$, we divide the domain of interest $[0, L]$ into 40 bins of length $h = L/40 = 25 \, \mu\text{m}$ and we calculate the number of molecules in each bin $[(i - 1)h, ih)$, $i = 1, 2, \ldots, 40$. The results are plotted in Figure 4.6(b) as the grey histogram. The deterministic counterpart to the SSA (a13)–(e13) is the solution of the diffusion equation (4.4) with boundary conditions

$$p_x(0, t) = 0, \qquad \frac{\partial p_x}{\partial x}(L, t) = 0. \qquad (4.73)$$

The solution of (4.4), with the boundary conditions (4.73) and initial density equal to a constant function, is plotted as the red solid line in Figure 4.6(b) for comparison.

The compartment-based diffusion model was studied in Section 4.2. Considering a one-dimensional interval $[0, L]$ with reflective boundary conditions, the compartment-based model can be formulated as the chain of "chemical reactions" (4.13). To implement the fully adsorbing boundary condition at $x = 0$, we simply add another reaction (a jump from the first compartment to the left where the molecules are removed from the solution)

$$A_1 \xrightarrow{d} \emptyset, \qquad (4.74)$$

where d is given as usual by (4.14). Equation (4.74) states that all diffusion jumps to the left from the boundary compartment lead to the removal of the jumping molecule. In a similar way, one could also model the reflective boundary condition as jumps to the right from the last compartment. Every molecule leaving the last compartment would jump back to the last compartment. The resulting "reaction" would be $A_K \longrightarrow A_K$. Since this "reaction" does not change the system it was not included previously, and we do not need to include it here. Thus a reflective boundary condition is modelled by the chain of reactions (4.13) only, while a fully adsorbing boundary (at $x = 0$) is modelled by including the additional reaction (4.74).

In Figure 4.6(c), we present results obtained by the Gillespie SSA (a5)–(d5) applied to the system of chemical reactions (4.13) and (4.74). We use the same parameters as in Figure 4.6(b) to enable direct comparison. The initial condition is chosen as 50 molecules per compartment and the number of compartments is $K = 40$. The solution of (4.4) with the boundary conditions (4.73) is plotted as the red solid line.

Finally let us consider the average time taken for a molecule to be adsorbed. Let $\tau(y)$ be the average time to adsorb a molecule given that it started at $X(0) = y$. Since the adsorbing boundary is at $x = 0$, we can compute the average time to adsorb a molecule by solving equation (3.60), which in this case reads

$$-1 = D\frac{d^2\tau}{dy^2}(y) \qquad \text{for } y \in (0, L), \qquad (4.75)$$

with the boundary conditions

$$\tau(0) = 0, \qquad \frac{d\tau}{dy}(L) = 0. \qquad (4.76)$$

Integrating (4.75), we obtain

$$\tau(y) = \frac{2Ly - y^2}{2D}. \qquad (4.77)$$

This function is plotted for $D = 10^{-4}$ mm^2 sec^{-1} and $L = 1$ mm in Figure 4.6(d) as the blue solid line. If we alter the problem so that the right boundary $y = 0$ is also adsorbing (i.e. $\tau(0) = \tau(L) = 0$ instead of the boundary conditions (4.76)) then the mean time to adsorb a molecule that started at $X(0) = y$ is given by

$$\tau_a(y) = \frac{Ly - y^2}{2D}. \qquad (4.78)$$

This curve is also plotted in Figure 4.6(d) as the red dashed line.

4.5 Reactive Boundary Conditions

A partially adsorbing boundary condition for SSAs can be formulated as follows: *whenever a molecule hits the boundary, it is adsorbed with some probability, and otherwise it is reflected.* So far we have seen the following special cases of this boundary condition: (a) the molecule is always reflected (studied in Sections 4.1, 4.2 and 4.3); and (b) the molecule is always adsorbed (introduced in Section 4.4). The reflecting boundary condition is often used when no adsorption of the diffusing molecules on the boundary takes place. On the other hand, if the molecule can chemically or physically attach to the boundary, then one has to assume that the boundary is (at least) partially adsorbing.

The key question is: What is the probability that the particle is adsorbed rather than reflected, and how does this relate to the reactive properties of the boundary for a given stochastic model? To answer this question, let us consider a chemical species A diffusing in the half space $\{x > 0\} = (0, \infty) \times \mathbb{R} \times \mathbb{R}$. We suppose that the boundary $\{x = 0\}$ is uniformly covered by receptors B. Molecules of A and B react according to the second-order chemical reaction

$$A + B \xrightarrow{k} \emptyset. \qquad (4.79)$$

Since the receptors are uniformly distributed on the boundary, there will be no concentration differences in the y- and z-directions. Thus we focus on the behaviour along the x-axis only. Assuming that we have a lot of molecules of A in the system, we can describe the system in terms of the density $\varrho(x, t)$ of molecules at the point $x \in [0, \infty)$ and time t, so that $\varrho(x, t)\, \delta x$ gives the number of molecules in the small box $[x, x + \delta x) \times [0, 1] \times [0, 1]$ at time t. The time evolution of $\varrho(x, t)$ is governed by the diffusion equation

$$\frac{\partial \varrho}{\partial t} = D \frac{\partial^2 \varrho}{\partial x^2}, \qquad (4.80)$$

where D is the diffusion constant. Let us now consider a small box $[0, \delta x) \times [0, 1] \times [0, 1]$ which lies just next to the boundary $x = 0$. It contains $\varrho(0, t)\, \delta x$ molecules of A on average. Let $b(t)$ be the number of available receptors B per unit of area of the boundary $\{x = 0\}$. Then, using Table 1.1, the probability that the chemical reaction (4.79) takes place in the time interval $[t, t + dt)$ is equal to

$$\left(\varrho(0, t)\, \delta x\, b(t)\, k/\delta x \right) dt = \sigma(t)\, \varrho(0, t)\, dt,$$

where we have written $\sigma(t) = kb(t)$. In what follows, we will assume that σ is independent of time, i.e. we consider the case where the number of available receptors does not change with time. This is a reasonable assumption when the number of receptors on the boundary is in excess of the number of diffusing molecules. The generalization to a time-dependent rate constant $\sigma(t)$ can be found in Erban and Chapman (2007b). The diffusive flux through the boundary (i.e. the average number of molecules that are removed by the boundary per unit area and per unit time) is thus

$$D \frac{\partial \varrho}{\partial x}(0, t) = \sigma \varrho(0, t), \qquad (4.81)$$

which is the so-called Robin boundary condition (or reactive boundary condition) at $x = 0$. The non-negative constant σ (which has units $\text{mm}\,\text{sec}^{-1}$) describes the reactivity of the boundary. A reflective boundary corresponds to $\sigma = 0$ and a fully adsorbing boundary corresponds to $\sigma = \infty$.

Our goal is to study the connections between the boundary conditions of the stochastic simulation and the Robin boundary condition (4.81) of the macroscopic diffusion equation (4.80). In particular, we will provide the relation between σ in (4.81) and the parameters of each SSA. The mathematical derivation of these relations is given in Erban and Chapman (2007a). For definiteness we consider the behaviour of molecules in a one-dimensional interval $[0, L]$ with a Robin boundary condition at $x = 0$ and a reflective boundary condition at $x = L$.

Let us consider first the compartment-based model of diffusion, given as the chain of "chemical reactions" (4.13). To implement the reactive boundary condition at $x = 0$, we simply add another reaction (a jump from the first compartment to the left where the molecules are removed from the solution)

$$A_1 \xrightarrow{\sigma/h} \emptyset. \qquad (4.82)$$

The key quantity here is the reaction rate σ/h, which is exactly what is needed to implement the Robin boundary condition (4.81). To see this, let $p(\mathbf{n}, t)$ be the joint probability that $A_i(t) = n_i$, $i = 1, 2, \dots, K$, where $\mathbf{n} = [n_1, n_2, \dots, n_K]$. Recalling the definition of the operators $R_i, L_i \colon \mathbb{N}^K \to \mathbb{N}^K$ by (4.20)–(4.21), the chemical master equation which corresponds to the system of chemical reactions given by (4.13) and (4.82) can be written as

$$\frac{\partial p(\mathbf{n})}{\partial t} = d \sum_{j=1}^{K-1} \left\{ (n_j + 1) p(R_j \mathbf{n}) - n_j p(\mathbf{n}) \right\}$$

$$+ d \sum_{j=2}^{K} \left\{ (n_j + 1) p(L_j \mathbf{n}) - n_j p(\mathbf{n}) \right\} \qquad (4.83)$$

$$+ \frac{\sigma}{h} \left\{ (n_1 + 1) p(n_1 + 1, n_2, n_3, \dots, n_K) - n_1 p(\mathbf{n}) \right\}.$$

Let us define the mean vector $\mathbf{M}(t) \equiv [M_1, M_2, \ldots, M_K]$ by (4.23). We have seen that away from the boundary $\mathbf{M}$ satisfies (4.24). To determine the boundary value M_1 we multiply (4.83) by n_1 and sum over $\mathbf{n}$ to give

$$\frac{\partial M_1}{\partial t} = d(M_2 - M_1) - \frac{\sigma M_1}{h}. \tag{4.84}$$

The mid-point of the kth compartment is $x_k = kh + h/2$, and by identifying $M_k(t)/h$ with $\varrho(x_k, t)$ we have seen that ϱ satisfies the diffusion equation. Using this same approximation in (4.84) and using (4.14) we obtain

$$h\frac{\partial \varrho}{\partial t}(h/2, t) = D\frac{\varrho(3h/2, t) - \varrho(h/2, t)}{h} - \sigma \varrho(h/2, t). \tag{4.85}$$

Taylor expanding $\varrho(3h/2, t)$ about the point $x = h/2$ gives

$$\varrho(3h/2, t) = \varrho(h/2, t) + h\frac{\partial \varrho}{\partial x}(h/2, t) + O(h^2),$$

so that (4.85) can be written

$$D\frac{\partial \varrho}{\partial x}(h/2, t) = \sigma \varrho(h/2, t) + O(h).$$

Passing to the limit $h \to 0$, we obtain the Robin boundary condition (4.81).

In Figure 4.7(a), we present the results obtained by the Gillespie SSA (a5)–(d5) applied to the system of chemical reactions (4.13) and (4.82). We use $D = 10^{-4}\,\mathrm{mm}^2\,\mathrm{sec}^{-1}$, $\sigma = 10^{-3}\,\mathrm{mm\,sec}^{-1}$ and $L = 1\,\mathrm{mm}$. We divide the computational domain $[0, L]$ into $K = 40$ compartments of length $h = L/K = 25\,\mu m$. Thus the diffusion jump rate is $d = D/h^2 = 0.16\,\mathrm{sec}^{-1}$. Starting with the uniform initial condition $A_i(0) = 400$, $i = 1, 2, \ldots, K$, we compute the number of molecules $A_i(t)$, $i = 1, 2, \ldots, K$, at time $t = 10\,\mathrm{min}$ by the SSA (a5)–(d5). They are plotted as a grey histogram in Figure 4.7(a). The red line is the solution of the diffusion equation (4.80) with the Robin boundary condition (4.81) at $x = 0$ and zero-flux boundary condition at $x = L$. The results compare well. The noise is smaller than in the previous sections because we simulated a system with more molecules than before. We started with $400K = 16\,000$ molecules, which is eight times more than in Figure 4.6(b) and (c).

We have seen that molecules are removed from the first compartment with the rate σ/h, while the diffusion jumps to the left occur with the rate $d = D/h^2$ (see (4.82)). Consequently, the probability of the removal of a molecule that has hit the boundary $x = 0$ is equal to

$$P_1(h) = \frac{\sigma/h}{d} = \frac{\sigma h}{D}, \tag{4.86}$$

where h has to be chosen sufficiently small to ensure that $P_1(h)$ is less than 1. Thus the boundary reaction (4.82) can be reformulated as the following

(a)   (b)

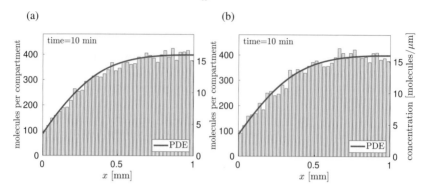

Figure 4.7 (a) *Numbers $A_i(t)$, $i = 1, 2, \ldots, K$, at time $t = 10\,min$ obtained by the SSA (a5)–(d5) applied to the system of chemical reactions (4.13) and (4.82). We use $D = 10^{-4}\,mm^2\,sec^{-1}$, $\sigma = 10^{-3}\,mm\,sec^{-1}$, $L = 1\,mm$, $K = 40$ and $h = L/K = 25\,\mu m$. At time $t = 0$, there are 16 000 molecules uniformly distributed in the domain $[0, L]$. (b) The results obtained by the SDE model (4.1) and the partially adsorbing boundary condition (a_p)–(b_p). We choose the finite time step $\Delta t = 0.1\,sec$. The other parameters are chosen as in panel (a).*

partially adsorbing boundary condition at $x = 0$: *whenever a molecule hits the boundary, it is adsorbed with the probability $P_1(h)$, and otherwise it is reflected.* In this way we have rewritten the boundary condition using the general formulation introduced at the beginning of this section. This form will be particularly useful when comparing with the other models of diffusion.

The diffusion model based on SDEs was introduced as (4.1)–(4.3) in Section 4.1. To compute the trajectory of diffusing molecules, we chose the time step Δt and we used the computational definition (3.5) of SDEs, i.e. we computed the solution of (4.1)–(4.3) iteratively by (3.15)–(3.17). The reactive boundary condition at $x = 0$ can be implemented as follows: *whenever a molecule hits the boundary, it is adsorbed with probability $P_2(\Delta t)$, and otherwise it is reflected.* Obviously, if $X(t + \Delta t)$ computed by (3.15) is negative, a molecule has hit the boundary. However, as shown in (4.63), there is also a chance that a molecule hit the boundary during the finite time step even if $X(t + \Delta t)$ computed by (3.15) is positive; that is, during the time interval $(t, t + \Delta t)$ the molecule might have crossed to X negative and then crossed back to X positive again. This probability is equal to (4.63). Consequently, the partially adsorbing boundary condition is implemented using the following two steps:

(a_p) If $X(t + \Delta t)$ computed by (3.15) is negative then
$\qquad X(t + \Delta t) = -X(t) - \sqrt{2D\,\Delta t}\,\xi$ with probability $1 - P_2(\Delta t)$,
$\qquad$ otherwise we remove the molecule from the system.

(b$_p$) If $X(t + \Delta t)$ computed by (3.15) is positive then we remove the molecule from the system with probability $\exp[-X(t)X(t + \Delta t)/(D\Delta t)]P_2(\Delta t)$.

The partially adsorbing boundary condition (a$_p$)–(b$_p$) leads to the Robin boundary condition (4.81) if we choose

$$P_2(\Delta t) = \frac{\sigma \sqrt{\pi \Delta t}}{2 \sqrt{D}}. \tag{4.87}$$

Let us note that some authors use the case (a$_p$) only as the implementation of the partially reflective boundary condition, i.e. they do not take the correction (b$_p$) into account. Considering the random walk (3.15) with the boundary condition (a$_p$) only, the probability $P_2(\Delta t)$ can be computed as

$$P_2(\Delta t) = \frac{\sigma \sqrt{\pi \Delta t}}{\sqrt{D}}. \tag{4.88}$$

Comparing (4.87) and (4.88), we see that we lose a factor of 2 if we do not consider the correction (b$_p$); without this correction the molecule hits the boundary less frequently, so that the probability of reaction at each collision must be increased to compensate. Such a compensation is possible provided that P_2 given by (4.88) is not greater than 1. The mathematical justification of formulae (4.87) and (4.88) is a bit too complicated to include here; it is presented in Erban and Chapman (2007a).

In Figure 4.7(b), we redo the example from Figure 4.7(a) using the SDE model (4.1) and the boundary condition (a$_p$)–(b$_p$) introduced above. We choose the finite time step $\Delta t = 0.1$ sec. The other parameters are chosen as in Figure 4.7(a). We simulate a collection of 16 000 molecules that are initially uniformly distributed in $[0, L]$. Each trajectory follows (4.1) together with the partially adsorbing boundary condition (a$_p$)–(b$_p$) at $x = 0$ and the reflective boundary condition at $x = L$. At time $t = 10$ min, we divide the computational domain $[0, L]$ into $K = 40$ compartments of length $h = L/K = 25 \, \mu m$. We plot the number of molecules in compartments as a grey histogram in Figure 4.7(b). The red line is the solution of the diffusion equation (4.80) with the Robin boundary condition (4.81) at $x = 0$ and the zero-flux boundary condition at $x = L$.

Using our parameter values, the probabilities that the molecule is adsorbed rather than reflected are

$$P_1(h) = \frac{\sigma h}{D} = 25\%, \qquad P_2(\Delta t) = \frac{\sigma \sqrt{\pi \Delta t}}{2 \sqrt{D}} = 2.8\%.$$

Unless we are very lucky with the choice of parameters, the probabilities $P_1(h)$ and $P_2(\Delta t)$ differ. They depend on different (small) parameters h and

Δt, respectively, which must be small enough that $P_1(h) < 1$ and $P_2(\Delta t) < 1$. The convergence to the Robin boundary condition (4.81) can be proven in the limits $h \to 0$ and $\Delta t \to 0$, respectively.

The reactive boundary condition for the velocity-jump process given by the SSA (a11)–(c11) can be formulated as: *whenever a molecule hits the boundary, it is adsorbed with probability $P_3(s)$, and otherwise it is reflected.* Similarly the reactive boundary condition for the velocity-jump process (4.49)–(4.50) can be formulated as: *whenever a molecule hits the boundary, it is adsorbed with probability $P_4(\beta)$, and otherwise it is reflected.* We use

$$P_3(s) = \frac{2\sigma}{s} \quad \text{and} \quad P_4(\beta) = \frac{\sigma \sqrt{2\pi}}{\sqrt{D\beta}}. \tag{4.89}$$

Then the Robin boundary condition is recovered in the limit $s \to \infty$ and $\beta \to \infty$, respectively (Erban and Chapman, 2007a), which are the same limits we used to derive the diffusion equation from the corresponding SSAs in Section 4.3. Again, we see that the probability that the molecule is adsorbed rather than reflected depends on the choice of the diffusion model as well as on the reactivity of the boundary σ. The Robin boundary condition can also be derived for a variant of the velocity-jump process (4.49)–(4.50), which models a reactive boundary in terms of a boundary potential instead of the boundary probability $P_4(\beta)$; see Chapman et al. (2016) for details.

4.6 Einstein–Smoluchowski Relation

In Section 4.3, we formulated the diffusion model in the form of Newton's second law of motion. We coupled the ODE (4.33) with the SDE (4.51). The discretized form of the equations (4.33) and (4.51) was given by (4.49)–(4.50), namely

$$X(t + \Delta t) = X(t) + V(t)\,\Delta t, \tag{4.90}$$

$$V(t + \Delta t) = V(t) - \beta V(t)\,\Delta t + \beta\,\sqrt{2D\Delta t}\,\xi, \tag{4.91}$$

where Δt is a finite time step.

We now derive the Einstein–Smoluchowski relation for the diffusion coefficient by evaluating the mean-square displacement $\langle X^2 \rangle = \mathrm{E}[X^2]$ in two different ways.[*] First we multiply (4.91) by $X(t)$ to give

---

[*] Note that both forms $\mathrm{E}[\cdot]$ and $\langle \cdot \rangle$ are used to denote expected values (averages over many realizations) in this book. While notation $\mathrm{E}[\cdot]$ is common in the probability literature, symbol E is often reserved to denote energy in the physics literature, where notation $\langle \cdot \rangle$ originates. In this section and later in Chapter 8, we use $\langle \cdot \rangle$ to denote averages (and E to denote energy).

$$V(t + \Delta t)X(t) - V(t)X(t) = -\beta V(t)X(t)\,\Delta t + \beta\,\sqrt{2D\Delta t}\,X(t)\,\xi.$$

Using (4.90), we may write this as

$$V(t + \Delta t)X(t + \Delta t) - V(t)X(t) - V(t + \Delta t)V(t)\,\Delta t$$
$$= -\beta V(t)X(t)\,\Delta t + \beta\,\sqrt{2D\Delta t}\,X(t)\,\xi.$$

Using ODE (4.33) and dividing by Δt gives

$$\frac{1}{2\Delta t}\left(\frac{dX^2}{dt}(t + \Delta t) - \frac{dX^2}{dt}(t)\right) - V(t + \Delta t)V(t) = -\frac{\beta}{2}\frac{dX^2}{dt}(t) + \beta\,\sqrt{\frac{2D}{\Delta t}}\,X(t)\,\xi.$$

Now we take an average (keeping in mind that ξ is normally distributed with zero mean) to give

$$\frac{1}{\Delta t}\left(\frac{d\langle X^2\rangle}{dt}(t + \Delta t) - \frac{d\langle X^2\rangle}{dt}(t)\right) - 2\langle V(t + \Delta t)V(t)\rangle = -\beta\frac{d\langle X^2\rangle}{dt}(t),$$

where $\langle\cdot\rangle$ denotes the average. Passing to the limit $\Delta t \to 0$ gives

$$\frac{1}{\beta}\frac{d^2\langle X^2\rangle}{dt^2}(t) + \frac{d\langle X^2\rangle}{dt}(t) = \frac{2\langle V(t)^2\rangle}{\beta}. \tag{4.92}$$

As discussed in Section 4.3, the kinetic energy associated with the movement of a small particle along the x-axis is equal to $k_B T/2$, where T is the absolute temperature and $k_B = 1.38 \times 10^{-14}\,\mathrm{g\,mm^2\,sec^{-2}\,K^{-1}}$ is the Boltzmann constant. Thus $m_p\langle V(t)^2\rangle/2 = k_B T/2$, where m_p is the particle mass. Consequently, equation (4.92) implies

$$\frac{1}{\beta}\frac{d^2\langle X^2\rangle}{dt^2}(t) + \frac{d\langle X^2\rangle}{dt}(t) = \frac{2k_B T}{\beta m_p}.$$

Since β is a large parameter, we can (for times $t \gg 1/\beta$) drop the second-order time derivative to give our first expression for the evolution of $\langle X^2\rangle$,

$$\frac{d\langle X^2\rangle}{dt}(t) = \frac{2k_B T}{\beta m_p}. \tag{4.93}$$

To derive an alternative expression we take the square of the equation (4.11) to give

$$X(t + \Delta t)^2 = X(t)^2 + 2X(t)\,\sqrt{2D\,\Delta t}\,\xi + 2D\,\Delta t\,\xi^2.$$

Taking averages (again using the fact that ξ is normally distributed with zero mean and unit variance), and dividing by Δt, we get

$$\frac{\langle X(t + \Delta t)^2\rangle - \langle X(t)^2\rangle}{\Delta t} = 2D.$$

Passing to the limit $\Delta t \to 0$ gives our second expression for the evolution of $\langle X^2 \rangle$,

$$\frac{d\langle X^2 \rangle}{dt}(t) = 2D. \tag{4.94}$$

Comparing (4.93) and (4.94) we see that

$$D = \frac{k_B T}{\gamma}, \tag{4.95}$$

where we have written $\gamma = \beta m_p$. Equation (4.95) is the Einstein–Smoluchowski relation. The coefficient γ is called the frictional drag coefficient, and, unlike β, it is independent of the particle mass m_p. To see that, let us multiply the equation (4.51) by m_p, to obtain

$$m_p(V(t + dt) - V(t)) = -\gamma V(t)\, dt + \gamma \sqrt{2D}\, dW.$$

If we formally divide this equation by dt and pass to the limit $dt \to 0$, we get

$$m_p \frac{dV}{dt}(t) = -\gamma V(t) + \gamma \sqrt{2D}\, \frac{\text{``}dW\text{''}}{dt}, \tag{4.96}$$

which (apart from the noise term) is a more familiar way to write Newton's second law of motion. There are two forces on the right-hand side: the random force and frictional force $F_f(t) = -\gamma V(t)$; we see that γ is indeed the coefficient that relates the frictional drag force $F_f(t)$ to the particle velocity $V(t)$. The derivative of W is in quotes in equation (4.96) to highlight that this is not a derivative in the usual sense (W is not differentiable).

The value of γ depends on the geometry of the object and properties of the surrounding fluid. If the diffusing particle is a sphere of radius r diffusing in incompressible viscous liquid, Stokes law says that

$$\gamma = 6\pi\eta r,$$

where η is the coefficient of viscosity. Substituting into (4.95), we obtain the Einstein (or Stokes–Einstein) relation

$$D = \frac{k_B T}{6\pi\eta r}, \tag{4.97}$$

which relates the diffusion constant D to the radius of the diffusing sphere r, the absolute temperature T and the viscosity of the solution η.

We note that there are several alternative derivations of (4.97), all of which have to borrow some facts from statistical physics (Einstein, 1905; Chandrasekhar, 1943; Berg, 1983). In the derivation we have presented, all we needed from statistical physics was that the kinetic energy associated with the movement of a small particle along the x-axis is equal to $k_B T/2$. We will return to equation (4.96) in Chapter 8 where we consider more detailed models of the diffusion process and show how the stochastic models discussed in this chapter can be derived from the underlying detailed models. In particular, we also get a better description and understanding of the noise term in equation (4.96).

To close this chapter, we use the Stokes–Einstein relation (4.97) to determine the diffusion coefficient of a typical protein molecule. If we approximate the molecule as a sphere of radius $r = 2$ nm, in water ($\eta = 10^{-3}$ g mm^{-1} sec^{-1}) at room temperature ($T = 300$ K), then (4.97) gives $D \approx 10^{-4}$ mm^2 sec^{-1}, which is the value we used in all the examples presented in this chapter.

Exercises

4.1   Verify that (4.9) is a solution of (4.8).

4.2   Derive the system of equations (4.24)–(4.25).

4.3   Consider the stationary diffusion master equation corresponding to the chemical reaction system (4.13) which is given by (compare with (4.22))

$$0 = \sum_{j=1}^{K-1} \left\{ (n_j + 1)\, p_s(R_j\mathbf{n}) - n_j\, p_s(\mathbf{n}) \right\} + \sum_{j=2}^{K} \left\{ (n_j + 1)\, p_s(L_j\mathbf{n}) - n_j\, p_s(\mathbf{n}) \right\}.$$

Show that the solution of this equation can be written in the form

$$p_s(\mathbf{n}) \equiv p_s(n_1, n_2, \ldots, n_K) = \frac{C}{n_1!\, n_2! \cdots n_K!}, \qquad (4.98)$$

for $n_1 + n_2 + \cdots + n_K = N$, where N is the total number of diffusing molecules and C is the normalization constant. Find the value of C and use (4.98) to calculate the stationary values of the mean and variance vectors, $\mathbf{M}$ and $\mathbf{V}$, and covariance matrix $\{V_{i,j}\}$ defined by (4.23), (4.26) and (4.27), respectively.

4.4   Show that the SSA (a10)–(c10) and the SSA (a11)–(c11) are equivalent.

      Hint: Derive the formula (4.37) using the similar argument as in the derivation of the formula (1.5).

4.5   Derive the equation (4.41).

      Hint: Modify the derivation of (4.40) accordingly.

4.6   Derive the position–velocity Fokker–Planck equation (4.52).

Hint: Assuming that the change in velocity of the diffusing molecule during the time step is Δv (i.e. $\Delta v = V(t + \Delta t) - V(t)$) and the molecule is at the point x with the velocity v at time $t + \Delta t$, it had to be at the position $X(t) = x - (v - \Delta v)\Delta t$ with the velocity $V(t) = v - \Delta v$ at time t. Consequently, $f(x, v, t + \Delta t)$ can be computed from $f(\cdot, \cdot, t)$ by

$$f(x, v, t + \Delta t) = \int_{-\infty}^{\infty} f(x - (v - \Delta v)\Delta t, v - \Delta v, t)\, \psi(v - \Delta v; \Delta v)\, d\Delta v, \quad (4.99)$$

where $\psi(w; \Delta v)$ is a distribution function of the conditional probability that the change in velocity during the time step is Δv provided that $V(t) = w$. Using (4.50), we obtain

$$\psi(w; \Delta v) = \frac{1}{\beta \sqrt{4\pi D \Delta t}} \exp\left[-\frac{(\Delta v + \beta w \Delta t)^2}{4\beta^2 D \Delta t}\right]. \quad (4.100)$$

Using the Taylor expansion of f in (4.99), evaluating the integrals on the right-hand side of (4.99) with the help of (4.100), we obtain the position–velocity Fokker–Planck equation (4.52) in the limit $\Delta t \to 0$.

4.7   Use the velocity-jump process (4.49)–(4.50) together with the reflective boundary conditions to design an SSA for the computation of the diffusion example studied in Figures 4.2, 4.3, 4.4(c) and 4.4(d).

4.8   Find the conditional probability that the trajectory of SDE (4.1) crosses the line $x = 0$ provided that it starts at $X(t) = x_1 > 0$ and finishes at $X(t + \Delta t) = x_2 > 0$, i.e. verify the formula (4.63).

Hint: Use (4.9) or (3.33) to show that the probability distribution that the particle is found at the distance $x_2 - x_1$ after the time interval of length Δt is equal to

$$\frac{1}{\sqrt{4D\pi \Delta t}} \exp\left[-\frac{(x_2 - x_1)^2}{4D\Delta t}\right]. \quad (4.101)$$

The probability that the particle crossed the boundary and arrived at the point $x = x_2$ is equal to the probability that the particle arrived at the point $x = -x_2$ (to justify this statement, note that every trajectory which has reached the boundary $x = 0$ can be, after the hitting point, reflected around $x = 0$ to get the trajectory which ends up at $x = -x_2$). The probability distribution that the particle is found at the distance $-x_2 - x_1$ after the time interval of length Δt is equal to

$$\frac{1}{\sqrt{4D\pi \Delta t}} \exp\left[-\frac{(x_2 + x_1)^2}{4D\Delta t}\right]. \quad (4.102)$$

Dividing (4.102) by (4.101), we obtain (4.63).

4.9 Show that (4.65) is the solution of the diffusion equation (4.4) which satisfies the boundary conditions (4.64) and the initial condition $p_x(x, 0) \equiv$
p_{in} for $x \in (0, \infty)$.

4.10 Verify that (4.70) is the solution of the Laplace equation (4.67) satisfying
the boundary conditions (4.68)–(4.69).

4.11 Consider the velocity-jump process given by the algorithm (a12)–(d12)
in the interval $(0, \infty)$ where the boundary condition in the step (c12) is
modified as the following partially adsorbing boundary condition:

(c12)*   If $X(t + \Delta t)$ computed by (4.35) is less than 0, then
$$\begin{cases} X(t + \Delta t) = -X(t) - V(t)\,\Delta t \\ V(t + \Delta t) = -V(t) \end{cases} \text{with probability } 1 - 2\sigma/s.$$
Otherwise, the trajectory is terminated.

Consider this process in the limit $\Delta t \to 0$. Show that we can obtain the
Robin boundary condition (4.81) in the limit $s \to \infty$.

4.12 Consider the velocity-jump process from Exercise 4.11 in the finite interval $[0, L]$. The boundary condition at $x = 0$ is partially reflective, given
in (c12)*. The boundary condition at $x = L$ is fully reflective. Implement
this algorithm for the model presented in Figure 4.7. Compute the distribution of molecules at $t = 10$ minutes using this velocity-jump process.
Compare with results obtained in Figure 4.7.

4.13 Consider a fully adsorbing sphere (e.g. a cell membrane) with radius
R_1. Assume that a molecule diffuses according to (3.18)–(3.20) in the
exterior of this sphere, i.e. in the domain

$$\left\{ (x, y, z) \,\middle|\, R_1 < \sqrt{x^2 + y^2 + z^2} \right\}.$$

Whenever the molecule hits the sphere, it is adsorbed. Suppose that the
initial position of the molecule is $[X(0), Y(0), Z(0)] = (x_0, y_0, z_0)$. What
is the probability that the molecule will be adsorbed by the sphere rather
than wander away for good?

4.14 Suppose that the adsorbing sphere from Exercise 4.13 is inside a larger
sphere with radius R_2 (where $R_2 > R_1$) and the molecule can only diffuse
according to (3.18)–(3.20) in the finite domain:

$$\left\{ (x, y, z) \,\middle|\, R_1 < \sqrt{x^2 + y^2 + z^2} < R_2 \right\}.$$

Assume reflective boundary conditions on the larger sphere (with radius R_2). Calculate the average time until the molecule is adsorbed.

4.15 Suppose now that the sphere with radius R_2 in Exercise 4.14 is also adsorbing. If the initial position of the molecule is $[X(0), Y(0), Z(0)] = (x_0, y_0, z_0)$, what is the probability that it is eventually adsorbed by the inner sphere? By the outer sphere?

5

Efficient Stochastic Modelling of Chemical Reactions

To model any well-mixed chemical system, we can always use the Gillespie SSA (a5)–(d5) which was introduced in Section 1.5. This is a general algorithm which is equivalent to solving the chemical master equation (3.78). The Gillespie SSA can be used to design a computer code that has the potential to answer all questions of interest. If the computer code can be executed in a reasonably short time, there is nothing to worry about. Unfortunately, SSAs are sometimes computationally intensive, and do not run in a short time. If this is the case, then we have to look for a way to speed up the computation. In this section, we focus on such methods.

There can be several reasons why the Gillespie SSA is computationally intensive. The Gillespie SSA computes one reaction event per time step. Some reactions happen more often than others. A computer spends most of the time evaluating the (fast) reactions which happen often. On the other hand, we are often interested in the time evolution on the time scale of the slowest reactions in the system. In Section 5.1, we introduce a simple example involving fast and slow processes. In Section 5.2 we show that it is possible to significantly speed up the SSA by assuming that the fast processes are at equilibrium. Other approaches for problems with multiple time scales are presented in Section 5.3 and later in Chapter 9, which focuses on spatio-temporal processes occurring on multiple spatial and temporal scales.

The computational intensity of the Gillespie SSA increases with the number of chemical reactions and the number of chemical species in the model. In Section 4.2, we presented a compartment-based approach to diffusion. In order to use the Gillespie SSA, we considered the number of molecules in each compartment as a separate chemical species. In particular, we obtained a system with many chemical species and many reactions (e.g. the chain of "reactions" (4.13) in one dimension). Large systems also naturally appear when modelling gene regulatory networks. Tens or hundreds of genes and proteins might be involved in these networks. In the context of gene regulation, we have a system of many reactions with several different time scales. Thus some multiscale approaches from Sections 5.2 and 5.3 might be applied to speed up their computation. In the case of the compartment-based model of diffusion, all "reactions" in (4.13) have the same rate constant. However, we can still decrease the computational intensity of the Gillespie SSA by a careful programming of the algorithm. In Section 5.5, we present the so-called next-reaction SSA, which makes use of special data structures, and is an improvement on the first-reaction SSA introduced in Section 5.4. Both SSAs are equivalent to the Gillespie SSA, so that they are equivalent to solving the chemical master equation. Other ways to decrease the computational intensity of the exact stochastic simulation of chemical systems are also discussed in Section 5.5.

Which SSA should I use for modelling systems of chemical reactions? It is an interesting question. The easy answer is: use the Gillespie SSA (a5)–(d5) or any equivalent form presented in Sections 5.4 and 5.5. If you are not satisfied with the speed of the computation, try to use another exact method from Sections 5.4 and 5.5. All methods will yield the same answer. It is not a priori straightforward to decide which of the equivalent approaches is less computationally intensive. On the one hand, Gibson and Bruck (2000) theoretically showed that the next-reaction SSA should be superior to the Gillespie SSA (a5)–(d5) for systems with many chemical species and many reactions. However, on the other hand, the next-reaction SSA also involves non-trivial data structures whose maintenance cost depends on a particular computer and is not included in the theoretical analysis of computational complexity of the SSAs (Cao et al., 2004). It might be possible that for your computer and your model the optimized version of the Gillespie SSA (as discussed in Section 5.5) might perform better than the next-reaction SSA. The SSAs presented in Sections 5.2 and 5.3 can also be used to decrease the computational intensity of the problem. However, these methods are not exact, so there is an additional issue of accuracy of the computed results. One has to always check that the assumptions made to derive a particular approximation are indeed satisfied. If this is the case, then the multiscale approaches from Sections 5.2 and 5.3 are often helpful to compute results in a reasonable time.

In this section, we measure the accuracy of simulations by comparing the results computed by an SSA with the "exact results", which would be obtained by averaging over infinitely many realizations of the Gillespie SSA (a5)–(d5) or equivalently by solving the corresponding chemical master equation (3.78). In this way, we can define and study computational errors from the mathematical point of view. In applications, we also have to consider other errors caused by our modelling assumptions. Did we include all important processes in our simulation? Can we accurately describe the considered problem as a well-mixed chemical system? We will discuss models that incorporate more modelling details in Chapters 6, 7 and 8. Such models are even more computationally intensive and we conclude our book with Chapter 9, which discusses ways to speed up computations for more complex spatio-temporal models.

5.1 A Simple Multiscale Problem

We consider chemical species A, B and C in a container of volume v which are subject to the following chemical reactions:

$$\emptyset \xrightarrow{k_1} A \underset{k_3}{\overset{k_2}{\rightleftharpoons}} B \xrightarrow{k_4} C. \tag{5.1}$$

This set of reactions describes production of C through the intermediate species A and B. More precisely, A is produced from a source and reversibly transformed to B, which is converted to C. We choose the rate constants

$$k_1 v = 1 \sec^{-1}, \quad k_2 = 50 \sec^{-1}, \quad k_3 = 200 \sec^{-1}, \quad k_4 = 0.1 \sec^{-1} \tag{5.2}$$

and the initial conditions $A(0) = 0$, $B(0) = 0$ and $C(0) = 0$. In Figure 5.1(c), we present the time evolution of C computed by the Gillespie SSA (a5)–(d5) for the chemical system (5.1). Five realizations of the SSA (a5)–(d5) are plotted using five different colours. We observe that up to 25 molecules of C are produced in the first 50 seconds.

Let $p(a, b, c, t)$ be the probability that $A(t) = a$, $B(t) = b$ and $C(t) = c$, i.e. the probability that there are a molecules of the chemical species A, b molecules of the chemical species B and c molecules of the chemical species C at time t in the system. The chemical master equation for the model (5.1) reads as follows (compare with (3.78)):

$$\frac{\partial}{\partial t} p(a, b, c, t) = k_1 v \, p(a - 1, b, c, t) - k_1 v \, p(a, b, c, t)$$
$$+ k_2(a + 1) \, p(a + 1, b - 1, c, t) - k_2 a \, p(a, b, c, t)$$
$$+ k_3(b + 1) \, p(a - 1, b + 1, c, t) - k_3 b \, p(a, b, c, t)$$
$$+ k_4(b + 1) \, p(a, b + 1, c - 1, t) - k_4 b \, p(a, b, c, t). \tag{5.3}$$

Let $p_c(n, t)$ be the probability that $C(t) = n$, i.e. the probability that there are n molecules of C in the system at time t. To calculate $p_c(n, t)$ we need to sum $p(a, b, n, t)$ over all the possible values of a and b:

$$p_c(n, t) = \sum_{a=0}^{\infty} \sum_{b=0}^{\infty} p(a, b, n, t). \tag{5.4}$$

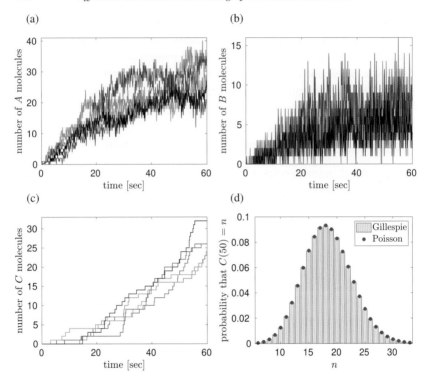

Figure 5.1 *Stochastic simulation of the system of chemical reactions (5.1) for $A(0) = 0$, $B(0) = 0$, $C(0) = 0$ and the rate constants given by (5.2). (a)–(c) Five realizations of the Gillespie SSA (a5)–(d5) showing the time evolution of A, B and C. The results corresponding to the same realization are plotted in the same colour. (d) Distribution $p_c(n, t)$ at time $t = 50$ sec obtained as an average of 10^6 realizations of the Gillespie SSA (a5)–(d5) (grey histogram) and as the Poisson distribution with mean $\mu(t)$ obtained by solving the corresponding deterministic ODE model (red circles).*

In principle, we could try to numerically solve the master equation (5.3) first and then compute $p_c(n, t)$ by (5.4). However, such an approach would be complicated by the fact that p is a function of four variables: even if we truncate the infinite set of ODEs we will still have a lot of ODEs to solve. An easier way to compute $p_c(n, t)$ is by averaging over many realizations of the SSA (a5)–(d5). In Figure 5.1(d), we present the histogram $p_c(n, t)$ at time $t = 50$ sec that was computed as an average over 10^6 realizations. For each realization, we run the simulation until the time $t = 50$ sec. We compute the histogram of the values of $C(50)$. We divide it by number of realizations to get the probability $p_c(n, 50)$, which is plotted in Figure 5.1(d) as the grey histogram.

Table 5.1 *Relative occurrences of the reactions in the model* (5.1) *during the time interval* [0, 50] *sec. We use* $A(0) = B(0) = C(0) = 0$ *and the rate constants given by* (5.2).

reaction	$\emptyset \rightarrow A$	$A \rightarrow B$	$B \rightarrow A$	$B \rightarrow C$
relative occurrence	0.068%	49.970%	49.937%	0.025%

Since our model (5.1) only includes zeroth-order and first-order chemical reactions and our initial condition is $A(0) = B(0) = C(0) = 0$, it is possible to show that $p_c(n, t)$ is the Poisson distribution with mean $\mu(t)$ given as the solution of the corresponding deterministic ODE model. We plot this Poisson distribution as the red circles in Figure 5.1(d). If we consider any chemical reaction network which only includes zeroth-order and first-order chemical reactions with arbitrary initial conditions, it is possible to find an exact solution formula of the chemical master equation in terms of the convolution of multinomial and product Poisson distributions with time-dependent parameters evolving according to the deterministic ODEs (Jahnke and Huisinga, 2007). However, such solutions are not available for general chemical reaction systems. Thus, in what follows, we will consider many realizations of the Gillespie SSA as the exact result to which that we want to compare our solutions. The discussed computational approaches are not just applicable to our simple example (5.1), but can be used to speed up computations of general chemical reaction systems. First, we need to identify the bottleneck that limits the speed of our simulations.

In Table 5.1, we present the relative occurrence of each of the four reactions of the model (5.1) during the time interval [0, 50] sec. We observe that the following two reactions happen most of the time:

$$A \underset{k_3}{\overset{k_2}{\rightleftarrows}} B. \tag{5.5}$$

In fact, 99.9% of all reaction events are given by (5.5). The remaining two reactions $\emptyset \rightarrow A$ and $B \rightarrow C$ happen rarely (only 0.1% of the total reaction events). Thus we shall call the chemical reactions (5.5) *fast reactions*. The fast reactions involve only the chemical species A and B. They do not change the number of molecules of C. We shall call the chemical species A and B *fast* and the chemical species C *slow*. The time evolution of the fast chemical species A and B is shown in Figure 5.1(a) and (b). We confirm that the changes of the fast chemical species A and B happen more frequently than the changes of the slow chemical species C.

In this section, we have introduced a simple multiscale problem (5.1) and studied it using the Gillespie SSA (a5)–(d5). Using a standard desktop computer, it took about 1 hour and 13 minutes to compute the histogram presented in Figure 5.1(d). In the following section, we present a multiscale SSA which is able to compute the same results about one thousand times faster (in less than 4 sec using the same desktop computer). This is a significant improvement.

Of course the computational cost can also be decreased simply by using a more powerful computer. Our illustrative computer codes for this example were written ten years ago, when the first draft of this chapter was developed. At that time, our Gillespie SSA took about 2 hours and 23 minutes to run and the multiscale SSA results were obtained in about 8 sec, which suggests that the computer used then was about twice slower. The reported computational times could only be meaningful and reproducible if we provided additional details of the computers used. However, such information would be quickly outdated. A more important piece of information is the ratio of the computational times, a speed-up, which has been about one thousand times on each computer. We can expect to get similar conclusions on other current or future generation of computers.

Another option to decrease the computational cost is to observe that the histogram in Figure 5.1(d) was computed as the average of 10^6 realizations of the Gillespie SSA. Such a computation is trivially parallelizable, e.g. we can run 10^5 realizations on 10 computers (instead of running 10^6 realizations on one computer) to get the same results in a shorter time. This observation can also be used to write more efficient computer codes for hardware systems designed for parallel computing (from graphics processsing units to supercomputers). However, if our hardware resources fail to decrease enough the overall computational cost of the Gillespie SSA, suitable multiscale methods must be considered. We will present a number of such methods in the following two sections. Finally, let us note that different realizations of the multiscale SSA can also be run on different computers to further decrease the time needed to compute our results.

5.2 Multiscale SSA with Partial Equilibrium Assumption

In this section, we present an approximate SSA for chemical systems with fast and slow reactions which was introduced by Cao et al. (2005*a,b*). It can be applied to any system that has reactions happening on disparate time scales. We will explain the algorithm on the multiscale model (5.1) with the rate constants given by (5.2). In the previous section, we identified its fast reactions as (5.5) and its slow reactions as

$$\emptyset \xrightarrow{k_1} A, \tag{5.6}$$

$$B \xrightarrow{k_4} C. \tag{5.7}$$

Considering a general chemical system with fast and slow reactions, we define the fast chemical species as those chemical species which are changed by at least one fast chemical reaction. The chemical species which are not changed by the fast chemical reactions will be called the slow chemical species. In the case of the multiscale model (5.1), the fast chemical species are A and B and the slow chemical species is C. Note that this definition does not preclude the number of molecules of the fast chemical species being changed by the slow chemical reactions. For example, in the model (5.1), a molecule of the fast chemical species B is lost by one occurrence of the slow chemical reaction (5.7). On the other hand, the numbers of molecules of the slow chemical species cannot be changed by the fast chemical reactions.

The main assumption of the approximate SSA is that the fast reactions are effectively at equilibrium, which means that the distributions of fast species are steady on a time scale that is larger than the typical time scale of the fast reactions and shorter than the time scale of the slow reactions. It is often the case that we are interested only in the time evolution of the slow chemical species, and this is what the approximate SSA simulates. Thus we focus on the computation of the time evolution of C in the multiscale model (5.1). Our goal is to get the same results as in Figure 5.1(d) using a more efficient algorithm. Although we focus here on C, we note that it is possible to get the distributions of the fast chemical species A and B as well; we will present the corresponding method in Section 5.3. In general, the slow reactions can be divided into two categories:

(i) slow reactions whose propensities do not depend on the fast chemical species, e.g. the slow reaction (5.6) of the model (5.1) which has the propensity $\alpha_1(t) = k_1 v$;

(ii) slow reactions whose propensities do depend on the fast chemical species, e.g. the slow reaction (5.7) of the model (5.1) which has the propensity $\alpha_4(t) = k_4 B(t)$ that depends on the fast chemical species B.

Let δt be a time interval that is much smaller than the time scale of the slow reactions T_{slow} and much larger than the time scale of the fast reactions T_{fast}, i.e. $T_{\text{fast}} \ll \delta t \ll T_{\text{slow}}$. We define the *effective propensity function* $\overline{\alpha}(t)$ of a slow chemical reaction so that the product $\overline{\alpha}(t)\delta t$ gives approximately the probability that the slow reaction occurs in the time interval $[t, t + \delta t)$. If the slow reaction is of the type (i), then its effective propensity function $\bar{\alpha}(t)$ is equal to its propensity function at time t. To compute the effective propensity function for slow reactions of the type (ii), we take into account the stochastic partial equilibrium assumption, which says that fast processes are effectively at equilibrium on the time scale δt. We explain the corresponding derivation of the effective propensity function below, using the slow chemical reaction (5.7) as an example.

So, to put these ideas into practice, let us consider the multiscale model (5.1). Let us assume that there are $A(t)$, $B(t)$ and $C(t)$ molecules of the chemical species A, B and C at time t, respectively. We want to find the effective propensity functions $\bar{\alpha}_1(t)$ and $\bar{\alpha}_4(t)$ of the slow reactions (5.6) and (5.7). Since the slow reaction (5.6) is of type (i), its effective propensity function is equal to its propensity function, i.e.

$$\bar{\alpha}_1(t) = \alpha_1(t) = k_1 \nu. \tag{5.8}$$

To compute the effective propensity functions of slow reactions of type (ii), we first study only the fast system under the condition that the numbers of molecules of the slow chemical species do not change. Thus, in the case of the slow reaction (5.7), we study the chemical system (5.5) for the chemical species A and B. The chemical system (5.5) has the conservation relation

$$A(\tilde{t}) + B(\tilde{t}) = A(t) + B(t), \qquad \text{for } \tilde{t} \geq t. \tag{5.9}$$

To simplify the following formulae, we define the auxiliary variable

$$Q(t) = A(t) + B(t). \tag{5.10}$$

Using (5.9) and (5.10), we can write $A(\tilde{t})$ in terms of $B(\tilde{t})$ as

$$A(\tilde{t}) = Q(t) - B(\tilde{t}), \qquad \text{for } \tilde{t} \geq t. \tag{5.11}$$

Now let $p(\tilde{b}, \tilde{t})$ be the probability that $B(\tilde{t}) = \tilde{b}$ given that $A(t) = a$, $B(t) = b$ and $\tilde{t} \geq t$. Using (5.11), the distribution $p(\tilde{b}, \tilde{t})$ satisfies the chemical master equation

$$\frac{\partial}{\partial \tilde{t}} p(\tilde{b}, \tilde{t}) = k_2(Q(t) - \tilde{b} + 1) p(\tilde{b} - 1, \tilde{t}) - k_2(Q(t) - \tilde{b}) p(\tilde{b}, \tilde{t})$$
$$+ k_3(\tilde{b} + 1) p(\tilde{b} + 1, \tilde{t}) - k_3 \tilde{b} p(\tilde{b}, \tilde{t}). \tag{5.12}$$

Since $T_{\text{fast}} \ll \delta t$, we can assume that the fast reactions reach the steady distribution

$$p_s(\tilde{b}) = \lim_{\tilde{t} \to \infty} p(\tilde{b}, \tilde{t})$$

on a time scale shorter than δt. Using (5.12), the steady distribution $p_s(\tilde{b})$ satisfies the stationary chemical master equation

$$0 = k_2(Q(t) - \tilde{b} + 1)\, p_s(\tilde{b} - 1) - k_2(Q(t) - \tilde{b})\, p_s(\tilde{b})$$
$$+ k_3(\tilde{b} + 1)\, p_s(\tilde{b} + 1) - k_3\, \tilde{b}\, p_s(\tilde{b}). \tag{5.13}$$

Multiplying this equation by $\tilde{b}$ and summing the resulting equation over $\tilde{b}$, we obtain (see Exercise 5.1)

$$\sum_{\tilde{b}=0}^{Q(t)} \tilde{b}\, p_s(\tilde{b}) = \frac{k_2 Q(t)}{k_2 + k_3}. \tag{5.14}$$

Now for $\tilde{t} \in (t, t + \delta t)$, we know that if $B(\tilde{t}) = \tilde{b}$ then the probability that the slow reaction (5.7) occurs in the infinitesimally small time interval $[\tilde{t}, \tilde{t} + dt)$ is equal to

$$\alpha_4(\tilde{b})\, dt = k_4\, B(\tilde{t})\, dt = k_4\, \tilde{b}\, dt.$$

Assuming that the fast reactions (5.5) are at equilibrium, the variable $B(\tilde{t})$ is distributed according to the stationary distribution $p_s(\tilde{b})$. Thus the probability that the slow reaction (5.7) occurs in the infinitesimally small time interval $[\tilde{t}, \tilde{t} + dt)$ is equal to

$$\sum_{\tilde{b}=0}^{Q(t)} p_s(\tilde{b})\, k_4\, \tilde{b}\, dt = k_4 \left(\sum_{\tilde{b}=0}^{Q(t)} \tilde{b}\, p_s(\tilde{b}) \right) dt = \frac{k_2 k_4 Q(t)}{k_2 + k_3}\, dt,$$

where we used the formula (5.14). Thus, using the partial equilibrium assumption in the whole interval $[t, t + \delta t)$, we obtain the effective propensity function of the slow reaction (5.7) as

$$\overline{\alpha}_4(t) = \frac{k_2 k_4 Q(t)}{k_2 + k_3}, \tag{5.15}$$

where $Q(t) = A(t) + B(t)$. The effective propensity function $\overline{\alpha}_4(t)$ depends on $Q(t)$, which is a conserved quantity of the fast reactions – see the conservation relation (5.9) – and thus is itself a slow variable. Thus we have successfully evaluated the effective propensity functions $\overline{\alpha}_1(t)$ and $\overline{\alpha}_4(t)$ of the slow reactions in terms of quantities that do not change during the (many) fast reactions which happen between each slow reaction. In particular, this means that the fast reactions do not have to be simulated; all we need to keep track of is the slow variable $C(t)$ and the auxiliary quantity $Q(t)$ which was defined by (5.10). This observation leads to an (approximate) multiscale SSA which is more efficient than the (exact) Gillespie SSA (a5)–(d5). The approximate SSA can be formulated for the multiscale model (5.1) as follows. Starting with the initial condition $A(0) = a_0$, $B(0) = b_0$ and $C(0) = c_0$ at time $t = 0$, we initialize Q as $Q(0) = a_0 + b_0$ and we perform the following five steps at time t:

(a14) Generate two random numbers r_1, r_2 uniformly distributed in $(0, 1)$.

(b14) Compute the effective propensity functions $\overline{\alpha}_1(t)$ and $\overline{\alpha}_4(t)$ by (5.8) and (5.15). Compute $\overline{\alpha}_0 = \overline{\alpha}_1 + \overline{\alpha}_4$.

(c14) Compute the time when the next slow chemical reaction takes place as $t + \tau$, where

$$\tau = \frac{1}{\overline{\alpha}_0} \ln \left[\frac{1}{r_1} \right]. \qquad (5.16)$$

(d14) Compute the number of molecules of the slow chemical species C at time $t + \tau$ by

$$C(t + \tau) = \begin{cases} C(t) & \text{if } r_2 < \overline{\alpha}_1 / \overline{\alpha}_0, \\ C(t) + 1 & \text{if } r_2 \geq \overline{\alpha}_1 / \overline{\alpha}_0. \end{cases}$$

(e14) Update the auxiliary variable Q by

$$Q(t + \tau) = \begin{cases} Q(t) + 1 & \text{if } r_2 < \overline{\alpha}_1 / \overline{\alpha}_0, \\ Q(t) - 1 & \text{if } r_2 \geq \overline{\alpha}_1 / \overline{\alpha}_0. \end{cases}$$

Then continue with step (a14) for time $t + \tau$.

The time of the next slow chemical reaction is computed by (5.16), which can be derived in a similar way as the formula (1.54) in the Gillespie SSA. The step (d14) specifies the changes of the slow variable C by the slow reactions (5.6)–(5.7). If $r_2 < \overline{\alpha}_1 / \overline{\alpha}_0$, then the slow reaction (5.6) occurs and C does not change. Otherwise, one molecule of C is produced by the slow reaction (5.7). The auxiliary variable $Q(t)$ is updated in the step (e14). The slow reaction (5.6) increases $A(t)$ by 1 and does not change $B(t)$. Consequently, $Q(t) = A(t) + B(t)$ is increased by 1 whenever the slow reaction (5.6) occurs. If the slow reaction (5.7) occurs, then $Q(t)$ is decreased by 1.

Starting with the initial condition $A(0) = 0$, $B(0) = 0$ and $C(0) = 0$, we present the time evolution of C computed by the multiscale SSA (a14)–(e14) in Figure 5.2(a). Five realizations of the SSA (a14)–(e14) are plotted using five different colours. This plot can be compared with Figure 5.1(c). We see that the Gillespie SSA (a5)–(d5) and the multiscale SSA (a14)–(e14) give comparable results.

To get a better idea of the accuracy of the multiscale SSA (a14)–(e14), we compute the distribution $p_c(n, t)$ at time $t = 50$ sec, i.e. we redo the plot from Figure 5.1(d). We compute the distribution $p_c(n, 50)$ as an average of 10^6

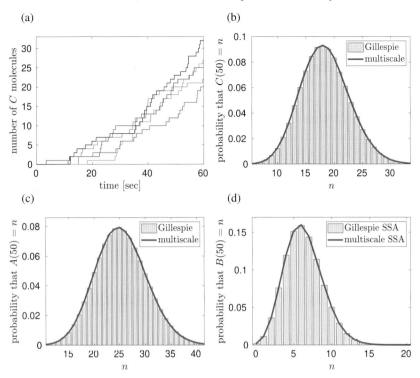

Figure 5.2 *Stochastic simulation of the system of chemical reactions* (5.1) *for* $A(0) = 0$, $B(0) = 0$, $C(0) = 0$ *and the rate constants given by* (5.2). (a) $C(t)$ *given by five realizations of the multiscale SSA (a14)–(e14).* (b) *Distribution* $p_c(n, t)$ *at time* $t = 50\,sec$ *obtained as an average of* 10^6 *realizations of the Gillespie SSA (a5)–(d5) (grey histogram) and as an average of* 10^6 *realizations of the multiscale SSA (a14)–(e14) (red line).* (c), (d) *Distributions* $p_a(n, t)$ (c) *and* $p_b(n, t)$ (d) *at time* $t = 50\,sec$ *obtained as averages of* 10^6 *realizations of the Gillespie SSA (a5)–(d5) (grey histograms) and as averages of* 10^6 *realizations of the multiscale SSA (a14)–(e14) in the time interval* $(0, 49.99)\,sec$ *followed by the Gillespie SSA (a5)–(d5) in the time interval* $(49.99, 50)\,sec$ *(red lines).*

realizations of the multiscale SSA (a14)–(e14). The result is plotted as the red line in Figure 5.2(b) together with the result presented in Figure 5.1(d). The comparison is excellent. The multiscale SSA (a14)–(e14) gives approximately the same results as the Gillespie SSA (a5)–(d5) applied to the full model (5.1). Moreover, the computation of 10^6 realizations of the multiscale SSA (a14)–(e14) took about 3.4 sec, which was about one thousand times faster than our original computation of 10^6 realizations of the Gillespie SSA (a5)–(d5), using the same computer – see Section 5.1 for further discussion.

5.3 Multiscale Modelling

In the previous section, we presented the multiscale SSA (a14)–(e14) which was used to compute the time evolution of the slow variable C. We showed in Figure 5.2(b) that the multiscale SSA (a14)–(e14) gives sufficiently accurate results for the chemical system (5.1) with the rate constants given by (5.2). Moreover, the computation was about a thousand times faster than by the Gillespie SSA. The multiscale SSA (a14)–(e14) only simulates the time evolution of the slow chemical species C and the auxiliary variable Q. How do we compute the distributions of the fast variables? To answer this question, we modify the algorithm slightly.

Let $p_a(n, t)$ be the probability that $A(t) = n$ and $p_b(n, t)$ be the probability that $B(t) = n$, i.e. the probability that there are n molecules of A or n molecules of B in the system at time t. These are related to the probability $p(a, b, c, t)$ by (compare with (5.4))

$$p_a(n, t) = \sum_{b=0}^{\infty} \sum_{c=0}^{\infty} p(n, b, c, t), \qquad p_b(n, t) = \sum_{a=0}^{\infty} \sum_{c=0}^{\infty} p(a, n, c, t).$$

As in the case of $p_c(n, t)$ which was defined in (5.4), we can compute $p_a(n, t)$ and $p_b(n, t)$ by averaging over many realizations of the Gillespie SSA (a5)–(d5). In Figure 5.2(c) and (d), we present the histograms $p_a(n, t)$ and $p_b(n, t)$ at time $t = 50$ sec which were computed as averages over 10^6 realizations (grey histograms). For each realization, we run the Gillespie SSA until the time $t = 50$ sec. We compute the histogram of the values of $A(50)$ and $B(50)$, respectively. We divide it by the number of realizations to get the probabilities $p_a(n, 50)$ and $p_b(n, 50)$ which are plotted in Figure 5.2. As we mentioned in the previous section, the computation of 10^6 realizations of the Gillespie SSA took about 1 hour and 13 minutes on our computer. Next, we make use of the multiscale SSA (a14)–(e14) to decrease the computational intensity of the problem.

The stochastic partial equilibrium assumption tells us that the fast reactions are effectively at equilibrium, i.e. their steady distribution is reached in a short time. In the previous section we wrote down the stationary chemical master equation (5.13) for p_b. Although we did not solve that equation there (we calculated only the average value of p_b), we could have used it to find the stationary distribution. However, it is also possible to use an efficient stochastic simulation to find the distribution of p_a and p_b, as we now demonstrate. Let δt be a time step that is much larger than the time scale of the fast reactions T_{fast}, i.e. $T_{\text{fast}} \ll \delta t$. Then we can approximate the distributions $p_a(n, t)$ and $p_b(n, t)$ at time $t = 50$ sec as follows. We run many realizations of the multiscale

SSA (a14)–(e14) until the time $t = 50\,\text{sec} - \delta t$. Then we run the Gillespie SSA (a5)–(d5) for the remaining time interval $(50 - \delta t, 50)$ sec. We choose $\delta t = 10^{-2}$ sec. In Figure 5.2(c) and (d), we present the probability distributions $p_a(n, t)$ and $p_b(n, t)$ at time $t = 50$ sec which were computed as averages over 10^6 realizations (red lines). The comparison with the exact results is excellent. Since we run the full model over the time interval of length $\delta t = 10^{-2}$ sec only, we do not significantly increase the computational intensity of the multiscale SSA (a14)–(e14). In fact, the time interval $(49.99, 50)$ sec contains on average about 25 occurrences of the fast reactions. Thus the red lines in Figure 5.2(c) and (d) were computed in less than 5 seconds, which was still a significant improvement as compared to the 1 hour and 13 minutes needed by the Gillespie SSA (a5)–(d5).

The multiscale SSA (a14)–(e14) can be generalized to any problem with fast and slow reactions. However, there are some issues which need to be addressed. How do we compute the effective propensity functions of the slow reactions? This question can be reformulated as a computation of some averages of the form (5.14). In our case, the fast system was of the first order in fast variables. Consequently, we can find exact formulae for averages – e.g. (5.14). If our model included some higher-order fast reactions, then we might not be able to evaluate the effective propensity functions analytically and suitable approximations would have to be used; see Cao et al. (2005*a,b*).

How do we test the accuracy of multiscale SSAs? In the case of the illustrative example (5.1), we demonstrated the accuracy of the algorithm by comparison with the exact results computed by the Gillespie SSA. However, this is not a typical situation. Indeed, if we are able to solve a multiscale problem exactly, then there will be no reason to compute the results again by an approximate SSA. The multiscale SSA is useful for models for which the Gillespie SSA is so computationally intensive that it cannot compute the exact results in a reasonably short time. Then we have no exact results to compare with and we have to use suitable computational tests to determine the accuracy of the multiscale SSA. For example, the results in Figure 5.2(c) and (d) were computed for the parameter $\delta t = 10^{-2}$ sec of the method. The results depend on this parameter. How do we choose its value if we do not know the exact results? Well, we know that the accuracy of the method will increase with increasing δt. At least a few fast reactions should occur in the time interval δt. We can easily test that the time interval $\delta t = 10^{-2}$ sec contains on average about 25 fast reactions. Increasing δt to 10^{-1} sec, the steady distributions $p_a(n, 50)$ and $p_b(n, 50)$ do not change. Thus we can a posteriori confirm that the time interval $\delta t = 10^{-2}$ sec is long enough for computation of $p_a(n, 50)$ and $p_b(n, 50)$ with a reasonable accuracy. If the results of the multiscale approach for $\delta t = 10^{-2}$ sec and $\delta t = 10^{-1}$ sec differ significantly, it would indicate that δt was not chosen correctly. In such a case, we would have to increase δt to improve the accuracy of the computation.

How do we determine which reaction is fast and which is slow? Using the Gillespie SSA, we can compute the relative occurrences of the chemical reactions in the model – see Table 5.1 for the model (5.1). Such a table clearly indicates which reactions happen often (fast reactions) and which occur rarely (slow reactions). It is important to note that similar results as in Table 5.1 can be obtained by running the Gillespie SSA for a short time only. In particular, this auxiliary calculation does not increase the computational intensity of the multiscale computation. Similar approaches have been applied in the literature to classify reactions as slow and fast in much larger systems of chemical reactions, for example, in the eukaryotic cell cycle by Liu et al. (2012), where a mixed (hybrid) SSA is presented combining deterministic and stochastic modelling frameworks.

As discussed above, short bursts of appropriately initialized simulations of the full model (Gillespie SSA) can be used to test assumptions and accuracy of the multiscale computation. Short bursts of the Gillespie SSA can also be used to estimate the averages that appear in the formulae for effective propensity functions. This approach is useful if we cannot find the averages exactly as in equation (5.14) – see E et al. (2005) for details.

Another approach to multiscale modelling is to assume that the system is effectively described by the chemical Fokker–Planck equation (3.80) written in terms of slow variables. Even if this equation is not explicitly available, we can use short bursts of appropriately initialized stochastic simulation of the full model (Gillespie SSA) to estimate the drift and diffusion coefficients numerically. Once we know the drift and diffusion coefficients, the results of Section 3.7 can be applied – see Erban et al. (2006) and Cotter et al. (2011) for details. Such an approach is especially useful when estimating stationary distributions and switching times between favourable states of the multiscale system.

If we are interested in accelerating the time evolution of the Gillespie SSA, we can also use the so-called tau-leaping simulation method (Gillespie, 2001). This is similar in philosophy to the derivation of the chemical Langevin equation (Section 3.7). At the very beginning of Chapter 1 we introduced the naive SSA (a1)–(b1) which was based on taking small time steps Δt and determining whether any reaction had taken place. We quickly improved the efficiency of the method by switching to an event-driven philosophy in which we jumped forward to the next reaction event. Tau-leaping returns to the time-driven philosophy but considers a much larger time step, $\bar{\tau}$. The idea is that many reactions take place in the interval t to $t + \bar{\tau}$, but not so many that the propensity functions change significantly. The propensity functions are taken to be constant during the time interval, so that the number of reactions is Poisson distributed. Thus several reaction events are incorporated into every time step. Tau-leaping is an approximate method which is reasonably accurate as long as no propensity function changes its value significantly during any time step. A particular danger that needs to be guarded against is the small but non-zero probability that the number of molecules becomes negative: since the propensity is not updated between the reactions of a given time step, it is possible to remove more molecules than are present; see Gillespie (2001) and Cao et al. (2006) for details.

5.4 First-Reaction SSA

So far in this chapter we have discussed *approximate* SSAs, which were designed to decrease the computational intensity of the exact Gillespie SSA. In the remainder of the chapter we focus on *exact* SSAs, which might for some systems be less computationally intensive than the Gillespie SSA. We start with the so-called first-reaction SSA presented by Gillespie (1976). This method is often less efficient than the Gillespie SSA, but it can be transformed to the more efficient next-reaction SSA as we will show in Section 5.5.

Suppose we have a system of q chemical reactions. Let $\alpha_i(t)$ be the propensity function of the ith chemical reaction, $i = 1, 2, \ldots, q$, at time t, that is, $\alpha_i(t)\, dt$ is the probability that the ith reaction occurs in the time interval $[t, t + dt)$. Then the first-reaction SSA consists of the following four steps at time t:

(a15) Generate q random numbers $r_1, r_2, \ldots, r_q$ uniformly distributed in the interval $(0, 1)$.

(b15) Compute the propensity function $\alpha_i(t)$ of each reaction and the tentative reaction time τ_i by

$$\tau_i = \frac{1}{\alpha_i(t)} \ln\left[\frac{1}{r_i}\right], \qquad i = 1, 2, \ldots, q. \tag{5.17}$$

(c15) Find j which is the index of $\min\{\tau_1, \tau_2, \ldots, \tau_q\}$. Compute the time when the next chemical reaction takes place as $t + \tau_j$.

(d15) It is the jth reaction that occurs at time $t + \tau_j$, so update numbers of reactants and products of the jth reaction.
Continue with step (a15) for time $t + \tau_j$.

The first-reaction SSA (a15)–(d15) is equivalent to the Gillespie SSA (a5)–(d5). It is an exact method for stochastic simulation of well-stirred systems of chemical reactions. At each time step, we generate the tentative time τ_i, $i = 1, 2, \ldots, q$, to the next occurrence of the ith chemical reaction under the condition that no other reaction occurs first. This time is given in the formula (5.17), which can be justified using the same arguments as formulae (1.5), (1.10) or (1.54). The reaction that actually occurs is the reaction with the minimal τ_i, $i = 1, 2, \ldots, q$, which we denote as τ_j. The steps (c15) and (d15) are used to update the time and the numbers of reactants and products according to the jth reaction. Once the jth reaction occurs, some propensities might change. Consequently, the corresponding tentative times τ_i no longer indicate when the reactions occur and have to be recomputed. To do that, the algorithm generates another set of q random numbers in the next time step. Thus, the first-reaction SSA (a15)–(d15) needs q uniformly distributed random numbers at each time step. On the other hand, the Gillespie SSA requires only two random numbers per one reaction event. If we have a system of many reactions (i.e. $q \gg 2$), then the first-reaction SSA might be more computationally intensive than the Gillespie SSA.

In the following section, we show that $q - 1$ of the random numbers that are generated by the step (a15) can be actually "re-used" by the algorithm. This observation leads to a more efficient SSA which requires only one random number per one reaction event. Following Gibson and Bruck (2000), we will call it the next-reaction SSA.

5.5 Exact Efficient SSAs

In every time step of the first-reaction SSA (a15)–(d15), we use q uniformly distributed random numbers to compute the times τ_i by (5.17). Only the smallest one (τ_j) is actually used for updating the system. Let us consider the remaining times τ_i, $i \neq j$. By virtue of (c15), they all satisfy

$$\tau_j \leq \tau_i, \qquad \text{for } i = 1, 2, \ldots, q.$$

Using (5.17), we obtain

$$r_i = \exp\left[-\alpha_i(t)\tau_i\right].$$

Consequently, the r_i, $i \neq j$, satisfy

$$r_i = \exp\left[-\alpha_i(t)\tau_i\right] \leq \exp[-\alpha_i(t)\tau_j], \qquad \text{for } i = 1, 2, \ldots, q.$$

Thus r_i, $i \neq j$, is a random number uniformly distributed in the interval

$$(0, \exp[-\alpha_i(t)\tau_j]).$$

Consequently,

$$\bar{r}_i = r_i \exp[\alpha_i(t)\tau_j] = \exp[\alpha_i(t)(\tau_j - \tau_i)]$$

is a random number uniformly distributed in the interval $(0, 1)$. The random number $\bar{r}_i$ can be used in the next time step to evaluate the formula (5.17). Let us denote the tentative times (5.17) at the time $t_{\text{old}} \equiv t$ as $\tau_{i,\text{old}}$ and the tentative times (5.17) at the time $t_{\text{new}} \equiv t + \tau_j$ as $\tau_{i,\text{new}}$. Let $\alpha_{i,\text{old}}$ (resp. $\alpha_{i,\text{new}}$) be the propensity function of the ith chemical reaction at time t_{old} (resp. t_{new}). Using $\bar{r}_i$ in (5.17) at time t_{new}, we obtain

$$\tau_{i,\text{new}} = \frac{1}{\alpha_{i,\text{new}}} \ln\left[\frac{1}{\bar{r}_i}\right] = \frac{1}{\alpha_{i,\text{new}}} \ln\left[\frac{1}{\exp[\alpha_{i,\text{old}}(\tau_j - \tau_{i,\text{old}})]}\right]$$

$$= \frac{\alpha_{i,\text{old}}}{\alpha_{i,\text{new}}}(\tau_{i,\text{old}} - \tau_j), \qquad \text{for } i = 1, 2, \ldots, q, \ i \neq j. \qquad (5.18)$$

Consequently, instead of generating a new set of random numbers, we can transform the old set of tentative times $\tau_{i,\text{old}}$ to the new set of tentative times $\tau_{i,\text{new}}$ by the formula (5.18). To use this formula efficiently, we rewrite it in terms of the absolute tentative times t_i of the occurrence of the ith reaction. The absolute tentative times t_i are related to the relative tentative times τ_i by

$$t_i = t + \tau_i, \qquad i = 1, 2, \ldots, q,$$

where t is the time for which the relative tentative times were generated. In particular, we denote

$$t_{i,\text{new}} = t_{\text{new}} + \tau_{i,\text{new}} \qquad \text{and} \qquad t_{i,\text{old}} = t_{\text{old}} + \tau_{i,\text{old}}.$$

Using (5.18) and the relation $t_{\text{new}} = t_{\text{old}} + \tau_j$, we obtain the formula for transformation of the absolute tentative times:

$$t_{i,\text{new}} = t_{\text{new}} + \tau_{i,\text{new}} = t_{\text{new}} + \frac{\alpha_{i,\text{old}}}{\alpha_{i,\text{new}}}(\tau_{i,\text{old}} - \tau_j)$$

$$= t_{\text{new}} + \frac{\alpha_{i,\text{old}}}{\alpha_{i,\text{new}}}(t_{i,\text{old}} - t_{\text{new}}), \qquad \text{for } i = 1, 2, \ldots, q, \ i \neq j. \qquad (5.19)$$

If the propensity of the ith chemical reaction is not changed by the occurrence of the jth chemical reaction, we have $\alpha_{i,\text{old}} = \alpha_{i,\text{new}}$. Then, the relation (5.19) implies $t_{i,\text{new}} = t_{i,\text{old}}$. Thus we need to recompute the absolute tentative times by (5.19) only if the value of the propensity function changes as a consequence of the occurrence of the jth chemical reaction. To make use of this observation, Gibson and Bruck (2000) define the so-called *dependency graph*. This is a data structure that tells us precisely which propensity function α_i should change when a given reaction occurs. Each reaction is a node in the dependency graph. A directed edge $j \rightarrow i$ connects the jth node to the ith node if and only if the execution of the jth reaction affects the reactants of the ith reaction. The dependency graph can be constructed prior to the stochastic simulation. Then the next-reaction SSA can be formulated as follows. We generate q random numbers $r_1, r_2, \ldots, r_q$ uniformly distributed in $(0, 1)$ and we compute the initial absolute tentative times t_i by

$$t_i = \frac{1}{\alpha_i(0)} \ln\left[\frac{1}{r_i}\right], \qquad i = 1, 2, \ldots, q,$$

where $\alpha_i(0)$ is the initial propensity function of the ith chemical reaction. After this initial step, three steps are performed at each time step:

(a16) Find j such that $t_j = \min\{t_1, t_2, \ldots, t_q\}$. It is the jth reaction which occurs at time t_j, so set the current time as $t = t_j$ and update numbers of reactants and products of the jth reaction.

(b16) For each edge $j \to i$, $i = 1, 2, \ldots, q$, $i \neq j$, in the dependency graph, compute the new value of the propensity function $\alpha_i(t)$ and transform the absolute tentative time by the formula

$$t_i = t + \frac{\alpha_{i,\text{old}}}{\alpha_i(t)}(t_{i,\text{old}} - t), \qquad (5.20)$$

where $t_{i,\text{old}}$ and $\alpha_{i,\text{old}}$ are the old values of the absolute tentative time and propensity function of the ith reaction, respectively.

(c16) Generate one random number r uniformly distributed in $(0, 1)$. Compute the propensity function $\alpha_j(t)$ and the absolute tentative time t_j of the jth chemical reaction by

$$t_j = t + \frac{1}{\alpha_j(t)} \ln\left[\frac{1}{r}\right]. \qquad (5.21)$$

Continue with step (a16).

The dependency graph helps us to calculate in step (b16) only those propensity functions which have changed. The formula (5.20) is equivalent to the formula (5.19). It transforms the old tentative times to their new values. We have to be a little careful whenever $\alpha_i(t) = 0$, i.e. whenever there are no reactants of the ith chemical reaction present in the system. In this case, the formulae (5.20) or (5.21) imply that $t_i = \infty$. This is a correct answer. The ith reaction will never happen unless its propensity function becomes non-zero again (as a consequence of the occurrence of another chemical reaction in the model). Consequently, it makes sense to generate the new random number in the step (c16) only if the propensity function $\alpha_j(t)$ is non-zero. Otherwise, we put

$t_j = \infty$ by default. Similarly, the transformation formula (5.20) is used only if the propensity function $\alpha_i(t) \neq 0$ (see Exercise 5.7).

In step (a16) we are looking for the minimal absolute tentative time t_j. To find t_j efficiently, Gibson and Bruck (2000) suggest storing the tentative times $t_i, i = 1, 2, \ldots, q$, in a so-called *indexed priority queue*. This is a tree structure in which the nodes contain the tentative times t_i. They are partially ordered so that a parent node has smaller values of t_i than either of its children. As a consequence, the minimal time is always stored at the top node, which is easy to find. Once the value of a tentative time is updated, it bubbles up or down to restore the order of the indexed priority queue. It can be shown that the updates of the indexed priority queue take $\log(q)$ operations, where q is the number of chemical reactions in the model. This is superior to q operations needed in the Gillespie SSA (a5)–(d5) to find which reaction occurs in the step (d5). Thus the next-reaction SSA with tentative times stored in an indexed priority queue might be more efficient than the Gillespie SSA for chemical systems with many reactions.

The additional data structures of the next-reaction SSA require some computer time to maintain. Cao et al. (2004) performed computational tests of the next-reaction SSA and the Gillespie SSA and concluded that for some systems the optimized version of the Gillespie SSA might actually perform better than the next-reaction SSA. One way to optimize the Gillespie SSA is to recalculate the propensities only for those reactions which were affected by the last reaction. We already mentioned this improvement of the Gillespie SSA in Section 4.2 where we introduced the compartment-based model for diffusion. We note that the idea of the dependency graph might be borrowed from the next-reaction SSA and used in the step (b5) of the Gillespie SSA for this purpose, i.e. we only update the propensity functions of those reactions which are end nodes of the directed edges that start at the node corresponding to the last reaction that occurred. The dependency graph can be constructed prior to the simulation.

Another way to decrease the computational cost of the Gillespie SSA is to optimize the step (d5). In this step, we look for j such that

$$r_2 \geq \frac{1}{\alpha_0} \sum_{i=1}^{j-1} \alpha_i \qquad \text{and} \qquad r_2 < \frac{1}{\alpha_0} \sum_{i=1}^{j} \alpha_i, \qquad (5.22)$$

where r_2 is the random number generated in the step (a5). To find j, we first test the condition

$$r_2 < \frac{\alpha_1}{\alpha_0}.$$

If this condition is satisfied, then the search for j is finished because we can conclude that $j = 1$. Otherwise, we have to test the condition

$$r_2 < \frac{\alpha_1 + \alpha_2}{\alpha_0}.$$

If this condition is satisfied, then $j = 2$ and we are done. Otherwise we have to test the condition

$$r_2 < \frac{\alpha_1 + \alpha_2 + \alpha_3}{\alpha_0},$$

etc. Thus, the computational cost of the step (d5) depends on how far we have to go until the condition (5.22) is satisfied. In the worst possible case, we have to test $(q - 1)$ conditions to determine that $j = q$.

The above analysis confirms that the number of operations in the step (d5) of the Gillespie SSA (a5)–(d5) is proportional to the number of chemical reactions q in the model. This computational cost was one of the reasons why the next-reaction SSA (which performs its search in $\log(q)$ operations) is more efficient than the Gillespie SSA for large q. However, the above analysis also shows that we need $(q - 1)$ comparisons only if the last reaction in the list occurs. In particular, the computational cost depends on the order of the chemical reactions in the model. For example, let us consider the multiscale model (5.1) with the rate constants given by (5.2). Looking at Table 5.1, we see that the reactions $A \to B$ and $B \to A$ happen most of the time (99.9% of all reaction events). If we label these reactions as the first and the second reactions, we will need only one or two conditions to test in most of the time steps to determine which reaction happens in the step (d5). On the other hand, if the indices of the frequently occurring reactions are large, more conditions will have to be tested in the step (d5), making the simulation slower. Thus, a useful way to optimize the Gillespie SSA is to find the relative occurrences of chemical reactions (by a short burst of simulation) and then reorder the chemical reactions in the model. The first reaction in the new list should be the chemical reaction that occurs most frequently.

In this chapter we have presented several exact and approximate methods and tricks to decrease the computational intensity of the Gillespie SSA. It is often difficult to say a priori which method is the most efficient to use; it depends on the particular problem. Our general advice is to use an exact method first, e.g. the Gillespie SSA or the next-reaction SSA. If you are not happy with the computational intensity of the algorithm, try to find out where your computer code spends most of the time. Then you might try to speed up the computation by using some of the SSAs and tricks we have presented. If you decide to use an approximate method, bear in mind that its assumptions and accuracy should be also tested – see Section 5.3 for the corresponding discussion.

Exercises

5.1 Derive the formula (5.14).

5.2 Consider the chemical system

$$\emptyset \xrightarrow{k_1} A \xrightarrow{k_2} \emptyset, \qquad A + B \xrightarrow{k_3} \emptyset, \qquad \emptyset \xrightarrow{k_4} B,$$

with the initial conditions $A(0) = B(0) = 0$ and the rate constants

$$k_1 v = 20 \text{ sec}^{-1}, \quad k_2 = 2 \text{ sec}^{-1}, \quad k_3/v = 0.01 \text{ sec}^{-1}, \quad k_4 = 1 \text{ sec}^{-1}.$$

Show that the reactions $\emptyset \to A \to \emptyset$ are fast but that the others are slow. Show that the slow system may be approximated as

$$\emptyset \xrightarrow{k_4} B \xrightarrow{\bar{k}_3} \emptyset,$$

for some effective reaction rate $\bar{k}_3$, which you should determine.

5.3 Consider the chemical system

$$\emptyset \xrightarrow{k_1} A \underset{k_3}{\overset{k_2}{\rightleftarrows}} B, \qquad B + B \xrightarrow{k_4} C,$$

with the initial conditions $A(0) = B(0) = C(0) = 0$ and the rate constants

$$k_1 v = 1 \text{ sec}^{-1}, \quad k_2 = 50 \text{ sec}^{-1}, \quad k_3 = 200 \text{ sec}^{-1}, \quad k_4/v = 0.01 \text{ sec}^{-1}.$$

Show that the reactions $\emptyset \to A$ and $B+B \to C$ are slow. Find the effective propensity function of the reaction $B + B \to C$. Describe an approximate multiscale SSA that simulates the slow reactions only.

Hint: Use the stationary master equation of the fast system to evaluate the sum $(k_4/v) \sum_{b=2}^{\infty} b(b-1) p_s(b)$.

5.4 Consider the chemical system

$$\emptyset \xrightarrow{k_1} A \xrightarrow{k_2} B \xrightarrow{k_3} C \xrightarrow{k_4} D, \qquad C \xrightarrow{k_5} A,$$

where the rate constants (in units $[\text{sec}^{-1}]$) are

$$k_1 \nu = 1, \quad k_2 = 10, \quad k_3 = 100, \quad k_4 = 0.01, \quad k_5 = 10,$$

with the initial conditions $A(0) = B(0) = C(0) = D(0) = 0$. Show that the reactions $\emptyset \to A$ and $C \to D$ are slow, and that $Q = A + B + C$ is conserved during the fast reactions. Find the effective propensity function for the slow reaction $C \to D$ in terms of Q. Describe an approximate multiscale SSA that simulates the slow reactions only.

Hint: Use the stationary master equation of the fast system to evaluate the sums $\sum_{a=1}^{\infty} \sum_{b=1}^{\infty} a p_s(a, b)$ and $\sum_{a=1}^{\infty} \sum_{b=1}^{\infty} b p_s(a, b)$. Use the convention that $p_s(a, b) = 0$ if $a + b > Q$.

5.5 Consider the chemical system

$$\emptyset \xrightarrow{k_1} A \xrightarrow{k_2} B + B, \qquad B \xrightarrow{k_3} C \xrightarrow{k_4} D, \qquad C + C \xrightarrow{k_5} A,$$

where the rate constants (in units $[\text{sec}^{-1}]$) are

$$k_1 \nu = 1, \quad k_2 = 50, \quad k_3 = 100, \quad k_4 = 0.1, \quad k_5/\nu = 5,$$

with the initial conditions $A(0) = B(0) = C(0) = D(0) = 0$. Show that the reactions $\emptyset \to A$ and $C \to D$ are slow, and that during the fast reactions $Q = 2A + B + C$ is conserved. Describe how you could use this information to construct a multiscale SSA.

5.6 Write computer code using the Gillespie SSA (a5)–(d5), the first-reaction SSA (a15)–(d15) and the next-reaction SSA (a16)–(c16) to simulate the problem (5.1). Verify numerically that all three SSAs lead to the same results – compute the results of Figures 5.1(d), 5.2(c) and 5.2(d) using each algorithm.

5.7   Consider a well-mixed system of q chemical reactions which include the reactions $A \rightarrow B$, $A \rightarrow C$ and $\emptyset \rightarrow A$. Suppose that the reaction $A \rightarrow C$ has occurred at time t^1 and as a result no molecules of A are present in the system after time t^1. At time t^2 (where $t^2 > t^1$), a new molecule of A is produced by the chemical reaction $\emptyset \rightarrow A$. Note that some other reactions (not explicitly mentioned above) might occur in the time interval (t^1, t^2). How do you update the tentative times of the chemical reactions $A \rightarrow B$ and $A \rightarrow C$?

Hint: The chemical reactions $A \rightarrow B$ and $A \rightarrow C$ cannot occur in (t^1, t^2) because their propensity functions are zero. Consequently, their tentative times in the time interval (t^1, t^2) are infinity without computation. The tentative time of $A \rightarrow B$ at $t = t^2$ can be calculated from the tentative time at $t = t^1$. Follow the similar approach as in (5.19), where you use the tentative time at $t = t^1$ instead of $t_{i,old} = \infty$, i.e. use the last non-infinity value of the tentative time to derive the corresponding transformation formula. To compute the propensity function of the chemical reaction $A \rightarrow C$, use the formula (5.21) at $t = t^2$.

6

Stochastic Reaction–Diffusion Models

In Chapter 1, we presented methods for stochastic simulation of well-mixed systems of chemical reactions. In Chapter 4, we introduced models of molecular diffusion. In this chapter, we study approaches for modelling chemical reactions and molecular diffusion at the same time, i.e. we present methods for stochastic modelling of reaction–diffusion processes. We mostly focus on adding chemical reactions to the two position-jump models of molecular diffusion that were presented in Chapter 4. The first approach is based on the compartment-based diffusion model from Section 4.2, the second one on the SDE diffusion model from Section 4.1.

In Sections 6.1 and 6.2 we explain the basic principles of each approach, using an example that includes only zeroth- and first-order chemical reactions. Consequently, it is relatively straightforward to write, simulate and analyse this stochastic reaction–diffusion model.

The situation is more complicated whenever some chemical species are subject to higher-order chemical reactions. In Section 6.3, we present an illustrative example of a reaction–diffusion process incorporating the nonlinear chemical system (1.55)–(1.56) and apply the compartment-based framework that is introduced in Section 6.1. The SSA is based on dividing the computational domain into artificially well-mixed compartments and postulating that only molecules that are within the same compartment can react. Diffusion is then modelled as jumps between the neighbouring compartments. In order to use this method, we have to choose an appropriate compartment size. This question is addressed in Section 6.4.

The modelling of second-order reactions in the context of the SDE diffusion model from Section 4.1 is shown in Section 6.5. In Section 6.7, we discuss modelling reversible reactions, including reversible adsorption to a boundary and reversible bimolecular reactions. Finally, in Section 6.8, we conclude this chapter with discussion of applications to pattern formation in biology.

160

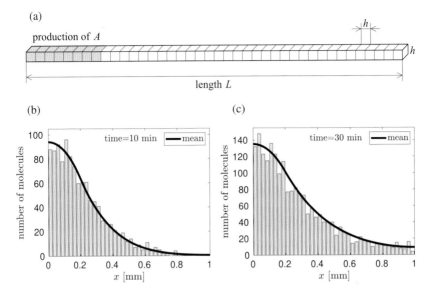

Figure 6.1   *Illustrative pseudo-one-dimensional reaction–diffusion example.*
(a) Schematic of the computational domain divided into compartments.
(b), (c) One realization of the Gillespie SSA (a5)–(d5) for the system of chemical
reactions (6.1)–(6.3). Grey histograms show numbers of molecules in compart-
ments at times t = 10 min and t = 30 min. The solution of (6.9)–(6.10) is plotted
as the black solid line.

6.1 A Compartment-Based Reaction–Diffusion Algorithm

In this section, we introduce chemical reactions to the compartment-based diffusion model that was studied in Section 4.2. Since we wanted to focus on ideas rather than technicalities, we concentrated in Section 4.2 on the one-dimensional case, where the compartment-based diffusion model can be formulated as the chain of "reactions" (4.13). The extension to three-dimensional (real) physical space is conceptually straightforward.

On the other hand, in Chapter 1 chemical reactions were always considered in a three-dimensional container of volume v. Although it might appear that this precludes the formulation of a one-dimensional reaction–diffusion problem, it is possible to design a problem that is effectively one-dimensional, and we study such a problem in this section.

We consider molecules of chemical species A that diffuse in the domain $[0, L] \times [0, h] \times [0, h]$, where $L = 1$ mm and $h = 25\,\mu$m. This domain is schematically shown in Figure 6.1(a). Although such an elongated domain looks like an artificial concept, it can be a useful simplification of real-world phenomena. Suppose, for example, that the chemical A really reacts and diffuses in a larger domain $[0, L] \times [0, L_y] \times [0, L_z]$ where $h \ll L_y$ and $h \ll L_z$. If we are mainly interested in the spatial distribution of A along the x-direction, and if we expect that there are no significant variations along the y- and z-directions, it is reasonable to restrict the simulation to the subdomain $[0, L] \times [0, h] \times [0, h]$ with periodic boundary conditions in the y- and z-directions. This restriction decreases the computational intensity of stochastic simulation significantly.

Our illustrative reaction–diffusion model can be formulated as follows. The molecules of chemical species A diffuse in the domain $[0, L] \times [0, h] \times [0, h]$ with the diffusion constant $D = 10^{-4} \, \text{mm}^2 \, \text{sec}^{-1}$. Initially, there are no molecules in the system. Molecules are produced uniformly in the part of the domain $[0, L/5] \times [0, h] \times [0, h]$ with rate $k_2 = 2 \times 10^{-5} \, \mu\text{m}^{-3} \, \text{sec}^{-1}$. This means that the probability that a molecule is created in a subvolume $v \subset [0, L/5] \times [0, h] \times [0, h]$ during the time interval $[t, t + dt)$ is equal to $k_2 v \, dt$. Consequently, the probability that a molecule is created somewhere in the subdomain $[0, L/5] \times [0, h] \times [0, h]$ during the time interval $[t, t+dt)$ is equal to $(k_2 h^2 L/5) \, dt$. Molecules are degraded at the rate $k_1 = 10^{-3} \, \text{sec}^{-1}$ according to the chemical reaction (1.1).

Following Section 4.2, we divide the computational domain $[0, L] \times [0, h] \times [0, h]$ into $K = L/h = 40$ compartments each of volume h^3 – see Figure 6.1(a). We denote the number of molecules of chemical species A in the ith compartment $[(i - 1)h, ih) \times [0, h] \times [0, h]$ by A_i, where i runs from 1 to K. Then our reaction–diffusion process is described by the system of chemical reactions:

$$A_1 \underset{d}{\overset{d}{\rightleftarrows}} A_2 \underset{d}{\overset{d}{\rightleftarrows}} A_3 \underset{d}{\overset{d}{\rightleftarrows}} \cdots \underset{d}{\overset{d}{\rightleftarrows}} A_K, \tag{6.1}$$

$$A_i \overset{k_1}{\longrightarrow} \emptyset, \qquad \text{for } i = 1, 2, \ldots, K, \tag{6.2}$$

$$\emptyset \overset{k_2}{\longrightarrow} A_i, \qquad \text{for } i = 1, 2, \ldots, K/5. \tag{6.3}$$

Reactions (6.1) describe diffusion and are identical to (4.13). In particular, the rate constant d is given by $d = D/h^2$. Reaction (6.2) describes the degradation of A and is, in fact, reaction (1.1) applied to every compartment. Reaction (6.3) describes the production of A in the first $K/5$ compartments, i.e. the production in the subdomain $[0, L/5] \times [0, h] \times [0, h]$, which is shaded grey in Figure 6.1(a). Since the compartment volume is equal to h^3, the rate of production per compartment is equal to $k_2 h^3$. Thus the propensity functions of chemical reactions in (6.3) are equal to $k_2 h^3$.

Thus the compartment-based modelling of reaction–diffusion processes is relatively straightforward. Since the diffusion is modelled as the chain of "reactions" (6.1), we end up with the system of chemical reactions (6.1)–(6.3) that can be simulated using the Gillespie SSA (a5)–(d5). In our case, the propensity functions of reactions in (6.1) are given as $A_i(t)d$, the propensity functions of reactions in (6.2) are given as $A_i(t)k_1$ and the propensity functions of reactions in (6.3) are equal to $k_2 h^3$. Starting with no molecules of A in the system, we compute one realization of the Gillespie SSA (a5)–(d5) for the system of reactions (6.1)–(6.3), and plot the numbers of molecules in compartments at two different times in Figure 6.1(b) and (c).

The model can be analysed using the reaction–diffusion master equation. Let $p(\mathbf{n}, t)$ be the joint probability that $A_i(t) = n_i$, $i = 1, 2, \ldots, K$, where we use the notation $\mathbf{n} = [n_1, n_2, \ldots, n_K]$. Let us define operators $R_i, L_i \colon \mathbb{N}^K \to \mathbb{N}^K$ by (4.20)–(4.21). Then the reaction–diffusion master equation, which corresponds to the system of reactions (6.1)–(6.3), can be written as follows:

$$\frac{\partial p(\mathbf{n})}{\partial t} = d \sum_{i=1}^{K-1} \left\{ (n_i + 1) \, p(R_i \mathbf{n}) - n_i \, p(\mathbf{n}) \right\} + d \sum_{i=2}^{K} \left\{ (n_i + 1) \, p(L_i \mathbf{n}) - n_i \, p(\mathbf{n}) \right\}$$

$$+ k_1 \sum_{i=1}^{K} \left\{ (n_i + 1) \, p(n_1, n_2, \ldots, n_i + 1, \ldots, n_K) - n_i \, p(\mathbf{n}) \right\}$$

$$+ k_2 h^3 \sum_{i=1}^{K/5} \left\{ p(n_1, n_2, \ldots, n_i - 1, \ldots, n_K) - p(\mathbf{n}) \right\}. \tag{6.4}$$

The first two sums correspond to diffusion (6.1), the third sum to degradation (6.2) and the fourth sum to production (6.3). The mean is defined as the vector $\mathbf{M}(t) \equiv [M_1, M_2, \ldots, M_K]$ where M_i is given by (4.23). This gives the mean number of molecules in the ith compartment, $i = 1, 2, \ldots, K$, at time t (averaged over many realizations of the Gillespie SSA (a5)–(d5)). To derive the evolution equation for the mean vector $\mathbf{M}(t)$, we can follow the method from Section 1.2 – see derivation of (1.15) from the chemical master equation (1.11). Multiplying (6.4) by n_i and summing over all n_j, $j = 1, 2, \ldots, K$, we obtain (leaving the details as Exercise 6.1)

$$\frac{\partial M_1}{\partial t} = d(M_2 - M_1) + k_2 h^3 - k_1 M_1, \tag{6.5}$$

$$\frac{\partial M_i}{\partial t} = d(M_{i+1} + M_{i-1} - 2M_i) + k_2 h^3 - k_1 M_i, \quad i = 2, \ldots, K/5, \tag{6.6}$$

$$\frac{\partial M_i}{\partial t} = d(M_{i+1} + M_{i-1} - 2M_i) - k_1 M_i, \quad i = K/5 + 1, \ldots, K - 1, \tag{6.7}$$

$$\frac{\partial M_K}{\partial t} = d(M_{K-1} - M_K) - k_1 M_K. \tag{6.8}$$

The concentration of molecules in the ith compartment is defined as $\overline{M}_i = M_i/h^3$, $i = 1, 2, \ldots, K$. Dividing (6.5)–(6.8) by h^3 and using $d = D/h^2$, we can write a system of ODEs for $\overline{M}_i$ in the following form:

$$\frac{\partial \overline{M}_1}{\partial t} = D \frac{\overline{M}_2 - \overline{M}_1}{h^2} + k_2 - k_1 \overline{M}_1,$$

$$\frac{\partial \overline{M}_i}{\partial t} = D \frac{\overline{M}_{i+1} + \overline{M}_{i-1} - 2\overline{M}_i}{h^2} + k_2 - k_1 \overline{M}_i, \qquad i = 2, \ldots, K/5,$$

$$\frac{\partial \overline{M}_i}{\partial t} = D \frac{\overline{M}_{i+1} + \overline{M}_{i-1} - 2\overline{M}_i}{h^2} - k_1 \overline{M}_i, \qquad i = K/5 + 1, \ldots, K - 1,$$

$$\frac{\partial \overline{M}_K}{\partial t} = D \frac{\overline{M}_{K-1} - \overline{M}_K}{h^2} - k_1 \overline{M}_K.$$

This is a discretized version of the reaction–diffusion PDE

$$\frac{\partial a}{\partial t} = D \frac{\partial^2 a}{\partial x^2} + k_2 \chi_{[0,L/5]} - k_1 a \tag{6.9}$$

with zero-flux boundary conditions

$$\frac{\partial a}{\partial x}(0) = \frac{\partial a}{\partial x}(L) = 0. \tag{6.10}$$

Here $a(x, t)$ is the concentration of molecules of A at point x and time t, and $\chi_{[0,L/5]}$ is the characteristic function of the interval $[0, L/5]$, so that $\chi_{[0,L/5]}(x) = 1$ if $x \in [0, L/5]$, and otherwise it is zero. Equation (6.9) provides a classic deterministic description of the reaction–diffusion process. Using the initial condition $a(x, 0) = 0$, we computed the solution of (6.9)–(6.10) numerically, and plotted it as the black solid line in Figure 6.1(b) and (c) for comparison. More precisely, we plot $a(x, t)h^3$, which expresses the results in terms of the number of molecules per compartment of volume h^3. As expected, the PDE (6.9) describes the time evolution of the mean well.

6.2 A Reaction–Diffusion SSA Based on the SDE Model of Diffusion

In this section, we present an SSA that implements the SDE model of diffusion from Section 4.1, that is, we follow the trajectories of individual molecules. The diffusion of each molecule is simulated according to the model (a8)–(c8). We explain the SSA using the reaction–diffusion example from Section 6.1. Since we are mostly interested in the spatial distribution of the chemical A along the x-direction, we formulate the algorithm for the time evolution of the x-coordinate only.

To model the reaction–diffusion system described in Figure 6.1(a), we choose a small time step Δt and perform the following three steps at time t:

(a17) For each molecule, compute its x-coordinate at time $t + \Delta t$ according to the steps (a8)–(c8).

(b17) For each molecule, generate a random number r_1 uniformly distributed in the interval $(0, 1)$. If $r_1 < k_1 \Delta t$, then remove the molecule from the system.

(c17) Generate a random number r_2 uniformly distributed in the interval $(0, 1)$. If $r_2 < (k_2 h^2 L/5) \Delta t$, then generate another random number r_3 uniformly distributed in the interval $(0, 1)$ and introduce a new molecule at the position with x-coordinate equal to $r_3 L/5$. Continue with step (a17) for time $t + \Delta t$.

Degradation of molecules is given by reaction (1.1). Thus $k_1 \, dt$ is the probability that a molecule is degraded in the time interval $[t, t + dt)$ for infinitesimally small dt. In the step (b17), we implement degradation by replacing dt by finite time step Δt (compare with the SSA (a1)–(b1)). In particular, Δt has to be chosen sufficiently small so that $k_1 \Delta t \ll 1$. Similarly, the probability that a molecule is created in $[0, L/5]$ in the time interval $[t, t + dt)$ is equal to $(k_2 h^2 L/5) \, dt$. Consequently, we have to choose Δt so small that $(k_2 h^2 L/5) \Delta t$ is significantly less than 1. We choose $\Delta t = 10^{-2}$ sec. Then $k_1 \Delta t = 10^{-5}$ and $(k_2 h^2 L/5) \Delta t = 2.5 \times 10^{-2}$ for our parameter values $k_1 = 10^{-3}$ sec^{-1}, $k_2 = 2 \times 10^{-5} \, \mu$m^{-3} sec^{-1} and $L = 1$ mm.

Starting with no molecules of A in the system, we compute one realization of the SSA (a17)–(c17). To visualize the results, we divide the interval $[0, L]$ into 40 bins and we plot the numbers of molecules in bins at time 10 minutes in Figure 6.2(a). The same plot at time 30 minutes is given in Figure 6.2(b). We use the same number of bins to visualize the results of the SSA (a17)–(c17) as we used previously in the compartment-based model. Thus Figure 6.2 is directly comparable with Figure 6.1. We also plot the solution of (6.9)–(6.10) as the black solid line for comparison.

Chemical reactions in our model have been implemented in steps (b17) and (c17) using a small time step Δt, following the approach of the "naive" SSA (a1)–(b1). Since each reaction in our model takes place at a randomly chosen time which is independent of the positions of molecules, it is relatively straightforward to design a more efficient event-based SSA, following the ideas behind the Gillespie SSA, which we leave to the reader as Exercise 6.2.

We conclude that zeroth- and first-order chemical reactions can be easily introduced to both the compartment-based and the SDE diffusion models

(a) (b)

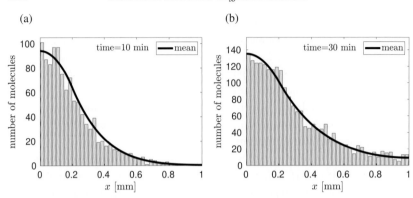

Figure 6.2 *One realization of the SSA (a17)–(c17). Dividing domain* [0, L] *into 40 bins, we plot the number of molecules in each bin at times:* (a) t = 10 *min; and* (b) t = 30 *min. Solution of (6.9)–(6.10) is plotted as the black solid line.*

which we studied in Chapter 4. The models can be further analysed to deduce that the classic deterministic PDE (6.9) describes the average behaviour of the model over many stochastic realizations. The SSA (a17)–(c17) can also be modified to incorporate a different individual-based model of diffusion, e.g. a velocity-jump process from Section 4.3 – see Exercise 6.3.

6.3 Compartment-Based SSA for Higher-Order Reactions

In Sections 6.1 and 6.2 we studied an example of a reaction–diffusion model which did not include second-order chemical reactions; that is, the reactions of the form (1.26). The compartment-based approach, introduced in Section 6.1, can be easily generalized to models that involve chemical reactions of any order. In this section, we explain this generalization using an illustrative reaction–diffusion system that incorporates the nonlinear model (1.55)–(1.56). The corresponding generalization of the SSA (a17)–(c17) to the case of the second-order chemical reactions (1.26) will be presented later in Section 6.5.

As in Figure 6.1(a), we consider the elongated domain $[0, L] \times [0, h] \times [0, h]$, where $L = 1$ mm and $h = 25\,\mu$m. We consider two chemical species A and B that diffuse in this domain and are subject to chemical reactions

$$A + A \xrightarrow{k_1} \emptyset, \qquad\qquad A + B \xrightarrow{k_2} \emptyset, \qquad\qquad (6.11)$$

$$A \xrightarrow{k_3} \emptyset, \qquad B \xrightarrow{k_4} \emptyset, \qquad \emptyset \xrightarrow{k_5} A, \qquad (6.12)$$

and (localized) production

$$\emptyset \xrightarrow{k_6} B \quad \text{in subdomain } [3L/5, L] \times [0, h] \times [0, h]. \tag{6.13}$$

Reactions (6.11) are identical to reactions (1.55). We couple them with the (linear) degradation and production of A and B. Notice that the production of B (reaction (6.13)) is restricted to a part of the computational domain. On the other hand, the reactions (6.11)–(6.12) take place in the whole domain.

To simulate the reaction–diffusion system (6.11)–(6.13), we use the compartment-based approach introduced in Section 6.1. We divide the computational domain $[0, L] \times [0, h] \times [0, h]$ into $K = L/h = 40$ compartments of volume h^3. We denote the number of molecules of chemical species A (resp. B) in the ith compartment $[(i - 1)h, ih] \times [0, h] \times [0, h]$ by A_i (resp. B_i), $i = 1, 2, \ldots, K$. We denote the diffusion coefficients of A and B by D_A and D_B respectively. Then diffusion corresponds to two chains of "chemical reactions":

$$A_1 \underset{d_A}{\overset{d_A}{\underset{\longleftarrow}{\longrightarrow}}} A_2 \underset{d_A}{\overset{d_A}{\underset{\longleftarrow}{\longrightarrow}}} A_3 \underset{d_A}{\overset{d_A}{\underset{\longleftarrow}{\longrightarrow}}} \cdots \underset{d_A}{\overset{d_A}{\underset{\longleftarrow}{\longrightarrow}}} A_K, \tag{6.14}$$

$$B_1 \underset{d_B}{\overset{d_B}{\underset{\longleftarrow}{\longrightarrow}}} B_2 \underset{d_B}{\overset{d_B}{\underset{\longleftarrow}{\longrightarrow}}} B_3 \underset{d_B}{\overset{d_B}{\underset{\longleftarrow}{\longrightarrow}}} \cdots \underset{d_B}{\overset{d_B}{\underset{\longleftarrow}{\longrightarrow}}} B_K, \tag{6.15}$$

where $d_A = D_A/h^2$ and $d_B = D_B/h^2$. Second-order (or higher-order) chemical reactions are localized in the compartment-based approach by postulating that molecules that are in the same compartment can react with each other and molecules in different compartments cannot. Thus reactions (6.11) are modelled as the following set of chemical reactions:

$$A_i + A_i \xrightarrow{k_1} \emptyset, \qquad A_i + B_i \xrightarrow{k_2} \emptyset, \qquad \text{for } i = 1, 2, \ldots, K. \tag{6.16}$$

Production and degradation have been already studied in Section 6.1. We again rewrite them as reactions for A_i and B_i applied in the corresponding compartments. Consequently, reactions (6.12)–(6.13) read as

$$A_i \xrightarrow{k_3} \emptyset, \qquad B_i \xrightarrow{k_4} \emptyset, \qquad \emptyset \xrightarrow{k_5} A_i, \qquad \text{for } i = 1, 2, \ldots, K, \tag{6.17}$$

$$\emptyset \xrightarrow{k_6} B_i, \qquad \text{for } i = 3K/5 + 1, \ldots, K, \tag{6.18}$$

where the last set of reactions takes into account the fact that molecules of chemical species B are only produced in subdomain $[3L/5, L] \times [0, h] \times [0, h]$ with the rate $k_6 h^3$ per compartment (of volume h^3).

Starting with no molecules in the system at time $t = 0$, we present one realization of the Gillespie SSA (a5)–(d5) applied to the chemical system

(a)   (b)

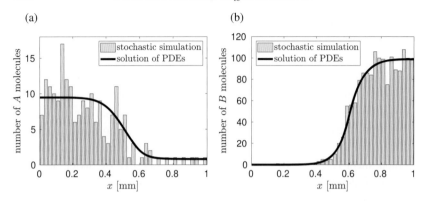

Figure 6.3 *One realization of the Gillespie SSA (a5)–(d5) for the system of chemical reactions (6.14)–(6.18). Numbers of molecules of chemical species A (in panel (a)) and B (in panel (b)) in compartments at time* 200 *minutes (grey histograms). Solution of (6.19)–(6.21) is plotted as the black solid line.*

(6.14)–(6.18) in Figure 6.3. We plot the numbers of molecules of A and B at time 200 minutes. We use the values of kinetic rate constants $k_1 = 3 \times 10^{-2} \, \mu\text{m}^3 \, \text{sec}^{-1}$, $k_2 = 3 \times 10^{-1} \, \mu\text{m}^3 \, \text{sec}^{-1}$, $k_3 = 10^{-4} \, \text{sec}^{-1}$, $k_4 = 10^{-4} \, \text{sec}^{-1}$, $k_5 = 10^{-7} \, \mu\text{m}^{-3} \, \text{sec}^{-1}$, $k_6 = 10^{-6} \, \mu\text{m}^{-3} \, \text{sec}^{-1}$ and the diffusion constants $D_A = D_B = 1 \, \mu\text{m}^2 \, \text{sec}^{-1}$.

In Section 1.5 we observed that the analysis of the master equation for chemical systems involving second-order reactions is not trivial. It is not possible to derive an equation for stochastic means as was done in Section 6.1 for the linear model. Hence, we will not attempt such an approach here. We also observed in Section 6.1 that the equation for the mean vector (6.5)–(6.8) was actually equal to a discretized version of the reaction–diffusion equation (6.9)–(6.10), which would be used as a traditional deterministic description. When considering the nonlinear chemical model (6.14)–(6.18), we cannot derive the equation for the mean vector but we can still write a deterministic system of PDEs for concentrations $a(x, t) \approx A_i(t)/h^3$ and $b(x, t) \approx B_i(t)/h^3$ where $x \approx ih$. We simply add diffusion and degradation to the system of ODEs (1.58)–(1.59) to obtain

$$\frac{\partial a}{\partial t} = D \frac{\partial^2 a}{\partial x^2} - 2k_1 a^2 - k_2 \, ab - k_3 \, a + k_5, \tag{6.19}$$

$$\frac{\partial b}{\partial t} = D \frac{\partial^2 b}{\partial x^2} - k_2 \, ab - k_4 \, b + k_6 \, \chi_{[3L/5, L]}, \tag{6.20}$$

and couple it with zero-flux boundary conditions

$$\frac{\partial a}{\partial x}(0) = \frac{\partial a}{\partial x}(L) = \frac{\partial b}{\partial x}(0) = \frac{\partial b}{\partial x}(L) = 0. \tag{6.21}$$

Using the initial condition $a(x,0) = 0$ and $b(x,0) = 0$, we can compute the solution of (6.19)–(6.21) numerically. It is plotted as the black solid line in Figure 6.3 for comparison. More precisely, we plot $a(x,t)h^3$ and $b(x,t)h^3$ to express the concentration profiles in terms of the number of molecules per compartment of volume h^3. We see that (6.19)–(6.21) gives a reasonable description of the system when comparing with one realization of the SSA (a5)–(d5). However, we emphasize again that the solution of (6.19)–(6.21) is not equal to the stochastic mean.

6.4 A Choice of Compartment Size *h*

An important question, which was not addressed in the previous sections, is: What is the appropriate choice of the compartment size h? In Sections 4.2 and 6.1, we considered linear models and we were able to derive the equations for the mean vectors (for example equations (6.5)–(6.8)). Dividing (6.5)–(6.8) by h and passing to the limit $h \to 0$, we derive the corresponding deterministic reaction–diffusion PDE (6.9) which can also be viewed (for linear models) as an equation for the probability distribution function of a single molecule (that is, the exact description that we are attempting to approximate using the compartment-based SSA). Consequently, for reaction–diffusion systems involving only zeroth- and first-order chemical reactions we can increase the accuracy of the SSA by decreasing h.

The situation is much more delicate when the system involves second- or higher-order reactions. In that case, although diffusion is modelled more accurately as h is decreased, the reactions are modelled less accurately as h is decreased, so that we lose accuracy if we choose h too small. We demonstrate this phenomenon using the following illustrative example.

We consider a chemical species A that diffuses in the cubic domain $[0, L] \times [0, L] \times [0, L]$ and is subject to the system of chemical reactions (1.27). In Section 1.4, we investigated this chemical system under the assumption that the reactor is well stirred. Then we can obtain an exact formula for the stationary value, M_s, of the mean number of molecules; see equation (1.48). Since the volume of the cubic domain $[0, L] \times [0, L] \times [0, L]$ is equal to $v = L^3$, equation (1.48) reads

$$M_s = \frac{1}{4} + \sqrt{\frac{k_2 L^6}{2k_1}}\, I_1'\!\left(2\sqrt{\frac{2k_2 L^6}{k_1}}\right)\left[I_1\!\left(2\sqrt{\frac{2k_2 L^6}{k_1}}\right)\right]^{-1}, \qquad (6.22)$$

where I_1 is the modified Bessel function of the first kind; see Figure 1.4(a). In this section, we apply the compartment-based reaction–diffusion approach to

(a) (b)

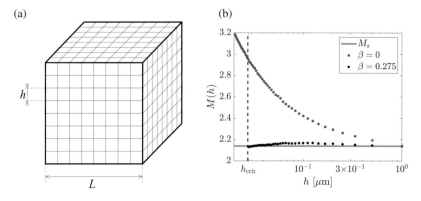

Figure 6.4 (a) *Domain* $[0, L] \times [0, L] \times [0, L]$ *is divided into* K^3 *compartments of volume* $h^3 = (L/K)^3$. *The division of the domain for* $K = 8$ *is shown on the picture.* (b) *The average value,* $M(h)$, *of the number of molecules of A as a function of the compartment length h, estimated from the long-time stochastic simulations of the chemical system* (6.23)–(6.24) *(red dots). Results computed for the modified propensity* (6.27) *with* $\beta = 0.275$ *are plotted as black dots. We use* $k_1 = 1\ \mu m^3\ sec^{-1}$, $k_2 = 8\ \mu m^{-3}\ sec^{-1}$, $D = 10\ \mu m^2\ sec^{-1}$ *and* $L = 1\ \mu m$. *The values* $M_s = 2.139$ *and* $h_{crit} = 2.8 \times 10^{-2} \mu m$ *are denoted by the green solid and blue dashed lines, respectively.*

the chemical system (1.27). Our goal is to illustrate the fact that a very small compartment size h leads to large computational errors.

We divide the cubic domain $[0, L] \times [0, L] \times [0, L]$ into K^3 cubic compartments of volume h^3 where $K \geq 1$ and $h = L/K$ (see Figure 6.4(a)). To formulate precisely the compartment-based SSA for the illustrative chemical system (1.27) in the reactor $[0, L] \times [0, L] \times [0, L]$, we denote the compartments by indices from the set

$$I_{all} = \{(i, j, k) \mid i, j, k \text{ are integers such that } 1 \leq i, j, k \leq K\}.$$

Let $A_{ijk}(t)$ denote the number of molecules of the chemical species A in the (i, j, k)th compartment at time t where $(i, j, k) \in I_{all}$. Diffusion is modelled as a jump process between neighbouring compartments. Let us define the set of possible directions of jumps

$$\mathbf{E} = \{[1, 0, 0], \ [-1, 0, 0], \ [0, 1, 0], \ [0, -1, 0], \ [0, 0, 1], \ [0, 0, -1]\}.$$

For every $(i, j, k) \in I_{all}$, we also define

$$\mathbf{E}_{ijk} = \{\mathbf{e} \in \mathbf{E} \mid ((i, j, k) + \mathbf{e}) \in I_{all}\},$$

i.e. $\mathbf{E}_{ijk}$ is the set of possible directions of jumps from the (i, j, k)th compartment. For most compartments $\mathbf{E}_{ijk}$ will be the full set of possible jumps $\mathbf{E}$, but for compartments on the boundary the set of jumps is restricted. The notation

$\mathbf{E}_{ijk}$ avoids us having to write down separate equations for each boundary compartment.

As we have seen, the main idea of the compartment-based approach is that the small compartments are assumed to be well mixed, and that only molecules in the same compartment can react according to bimolecular reactions. Thus the compartment-based reaction–diffusion model can be written using the chemical reaction formalism as follows. We have a system of K^3 "chemical species" A_{ijk} for $(i, j, k) \in I_{\text{all}}$ which are subject to the chemical reactions:

$$A_{ijk} + A_{ijk} \xrightarrow{k_1} \emptyset, \qquad \emptyset \xrightarrow{k_2} A_{ijk}, \qquad \text{for } (i, j, k) \in I_{\text{all}}, \qquad (6.23)$$

$$A_{ijk} \xrightarrow{D/h^2} A_{ijk+\mathbf{e}}, \qquad \text{for } (i, j, k) \in I_{\text{all}}, \ \mathbf{e} \in \mathbf{E}_{ijk}, \qquad (6.24)$$

where D is the diffusion constant. The chemical reactions (6.23) correspond to the chemical system (1.27) considered in each compartment; each compartment is well stirred and only molecules in the same compartment can react. The propensity functions of reactions (6.23) are

$$\alpha_{ijk,1}(t) = A_{ijk}(t)(A_{ijk}(t) - 1)\, k_1/h^3, \qquad \alpha_{ijk,2}(t) = k_2 h^3, \qquad (6.25)$$

where h^3 is the volume of the compartment (see Table 1.1). The reactions (6.24) correspond to diffusive jumps between neighbouring compartments. The propensity functions of these "reactions" are equal to $A_{ijk}(t)\, D/h^2$. There are $2K^3$ reactions in (6.23) and $6K^3 - 6K^2$ diffusion "reactions" in (6.24) (because there are six possible directions to jump from each inner compartment and some directions are missing for boundary compartments). Thus we are able to formulate the compartment-based reaction–diffusion model as the chemical system of K^3 chemical species A_{ijk} which are subject to $8K^3 - 6K^2$ reactions (6.23)–(6.24). The time evolution of the chemical system (6.23)–(6.24) can be simulated by the Gillespie SSA.

Now the number of molecules of A in the whole container $[0, L] \times [0, L] \times [0, L]$ is given by

$$A(t) = \sum_{(i,j,k) \in I_{\text{all}}} A_{ijk}(t).$$

Let $M(h)$ be the average number of molecules of A in the whole container at steady state (i.e. provided that the system is observed for a long time). It can be formally defined by

$$M(h) = \lim_{t \to \infty} \mathrm{E}\,[A(t)] = \lim_{t \to \infty} \mathrm{E}\left[\sum_{(i,j,k) \in I_{\text{all}}} A_{ijk}(t) \right], \qquad (6.26)$$

where notation $M(h)$ highlights that we are interested in the behaviour of M as a function of $h = L/K$. We have $M(L) = M_s$, where M_s is given by (6.22), because the case $h = L$ means that we simulate the whole domain as one compartment (i.e. $K = 1$). If we increase K, we decrease h through the formula $h = L/K$.

In the problems studied thus far in this chapter we have restricted the production of molecules to part of the computational domain so that we obtain a non-trivial spatial structure in the solution. In the current problem we have deliberately kept the production of A uniform throughout the whole domain. Thus, although we are simulating diffusion (for $K > 1$) and there will be spatial variations in any particular simulation due to random fluctuations, we do not expect any spatial variation in the probability distribution function. In particular this means that $M(h)$ should be equal to M_s given by (6.22), which is plotted as the green solid line in Figure 6.4(b) together with the values of $M(h)$ (red dots) estimated from 50 long-time simulations of the Gillespie SSA, with one simulation for each value of $h = L/K$ where $K = 1, 2, \ldots, 50$. We use the same parameter values in all simulations, namely $L = 1$ μm, $k_1 = 1$ μm^3 sec^{-1}, $k_2 = 8$ μm^{-3} sec^{-1} and $D = 10$ μm^2 sec^{-1}. Then, using (6.22) and recurrence formulae from Exercise 1.3, we obtain that $M_s = 2I_0(8)/I_1(8) \approx 2.139$.

In Figure 6.4(b), we observe that $M(h)$ increases as we decrease h. The values of $M(h)$ start at M_s for $h = L$, but they do not converge to M_s as $h \to 0$. In fact, the stationary distribution of A does not converge to any distribution as $h \to 0$; its average will increase further as h is decreased. This makes the assessment of the accuracy of computations more challenging than in the deterministic case. For example, if we solve numerically a PDE-based (deterministic) model using the cubic mesh shown in Figure 6.4(a) with mesh size h, it is often very informative to make another computation with a finer mesh where h is replaced by $h/2$. If the solutions computed do not significantly differ, we can conclude that the original mesh was chosen fine enough. On the other hand, the same trick cannot be applied to the compartment-based SSA because the results do not converge to anything in the limit $h \to 0$.

The increase of $M(h)$ for small values of h in Figure 6.4(b) is caused by the bimolecular reaction being lost in the limit $h \to 0$ (i.e. this reaction does not occur as frequently as it should whenever h is too small). In the present problem we coupled the bimolecular reaction with the production of A. The production rate in the whole domain is equal to

$$\sum_{(i,j,k)\in I_{\text{all}}} \alpha_{ijk,2}(t) = \sum_{(i,j,k)\in I_{\text{all}}} k_2 h^3 = K^3 k_2 h^3 = k_2 L^3,$$

i.e. it is independent of h. Thus, for small h, the slower removal of A by the bimolecular reaction and unchanged production rate of A result in more molecules of A in the system on average, corresponding to the increase of $M(h)$ in Figure 6.4(b).

The compartment-based SSA is generally considered valid only for a range of values of h. In particular, h must not be too small. For the second-order reaction (1.25) this constraint is usually stated in the form $h \gg k/(D_A + D_B)$, where k is the reaction rate constant and D_A and D_B are the diffusion constants of A and B respectively. In the case of the second-order reaction (1.27) we require $h \gg k_1/D_A$. To satisfy this condition in our particular example, we could simply choose $h = L$. However, the real importance of stochastic reaction–diffusion modelling is not in modelling of spatially homogeneous systems. If the system has some spatial variations (i.e. some parts of the computational domain are more preferred by molecules than others), then we obviously want to choose h small enough to capture the desired spatial resolution. This leads to a restriction on h from above, namely $L \gg h$. Thus it is often suggested to choose h small (to satisfy $L \gg h$) but not too small (to satisfy $h \gg k/(D_A + D_B)$). The optimal choice of h is a subject of current research – see, for example, papers by Erban and Chapman (2009), Isaacson (2009), Hellander et al. (2012), Kang et al. (2012), Isaacson (2013), Agbanusi and Isaacson (2014), Cao and Erban (2014), Hellander et al. (2015), Hellander and Petzold (2016) and Isaacson and Zhang (2018).

We have already said that the reason for the bound on h from below is that the second-order reaction does not happen as frequently as it should for small values of h. One way to fix this problem is to increase the probability of reaction for two molecules in the same compartment, i.e. to increase the propensity function (6.25) of the second-order reaction in each compartment. It is shown by Erban and Chapman (2009) that if the first propensity function in (6.25) is modified to

$$\alpha_{ijk,1}(t) = A_{ijk}(t)(A_{ijk}(t) - 1)\frac{Dk_1}{Dh^3 - \beta k_1 h^2}, \qquad (6.27)$$

where β is a suitable constant, we can recover correctly not only the value of mean $M(h)$ but the whole stationary distribution given by (1.50). In Figure 6.4(b), we present results of stochastic simulations which use the propensity function (6.27) with $\beta = 0.275$. We observe that $M(h)$ stays close to M_s. The exact value of β for this particular chemical system can be obtained analytically. It is slightly K-dependent: ranging from 0.33 for small values of K to 0.25 for large values of K. If we used such K-dependent β instead of $\beta = 0.275$ in our stochastic simulations, the black dots in Figure 6.4(b) would then exactly lie on the green line corresponding to M_s.

The presence of the additional term in the denominator of (6.27) increases the propensity function. Since this term is proportional to h^2 rather than h^3, it becomes more significant as h gets smaller. In fact, at the critical value

$$h_{\text{crit}} = \frac{\beta k_1}{D},\qquad (6.28)$$

the modified propensity function becomes infinite (providing $A_{ijk}(t) \geq 2$). For this value of h two molecules that are present in the same compartment immediately react. There is no way to correct the compartment-based algorithm for h smaller than h_{crit}. Thus we can effectively replace the condition $h \gg k_1/D$ by a sharp inequality $h \geq h_{\text{crit}}$ provided that the propensity function (6.25) is replaced by (6.27).

On the other hand, if $h \gg h_{\text{crit}}$, then (6.27) may be replaced by the original propensity function (6.25). In fact, the modified propensity function (6.27) can be rewritten as (6.25) multiplied by

$$\frac{h}{h - h_{\text{crit}}},\qquad (6.29)$$

which is equal to 1 for $h \gg h_{\text{crit}}$. Similar conclusions can be obtained for the second-order reactions involving two different molecular species, i.e. for reactions $A + B \rightarrow \emptyset$. We can again correct the compartment-based approach for $h \geq h_{\text{crit}}$ by multiplying the original propensity function by factor (6.29). Here, h_{crit} is given by (6.28), where $D = D_A + D_B$ is the sum of the diffusion constants of reactants A and B; see Erban and Chapman (2009) for details. The value of h_{crit} for our parameter values is denoted in Figure 6.4(b) as the blue dashed line.

6.5 Molecular-Based Approaches for Second-Order Reactions

In Section 6.2 we presented the reaction–diffusion SSA implementing the SDE model of diffusion from Section 4.1. The diffusive trajectory of each molecule was simulated according to the SSA (a8)–(c8), and it was relatively straightforward to add zeroth-order and first-order reactions to this model via the SSA (a17)–(c17). Second-order (bimolecular) reactions can be implemented in this framework using the idea of a *reaction radius*: we postulate that two molecules that are involved in a second-order reaction can react only if their separation is less than the reaction radius. We will illustrate this idea on the reaction–diffusion system that has been introduced in Section 6.3. We consider two chemical species A and B that diffuse in the elongated domain

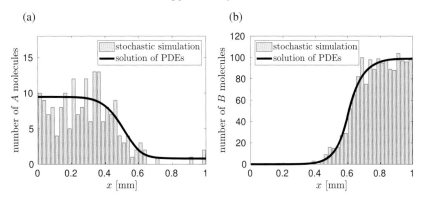

Figure 6.5 *One realization of the SSA (a18)–(e18). Numbers of molecules of chemical species A (in panel (a)) and B (in panel (b)) in bins of volume h^3 at time* 200 *minutes (grey histograms). Solution of (6.19)–(6.21) is plotted as the black solid line.*

$[0, L] \times [0, h] \times [0, h]$, where $L = 1$ mm and $h = 25 \,\mu$m, and are subject to the set of chemical reactions (6.11)–(6.13).

Since we now care about the y- and z-coordinates of each molecule, we have to generalize the SSA (a8)–(c8) to compute the time evolution of all three coordinates of diffusing molecules. We choose a small time step Δt. For each molecule, we generate three normally distributed (with zero mean and unit variance) random numbers ξ_x, ξ_y, ξ_z and we update the position of the molecule by (3.15)–(3.17), i.e.

$$X(t + \Delta t) = X(t) + \sqrt{2D\Delta t}\, \xi_x, \qquad (6.30)$$

$$Y(t + \Delta t) = Y(t) + \sqrt{2D\Delta t}\, \xi_y, \qquad (6.31)$$

$$Z(t + \Delta t) = Z(t) + \sqrt{2D\Delta t}\, \xi_z, \qquad (6.32)$$

where D is equal to D_A for molecules of A and to D_B for molecules of B. On the boundary, we implement reflective (zero-flux) boundary conditions, as we did in step (c8) of the SSA (a8)–(c8) for the x-coordinate. If the position of a molecule computed by (6.30)–(6.32) is outside the computational domain $[0, L] \times [0, h] \times [0, h]$, we return it back to the domain using the mirror reflection from the boundary.

To model the first reaction in (6.11) we introduce two parameters, the reaction radius ϱ_1 and the reaction probability P_1. If the distance between two molecules of A is less than ϱ_1 (at the end of the time step), the molecules will react with probability P_1. The appropriate choice of the parameters ϱ_1 and P_1 will be discussed in the following Section 6.6 where the relation between ϱ_1 and P_1 and the rate constant k_1 is presented. In this section, we simply choose the values of ϱ_1 and P_1 that will give us correct results, without discussing how the values of ϱ_1 and P_1 have been selected. Similarly, we model the second reaction in (6.11) with the help of two parameters: the reaction radius ϱ_2 and the reaction probability P_2. If the distance between a molecule of A and a molecule of B is less than ϱ_2 (at the end of the time step), the molecules will react with probability P_2. Again, the choice of ϱ_2 and P_2 is discussed in Section 6.6.

The first-order and zeroth-order chemical reactions in (6.12)–(6.13) are implemented in a similar way as in Section 6.2 (compare with the SSA (a17)–(c17)). Thus the reaction–diffusion system (6.11)–(6.13) is simulated by performing the following five steps at time t:

(a18) For each molecule of A (resp. B), compute its position at time $t + \Delta t$ by (6.30)–(6.32), where the diffusion constant D is equal to D_A (resp. D_B). If the position computed by (6.30)–(6.32) is outside the domain $[0, L] \times [0, h] \times [0, h]$, return it back to the domain using the mirror reflection from the boundary.

(b18) For each molecule of A (resp. B), generate a random number r uniformly distributed in the interval $(0, 1)$. If $r < k_3 \Delta t$ (resp. $r < k_4 \Delta t$), then remove the molecule from the system.

(c18) Generate a random number r_1 uniformly distributed in the interval $(0, 1)$. If $r_1 < k_5 h^2 L \Delta t$, then introduce a new molecule of A in a random position in the domain $[0, L] \times [0, h] \times [0, h]$.

(d18) Generate a random number r_2 uniformly distributed in the interval $(0, 1)$. If $r_2 < 2k_6 h^2 L \Delta t/5$, then introduce a new molecule of B in a random position in the subdomain $[3L/5, L] \times [0, h] \times [0, h]$.

(e18) If the distance between two molecules of A is less than ϱ_1, then remove them from the system with probability P_1. If the distance between a molecule of A and a molecule of B is less than ϱ_2, then remove them from the system with probability P_2.
Continue with step (a18) for time $t + \Delta t$.

The first step (a18) models diffusion of molecules. The step (b18) describes the (linear) degradation of A and B given by chemical reactions (6.12). The production of A molecules (last reaction in (6.12)) is implemented in the step (c18), the production of B molecules (reaction (6.13)) in the step (d18). The last step (e18) describes the second-order reactions (6.11). We have to choose the time step Δt small so that the probabilities of the zeroth-order and first-order reactions, i.e. the quantities $k_3 \Delta t$, $k_4 \Delta t$, $k_5 h^2 L \Delta t$ and $2k_6 h^2 L \Delta t/5$ that appear in the steps (b18)–(d18), are significantly less than 1. We use $\Delta t = 4 \times 10^{-4}$ sec.

Using the same parameter values as in Section 6.3, we present illustrative results computed by the SSA (a18)–(e18) in Figure 6.5. To enable direct comparison with compartment-based modelling (i.e. Figure 6.3), we divide the elongated domain $[0, L] \times [0, h] \times [0, h]$ into cubic bins of volume h^3 and we plot the numbers of molecules in each bin at time 200 minutes. We observe that the SSA (a18)–(e18) gives similar results to the compartment-based model. We used $\varrho_1 = \varrho_2 = 40$ nm, $P_1 = 9.5\%$ and $P_2 = 59.5\%$ in the step (e18). What remains is to show that these values of reaction radii and reaction probabilities indeed correspond to the rate constants $k_1 = 3 \times 10^{-2} \, \mu m^3 \, sec^{-1}$ and $k_2 = 3 \times 10^{-1} \, \mu m^3 \, sec^{-1}$. This will be the focus of the following section.

6.6 Reaction Radius and Reaction Probability

The position $[X(t), Y(t), Z(t)]$ of a diffusing molecule can be described by a system of three (uncoupled) SDEs (4.1)–(4.3). To solve them, we use the computational definition (3.5) of SDEs, i.e. we choose the time step Δt and compute the solution iteratively by (6.30)–(6.32). To model a bimolecular reaction, one approach is to postulate that two molecules (which are subject to a bimolecular reaction) *always* react whenever their separation is less than a given reaction radius ϱ. If the trajectories of molecules exactly follow the system of SDEs (4.1)–(4.3) it is then possible to find explicit formulae linking the reaction rate constant, the diffusion constant(s) of reactants and the reaction radius (Smoluchowski, 1917; Berg and Purcell, 1977). For example, let us consider the chemical reaction

$$A + B \xrightarrow{k} \emptyset. \tag{6.33}$$

If molecules of A and B are well mixed, then the average number of reactions per unit of time is given by the propensity function $\alpha(t) = A(t)B(t) \, k/v$, where $A(t)$ (resp. $B(t)$) is the number of molecules of A (resp. B) at time t and v is the volume of the reactor. Let us observe the system from the reference frame tied to a molecule of A. Then the molecule of A does not move and molecules of B diffuse with the diffusion constant $D_A + D_B$. They are removed from the system whenever their distance from the origin is less than the reaction radius ϱ. Thus we can view the system as the diffusion to an adsorbing sphere, which was studied in Section 4.4. We computed the average number of molecules that are removed by the adsorbing sphere per unit of time as $4\pi(D_A + D_B) \, c_\infty \, \varrho$ (compare with (4.71)), where ϱ is the radius of the sphere and c_∞ is the concentration of molecules of B at infinity, which can be approximated as $B(t)/v$. Multiplying the average number of molecules of B that interact with one molecule of A (per unit of time) by the number of molecules of A, we obtain the total number of reaction events per unit of time, namely $4\pi(D_A + D_B)\varrho\, A(t)B(t)/\nu$. Identifying this with the propensity function $\alpha(t) = A(t)B(t)\, k/\nu$, we obtain

$$A(t)B(t)\, k/\nu = 4\pi(D_A + D_B)\varrho\, A(t)B(t)/\nu.$$

Solving for ϱ, we get

$$\varrho = \frac{k}{4\pi(D_A + D_B)}. \tag{6.34}$$

This formula gives the reaction radius ϱ in terms of the reaction rate constant k and diffusion constants D_A and D_B of reacting molecules. Thus the chemical reaction (6.33) can be simulated in the stochastic reaction–diffusion modelling as follows: we simulate the trajectories of individual molecules by (6.30)–(6.32) and the reaction (6.33) takes place whenever the distance between a molecule of A and a molecule of B is less than ϱ given by (6.34). Although this idea is straightforward to implement, there are several problems that have to be taken into account. First of all, the molecules are modelled as points, while in reality they have a finite volume. Approximating the diffusing molecule as a sphere, we can estimate its radius by the Einstein relation (4.97). This gives

$$\varrho_m = \frac{k_B T}{6\pi\eta D}, \tag{6.35}$$

where $k_B = 1.38 \times 10^{-14}$ g mm^2 sec^{-2} K^{-1} is the Boltzmann constant, T is the absolute temperature, η is the coefficient of viscosity and D is the diffusion constant. Let us investigate how the reaction radius ϱ compares to the molecular radius ϱ_m. Using (6.34), (6.35) and $D = D_A = D_B$, the inequality $\varrho > \varrho_m$ is equivalent to

$$k > \frac{4k_B T}{3\eta}.$$

Considering a solution in water ($\eta = 10^{-3}$ g mm^{-1} sec^{-1}) at room temperature ($T = 300$ K), we find that the reaction radius is larger than the molecular radius only for reaction rate constants k greater than about 10^9 M^{-1} sec^{-1}. On the other hand, typical values of k for interactions between proteins are of the order of 10^6 M^{-1} sec^{-1}. Thus, for typical values of the reaction rate constant k, the model requires the reaction radius ϱ to be chosen much smaller than the molecular radius ϱ_m. In particular, this model cannot be correct on a molecular level – it does not correctly take into account the finite size of molecules.

Perhaps more importantly, a small reaction radius also provides restrictions on the simulation time step Δt. For the formula (6.34) to be valid we have to make sure that the average change in the distance between molecules during one time step is much less than the reaction radius ϱ. We formulate this condition as $s \ll \varrho$, where s is the root mean-square step length (in each coordinate) given by

$$s = \sqrt{2(D_A + D_B)\Delta t}. \tag{6.36}$$

Using $D_A = D_B = 10\,\mu\text{m}^2\,\text{sec}^{-1}$ and $k = 10^6\,\text{M}^{-1}\,\text{sec}^{-1}$, we obtain that Δt has to be significantly less than a nanosecond. This limitation is even more severe for faster-diffusing molecules.

Considering the bimolecular reaction only, it can, in principle, be simulated with a much larger time step Δt, but the formula for ϱ has to be modified accordingly. If $s \gg \varrho$, then the probability that a given pair of molecules interacts during the time step $(t, t + \Delta t)$ is proportional to the volume fraction $4\pi\varrho_\infty^3/(3\nu)$, where ϱ_∞ is the modified reaction radius. Comparing with $k\,\Delta t/\nu$, we obtain

$$\varrho_\infty = \left(\frac{3k\,\Delta t}{4\pi}\right)^{1/3}. \tag{6.37}$$

This formula gives a larger (Δt-dependent) reaction radius, but it does not have the potential to provide a spatial resolution close to the size of individual molecules. Andrews and Bray (2004) designed a computational algorithm for intermediate values of Δt that satisfy $s \approx \varrho$. In this case, it is not possible to derive an explicit formula relating ϱ and k (as was done in (6.34) for $s \ll \varrho$ and in (6.37) for $s \gg \varrho$). Instead their algorithm, called Smoldyn (Andrews and Bray, 2004; Andrews et al., 2010; Robinson et al., 2015), uses a look-up table relating the (scaled) reaction rate constant k and reaction radius ϱ. However, the reaction radius can still be smaller than the molecular radius ϱ_m in their algorithm.

The major assumption of the molecular-based models discussed above is that molecules always react whenever the distance between them is less than the reaction radius ϱ. The reaction radius ϱ is related to the rate constant of the bimolecular reaction by a simple formula (for example, (6.34) for the Smoluchowski model) or by a look-up table for the Andrews–Bray model. In the rest of this section, we present models that implement bimolecular reactions with the help of two parameters: the reaction radius $\bar{\varrho}$ and the reaction rate λ. We postulate that the bimolecular reaction can take place only when the separation between molecules is less than $\bar{\varrho}$, and that in this case the bimolecular reaction events happen at a rate λ (thus, as far as reactions are concerned, the reaction radius is playing the role that a compartment plays in the compartment-based model). In what follows, this model is called *the λ–$\bar{\varrho}$ model*. It has also been called the Doi model in the literature. One advantage of the λ–$\bar{\varrho}$ model is that the reaction radius $\bar{\varrho}$ can be chosen as large as the molecular radius ϱ_m. Unlike the previous models, which consider all "collisions" of reactants as reactive, the λ–$\bar{\varrho}$ model takes into account that in reality many non-reactive collisions happen before the reaction takes place, i.e. it better captures the processes on a molecular level.

We explain the λ–$\bar{\varrho}$ model on the second-order reaction (6.33). Molecular diffusion is simulated as in Section 6.5. We choose a small time step Δt. The trajectory of every molecule is computed by (6.30)–(6.32), where $D = D_A$ for molecules of A and $D = D_B$ for molecules of B. Let s be given by (6.36). We distinguish two cases of the value of the time step: (i) time step Δt is chosen so small that $s \ll \bar{\varrho}$; and (ii) time step Δt is larger so that $s \approx \bar{\varrho}$.

(i) Small time step Δt. To model the second-order reaction (6.33), we introduce two parameters: the reaction radius $\bar{\varrho}$ and the rate λ. The reaction radius is expressed in units of length and rate λ in units per time. Whenever the distance of a molecule of A from a molecule of B is less than the reaction radius $\bar{\varrho}$, then the reaction (6.33) takes place with the rate λ. The relation between $\bar{\varrho}$, λ and the rate constant k of (6.33) is (see Exercise 6.4)

$$k = 4\pi(D_A + D_B)\left(\bar{\varrho} - \sqrt{\frac{D_A + D_B}{\lambda}} \, \tanh\left(\bar{\varrho} \, \sqrt{\frac{\lambda}{D_A + D_B}}\right)\right). \qquad (6.38)$$

This is one condition for two unknowns $\bar{\varrho}$ and λ. In particular, we can choose $\bar{\varrho}$ comparable to the radii of reacting molecules and use (6.38) to compute the corresponding λ. Notice that (6.38) is a simple nonlinear equation which can be solved by any numerical method for finding roots of a real-valued function (for example, Newton's method or the bisection method).

If $\lambda = \infty$ (that is, if molecules react immediately whenever they are within the reaction radius), then (6.38) simplifies to (6.34) as desired. On the other hand, if λ is small so that $\lambda \ll (D_A + D_B)/\bar{\varrho}^2$, then we can use Taylor expansion in (6.38) to approximate

$$\tanh\left(\bar{\varrho} \, \sqrt{\lambda/(D_A + D_B)}\right) \approx \bar{\varrho} \, \sqrt{\lambda/(D_A + D_B)} - \tfrac{1}{3}\left(\bar{\varrho} \, \sqrt{\lambda/(D_A + D_B)}\right)^3.$$

Consequently, (6.38) simplifies to $k \approx 4\pi\bar{\varrho}^3\lambda/3$, which can be equivalently rewritten as

$$\lambda \approx \frac{k}{4\pi\bar{\varrho}^3/3}, \qquad (6.39)$$

i.e. the reaction rate λ is given as the reaction rate constant k divided by the volume, $4\pi\bar{\varrho}^3/3$, of the ball in which the reaction takes place. Formula (6.39) is analogous to the formula for the reaction rate per compartment in the compartment-based approach for large compartment size h. If h is large satisfying $h \gg h_{\text{crit}}$, then the reaction rate per compartment is given as k/h^3, which is the reaction rate constant k divided by the volume, h^3, of the compartment – see (6.25) and (6.27).

(ii) SSA for larger time steps. In this case we introduce two slightly different parameters: the reaction radius $\bar{\varrho}$ and the probability P_λ. The second-order reaction (6.33) is modelled as follows. Whenever the distance between a molecule of A and a molecule of B (at the end of a time step) is less than the reaction radius $\bar{\varrho}$, then the reaction takes place with probability P_λ; that is, we generate a random number r uniformly distributed in $(0, 1)$ and the reaction (removal/addition of molecules) is performed whenever $r < P_\lambda$. Notice that the previous algorithm (for small time step Δt) can also be formulated in terms of parameters $\bar{\varrho}$ and P_λ rather than $\bar{\varrho}$ and λ. Indeed, if $\lambda \Delta t \ll 1$, we have $P_\lambda \approx \lambda \Delta t$. If Δt is larger, then the relation between P_λ and λ is more complicated. However, from the practical point of view, there is no need to know the rate λ: it is sufficient to formulate the algorithms in terms of $\bar{\varrho}$ and P_λ.

We need to present the condition relating $\bar{\varrho}$ and P_λ with the rate constant k of (6.33). To do this we define the dimensionless parameters

$$\gamma = \frac{s}{\bar{\varrho}} = \frac{\sqrt{2(D_A + D_B)\Delta t}}{\bar{\varrho}}, \qquad \kappa = \frac{k\,\Delta t}{\bar{\varrho}^3}. \qquad (6.40)$$

Here γ is the ratio of a typical jump (in one coordinate) of a molecule to the reaction radius, while κ is a measure of the number of bimolecular reaction events per pair of reactants in the time Δt in a well-mixed compartment of volume $\bar{\varrho}^3$ (compare Table 1.1). In applications, we first specify the time step Δt. We also want to specify $\bar{\varrho}$ in a realistic parameter range. Consequently, γ and κ can be considered as given numbers in what follows. For example, we can choose $s = \bar{\varrho}$. Then (6.40) gives $\gamma = 1$. The key modelling question is: What is the appropriate value of the probability P_λ?

To formulate the equation for P_λ, it is useful to define an (auxiliary) function $g(r): [0, \infty) \to [0, 1]$ as the solution of the integral equation

$$g(r) = (1 - P_\lambda) \int_0^1 K(r, r'; \gamma)\, g(r')\, dr' + \int_1^\infty K(r, r'; \gamma)\, g(r')\, dr', \qquad (6.41)$$

satisfying $g(r) \to 1$ as $r \to \infty$, where

$$K(r, r'; \gamma) = \frac{r'}{r\gamma \sqrt{2\pi}} \left(\exp\left[-\frac{(r - r')^2}{2\gamma^2} \right] - \exp\left[-\frac{(r + r')^2}{2\gamma^2} \right] \right). \qquad (6.42)$$

(a)   (b)

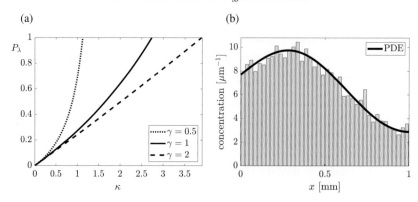

Figure 6.6 (a) *The λ–ρ̄ model. Dependence of P_λ on κ for three different values of γ. Dimensionless parameters κ and γ are given by (6.40). (b) Diffusion of molecules with the reversible reactive boundary at $x = 0$. The distribution of molecules above the boundary computed by the molecular-based SSA (grey histogram) is compared with the solution of the diffusion equation (4.80) with the boundary conditions (6.48)–(6.49) (black line).*

The function $g(r)$ depends on dimensionless parameters P_λ and γ; we make this explicit by writing $g(r; P_\lambda, \gamma) \equiv g(r)$. Then the model parameters $\bar{\varrho}$, P_λ, Δt are related to rate constant k and diffusion constants D_A and D_B by

$$\kappa = P_\lambda \int_0^1 4\pi r^2 g(r; P_\lambda, \gamma) \, dr. \qquad (6.43)$$

The derivation of (6.43) is left to the reader as Exercise 6.5. Since k, D_A and D_B are known and the parameters Δt and $\bar{\varrho}$ are chosen, the parameters γ and κ are in applications given numbers. Thus (6.43) is one equation for one unknown P_λ which can be solved numerically, using, for example, a numerical approach presented by Erban and Chapman (2009). In Figure 6.6(a), we present the dependence of P_λ on κ for three different values of γ. Note that, for small Δt, we can interpret equation (6.39) as $P_\lambda \approx \lambda \Delta t \approx 3\kappa/(4\pi)$. Indeed, all three curves in Figure 6.6(a) initially grow as $3\kappa/(4\pi)$.

Equations (6.41)–(6.43) were used to obtain appropriate values of reaction probabilities for the step (e18) of the SSA (a18)–(e18). This step models the two second-order reactions (6.11). The second reaction in (6.11) is the same as the reaction (6.33). The rate constant of this reaction is $k_2 = 3 \times 10^{-1} \, \mu m^3 \, \text{sec}^{-1}$. In Section 6.5, we use the reaction radius $\bar{\varrho} \equiv \varrho_2 = 40$ nm, which is equal to s given by equation (6.36). Since $\Delta t = 4 \times 10^{-4}$ sec and $D_A = D_B = 1 \, \mu m^2 \, \text{sec}^{-1}$, formulae (6.40) imply that $\gamma = 1$ and $\kappa = 1.875$. Solving (6.41)–(6.43) for P_λ, we obtain that the corresponding reaction probability is $P_\lambda \equiv P_2 = 59.5\%$. This can also be confirmed by looking at the line corresponding to $\gamma = 1$ in Figure 6.6(a).

The first reaction in (6.11) is an example of the bimolecular reaction $A + A \to \emptyset$. Since two molecules of A are removed from the system whenever the reaction takes place, we have to replace k_1 by $2k_1$ in the above formulae. In particular, we replace κ by 2κ in (6.43). Moreover, $D_A + D_B$ has to be replaced by $2D_A$ in all formulae. Otherwise, the method (6.41)–(6.43) for computing P_λ stays the same. Using $k_1 = 3 \times 10^{-2} \, \mu\text{m}^3 \, \text{sec}^{-1}$, we obtain $P_\lambda \equiv P_1 = 9.5\%$. This value has been used to compute the illustrative example in Figure 6.5. Since $P_\lambda \equiv P_1$ is small, we could also obtain a comparable value for P_λ from (6.39), i.e. by using the linear approximation $P_\lambda \approx 3\kappa/(4\pi)$.

6.7 Modelling Reversible Reactions

Thus far in this chapter, whenever two molecules have been involved in a reaction they have been removed from the system; the reactions we have considered have all been irreversible, and the product has not been of interest. In this section, we will discuss modelling reversible reactions. An example is the reversible second-order reaction

$$A + B \underset{k_2}{\overset{k_1}{\rightleftarrows}} C. \tag{6.44}$$

If the molecules of A, B and C are well mixed, we can simulate the time evolution of the reversible reaction (6.44) by the Gillespie SSA applied to the system of two chemical reactions:

$$A + B \xrightarrow{k_1} C \qquad \text{and} \qquad C \xrightarrow{k_2} A + B. \tag{6.45}$$

In particular, reversible reactions do not present any special challenge if the chemical system is well stirred.

Since the compartment-based model of diffusion is based on the Gillespie SSA, reversible reactions can be again simulated without any difficulty in the compartment-based approach. A reversible reaction simply means two chemical reactions in the corresponding compartment. For example, the production and degradation of molecules of A in (6.12) can be considered as two reactions forming a reversible reaction (we have used this notation in (2.9)). Thus we have already seen the implementation of a reversible reaction in the compartment-based model in Section 6.3. The bimolecular reversible reaction (6.44) can also be easily simulated as two reactions (6.45) in each compartment.

The situation is less trivial when the molecular-based model for diffusion is considered. We saw in Section 6.6 that one way to model bimolecular reactions is to postulate that two molecules (which are subject to the bimolecular reaction) always react whenever their distance is less than a given reaction radius ϱ. This idea can be implemented to model the first reaction in (6.45). The second reaction in (6.45) is a reaction of the first order, that is, the probability that the complex C dissociates during the time interval $(t, t + \Delta t)$ is equal to $k_2 \, \Delta t$ provided that Δt is small enough that $k_2 \, \Delta t \ll 1$. When the complex C dissociates, a molecule of A and a molecule of B are introduced to the system. What should be their positions? A natural way to initialize the positions of new molecules of A and B (modelled as points) is by the position of the complex C that was destroyed during this reaction. However, if the next step of the computer code describes the first reaction in (6.45), then the new molecules of A and B will immediately create the complex C again: they have to react because their separation is zero. In particular, this means that we might obtain different results by different orderings of the subroutines in the computer code, which is clearly an unsatisfactory situation. Some authors (Andrews and Bray, 2004) solve this problem by introducing the so-called unbinding radius, which is greater than the reaction radius ϱ. When the complex C dissociates they introduce new molecules of A and B in such a way that their initial separation is equal to the unbinding radius. The unbinding radius has to be chosen carefully to get the right rate constants for both the forward and back reactions. If instead we adopt the λ–$\bar{\varrho}$ model discussed in Section 6.6, then it is safe to initialize the positions of new molecules of A and B at the position of the complex C, since there is only a finite (and often small) probability that A and B will react in the next time step. In particular, we do not have to introduce an unbinding radius (Lipkova et al., 2011). We obtain very similar results by different ordering of subroutines in the computer code even if the initial distance of new molecules is zero (which is less than the reaction radius $\bar{\varrho}$) because the molecules rarely react.

We conclude this section by discussing the modelling of reversible reactions with receptors on the domain boundary. In Section 4.5, we presented methods for modelling reactive boundaries. There it was assumed that the molecules of A bound irreversibly to the boundary according to the reaction (4.79). In some applications, molecules can also detach from the boundary, i.e. molecules of A and receptor molecules B (on the boundary) react according to the reversible second-order chemical reaction:

$$A + B \underset{k_2}{\overset{k_1}{\rightleftharpoons}} \overline{AB}. \tag{6.46}$$

This reaction looks similar to (6.44). The main difference is that the receptor molecules B and the complexes $\overline{AB}$ are localized on the boundary. The reaction (6.46) can be equivalently written as two reactions (compare with (6.45)):

$$A + B \xrightarrow{k_1} \overline{AB} \qquad \text{and} \qquad \overline{AB} \xrightarrow{k_2} A + B. \qquad (6.47)$$

The first reaction in (6.47) is the same as (4.79). Therefore we can apply all algorithms in Section 4.5 to simulate it. In addition, we need to keep track of complexes $\overline{AB}$ and their positions on the boundary. Whenever a complex $\overline{AB}$ dissociates, a molecule of A is re-introduced to the solution. Here, we present a molecular-based algorithm for modelling the reversible boundary reactions. We leave the implementation of the compartment-based model to the reader as Exercise 6.6.

We consider the chemical species A diffusing in the domain $[0, L] \times [0, \overline{L}] \times [0, \overline{L}]$. The boundary $x = 0$ is uniformly covered by receptors B. Molecules of A and B react according to the reversible second-order chemical reaction (6.46). We consider reflective boundary conditions on other boundaries. Since the receptors are uniformly distributed on the boundary, there will be no concentration differences (on average) in the y- and z-directions. As in Section 4.5, we focus only on the behaviour along the x-axis. That is, we study the behaviour of molecules in the one-dimensional interval $[0, L]$ with the reversible reactive boundary condition at $x = 0$ and the reflective boundary condition at $x = L$. To compute the trajectory of diffusing molecules, we choose the time step Δt and we use (6.30). The first reaction in (6.47) can be implemented as in Section 4.5. Roughly speaking, whenever a molecule hits the boundary, it is adsorbed with probability $P_2(\Delta t)$ given by formula (4.87), and otherwise it is reflected. As discussed in Section 4.5, the precise implementation of this boundary condition is given by the following two steps:

(a_p) If $X(t + \Delta t)$ computed by (6.30) is negative then
$X(t + \Delta t) = -X(t) - \sqrt{2D\Delta t}\, \xi$ with probability $1 - P_2(\Delta t)$,
otherwise the molecule reacts with the boundary.

(b_p) If $X(t + \Delta t)$ computed by (6.30) is positive then the molecule reacts with the boundary with probability $\exp[-X(t)X(t + \Delta t)/(D\Delta t)]P_2(\Delta t)$.

At the end of each time step, we also have to generate one random number r uniformly distributed in $(0,1)$ for each complex $\overline{AB}$. If $r < k_2 \Delta t$, then the corresponding complex dissociates and we introduce a new molecule of A in position $x = 0$ to the solution.[*] We note that this algorithm will work correctly provided that $P_2(\Delta t)$ is small. Otherwise, the molecule of A at the position $x = 0$ might react immediately with the receptor B (depending on the order of the subroutines in the computer program).

---

[*] In fact, since all the boundary complexes are the same, we could combine these independent trials into one trial sampling from a binomial distribution.

In Figure 6.6(b), we present results of an illustrative computation. As in Section 4.5, we assume that the number of available receptors on the boundary is independent of time (i.e. the receptors are in excess). We use $\sigma = k_1 b = 2 \times 10^{-3}$ mm sec^{-1}, where b is the number of available receptors per unit of area of the boundary $x = 0$. Other parameters are $k_2 = 5 \times 10^{-3}$ sec, $D = 10^{-4}$ mm^2 sec^{-1} and $\Delta t = 0.1$ sec. We simulate a collection of 10 000 molecules, which are initially uniformly distributed in $[0, L/2]$. At time $t = 10$ min, we divide the computational domain $[0, L]$ into $K = 40$ bins of length $h = L/K = 25 \,\mu$m. We plot the number of molecules in bins as the grey histogram in Figure 6.6(b). The black line is the solution of the diffusion equation (4.80) with the boundary conditions

$$D \frac{\partial \varrho}{\partial x}(0, t) = \sigma \varrho(0, t) - k_2 \, \overline{ab}, \qquad \frac{\partial \varrho}{\partial x}(L, t) = 0, \qquad (6.48)$$

where $\overline{ab}$ is the concentration of the boundary complex $\overline{AB}$ evolving according to the ODE

$$\frac{d \overline{ab}}{dt} = \sigma \varrho(0, t) - k_2 \, \overline{ab}. \qquad (6.49)$$

The solution of the diffusion equation compares well with the results of the SSA. It can be shown that the diffusion equation (4.80) with the boundary conditions (6.48)–(6.49) is an exact description of this algorithm in the limit $\Delta t \to 0$. If we want to use it for a finite time step Δt, we have to make sure that the time step is small enough that $P_2(\Delta t) \ll 1$. We have $P_2(\Delta t) = 5.6\%$ for our parameter values.

6.8 Biological Pattern Formation

There is plenty of evidence that stochasticity (noise) plays an important role in many biological systems (Rao et al., 2002; Elowitz et al., 2002). Reaction-diffusion processes are key elements of pattern-forming mechanisms in developmental biology. In this section, we present stochastic analogues of two classic pattern-forming models. The first one is the so-called French flag model where we re-interpret the illustrative example from Sections 6.1 and 6.2. Then we present a reaction–diffusion pattern-forming model based on the so-called Turing instability.

The illustrative example from Sections 6.1 and 6.2 was a caricature of more complicated morphogenesis applications (Shimmi et al., 2005; Reeves et al., 2005; Rogers and Schier, 2011) where one assumes that some prepatterning

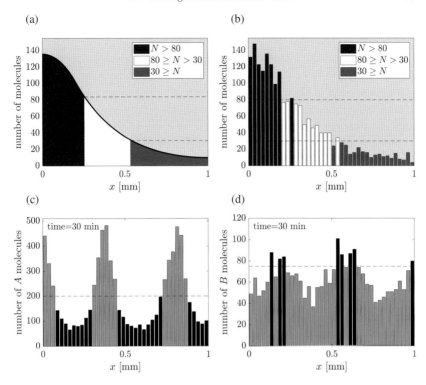

Figure 6.7 *Examples of reaction–diffusion models for pattern formation.*
(a) *French flag model – deterministic description.* (b) *French flag model – stochastic description.* (c), (d) *Turing patterns. Numbers of molecules of chemical species A and B in each compartment at time 30 minutes.*

in the domain exists and one wants to validate the reaction–diffusion mechanism of the next stage of the patterning of the embryo. In our example, we considered a chemical A that is produced in only part of the domain $[0, L]$ (specifically, in $[0, L/5]$). Thus the domain was divided into two different regions $[0, L/5]$ and $[L/5, L]$; this is the prepatterning. The simplest idea of further patterning is the so-called French flag model (Wolpert et al., 2002). We assume that the interval $[0, L]$ describes a layer of cells which are sensitive to the concentration of the chemical A. In particular, we suppose that a cell can have three different fates (e.g. different genes are switched on or off) depending on the concentration of A. Then the concentration gradient of A can help to distinguish three different regions in $[0, L]$; see Figure 6.7(a) and (b). If the concentration of A is high enough (above a certain threshold), a cell follows the first programme (denoted blue in Figure 6.7(a) and (b)). The "white programme" is followed for medium concentrations of A, and the "red programme" is followed for low concentrations.

The deterministic version of the French flag model is presented in Figure 6.7(a). We consider a solution of (6.9)–(6.10) at time 30 minutes, which is the black curve in Figure 6.1(c) or Figure 6.2(b). The solution of (6.9)–(6.10) is decreasing in space. Introducing two thresholds we obtain three well-defined regions as seen in Figure 6.7(a).

The stochastic version of the French flag model is shown in Figure 6.7(b). We take the spatial histogram presented in Figure 6.1(c). We introduce two thresholds (80 and 30 molecules) as before and replot the histogram using the corresponding colours. Clearly, the resulting "French flag" is noisy. Different realizations of the SSA would lead to different noisy French flags. The same is true for the SSA from Figure 6.2(b). Not only do concentration gradients regulate patterning in developmental biology, but intracellular concentration gradients also contribute to spatial organization inside single cells (Saunders et al., 2012). An important research question in these applications is how such noisy concentration gradients can robustly determine patterning and positional information (Howard, 2012). In the case of Figure 6.1(c), we would clearly get a less noisy pattern if we averaged observed concentration profiles over a sufficiently long time window. Time averaging of input concentration profiles can be done by a signal-processing biochemical network (Tostevin et al., 2007). In Section 7.3, we will discuss a related "time-averaging" mechanism that helps bacteria to navigate in noisy gradients of extracellular signals.

Our second example of patterning in developmental biology concerns so-called Turing patterns (Turing, 1952; Gierer and Meinhardt, 1972; Murray, 2002; Sick et al., 2006). These do not require any prepatterning: molecules are subject to the same chemical reactions in the whole domain of interest. For example, let us consider a system of two chemical species A and B in the elongated computational domain $[0, L] \times [0, h] \times [0, h]$, where $L = 1$ mm and $h = 25\,\mu$m, which react according to the Schnakenberg system of chemical reactions (2.9). Let us choose the values of rate constants as $k_1/h^6 = 10^{-6}\,\text{sec}^{-1}$, $k_2 h^3 = 1\,\text{sec}^{-1}$, $k_3 = 0.02\,\text{sec}^{-1}$ and $k_4 h^3 = 3\,\text{sec}^{-1}$. If molecules of A and B are well mixed, then the corresponding deterministic system of ODEs for (2.9) has one non-negative stable steady state equal to $a_s = 200$ and $b_s = 75$ molecules per volume h^3. Introducing diffusion to the model, one steady-state solution of the spatial problem is the constant one (a_s, b_s) everywhere. However, this solution might not be stable (so might not be seen in reality) if the diffusion constants of A and B differ significantly.

We choose $D_A = 10^{-5}\,\text{mm}^2\,\text{sec}^{-1}$ and $D_B = 10^{-3}\,\text{mm}^2\,\text{sec}^{-1}$, which gives $D_B/D_A = 100$. To simulate the reaction–diffusion problem with the

Schnakenberg system of chemical reactions (2.9), we follow the compartment-based method of Section 6.1. We divide the domain $[0, L] \times [0, h] \times [0, h]$ into $K = 40$ compartments of length $h = L/K = 25 \, \mu m$. We denote the number of molecules of chemical species A (resp. B) in the ith compartment $[(i - 1)h, ih)$ by A_i (resp. B_i), $i = 1, 2, \ldots, K$. Diffusion is described by two chains of chemical reactions (6.14)–(6.15), where the rates of "chemical reactions" are equal to $d_A = D_A/h^2$ for chemical species A and $d_B = D_B/h^2$ for chemical species B. Chemical reactions (2.9) are considered in every compartment (the values of rate constants in (2.9) were already expressed in units per compartment).

Starting with a uniform distribution of chemicals $A_i(0) = a_s = 200$ and $B_i(0) = b_s = 75$, $i = 1, 2, \ldots, K$, at time $t = 0$, we plot the numbers of molecules in each compartment at time $t = 30$ minutes computed by the SSA (a5)–(d5) in Figure 6.7(c) and (d). To demonstrate the idea of patterning, compartments with many molecules of A (above steady-state value a_s) are plotted as green; other compartments are plotted as black in Figure 6.7(c). We see that chemical A can be clearly used to divide our computational domain into several regions. There are two and a half green peaks in this figure. The number of green peaks depends on the size of the computational domain $[0, L]$ and it is not a unique number in general. The reaction–diffusion system has several favourable states, each with a different number of green peaks.

In Figure 6.7(d), compartments with many molecules of B (above steady-state value b_s) are plotted as black; other compartments are plotted as green. Although we are able to identify similarly positioned green regions as in Figure 6.7(c), the resulting pattern is more noisy. The compartment-based model helps us to investigate the role of noise in pattern formation and has been used for the analysis of stochastic Turing patterns by Biancalani et al. (2010). However, we have already seen in Section 6.4 that some compartment-based results can depend on the compartment size h if it is not carefully chosen. In the case of Turing patterns presented in this section, we have two chemical species with very different diffusion constants, i.e. $D_B = 100 D_A$. Since the conditions on the compartment size, discussed in Section 6.4, depend on the diffusion constants, we could also consider a compartment-based approach that uses two different sizes of compartments, one for the chemical species A and one for B. Unfortunately, this alternative compartment-based approach can modify the parameter regimes where Turing patterns can be observed (Cao and Erban, 2014). To get more confidence in the conclusions obtained for stochastic Turing patterns by the compartment-based models, we could also consider the molecular-based approach studied in Section 6.5. To apply it to the Schnakenberg system of chemical reactions (2.9), we would need to decide how the trimolecular reaction in (2.9) is modelled; see (Flegg, 2016) for discussion of such methods.

In this section, we discussed stochastic analogues of two classic spatio-temporal biological models. In recent years, stochastic reaction–diffusion algorithms have been successfully applied to more complicated biological systems, including models of an MAPK pathway (Takahashi et al., 2010), signal transduction in *Escherichia coli* chemotaxis (Lipkow et al., 2005), oscillation of Min proteins in cell division (Fange and Elf, 2006), intracellular and morphogen gradients (Howard, 2012), actin dynamics in filopodia (Zhuravlev and Papoian, 2009), ribosome biogenesis (Earnest et al., 2015) and intracellular calcium dynamics (Dobramysl et al., 2016). Both molecular-based and compartment-based models have been used. They have been implemented in several software packages, including Smoldyn (Andrews and Bray, 2004; Andrews et al., 2010; Robinson et al., 2015), Lattice Microbes (Roberts et al., 2013), MCell (Stiles and Bartol, 2001), GFRD (van Zon and ten Wolde, 2005), MesoRD (Hattne et al., 2005), URDME (Engblom et al., 2009) and STEPS (Wils and De Schutter, 2009). In some applications, the computational domain is also divided into subdomains, some of which use molecular-based models and some of which use compartment-based models. Such approaches will be discussed in Chapter 9.

Exercises

6.1   Derive the system of equations (6.5)–(6.8) for the mean vector $\mathbf{M}(t)$.

6.2   Reformulate the SSA (a17)–(c17) in such a way that it does not use a finite time step Δt in steps (b17) and (c17), but computes the times when the next production and degradation events happen using exponentially distributed random numbers given by formula (5.17). Verify your algorithm numerically by recomputing the results from Figures 6.1 and 6.2. Can you further modify this SSA to a fully event-based algorithm which does not use the finite time step at all?

6.3   Design an SSA for modelling the reaction–diffusion system from Sections 6.1 and 6.2 which has the velocity-jump process (a10)–(c10) as the underlying diffusion model. Verify your algorithm numerically by recomputing the results from Figures 6.1 and 6.2.

Hint: Modify the SSA (a17)–(b17). Replace (a8)–(c8) in the step (a17) by the velocity-jump process (a12)–(c12).

6.4 Derive the formula (6.38) relating the reaction rate constant k and the parameters of the $\lambda-\bar{\varrho}$ model.

Hint: Consider the diffusion to the ball of radius $\bar{\varrho}$ centred at the origin which removes molecules with the rate λ. Let $c(r)$ be the equilibrium concentration of molecules at distance r from the origin. Then c is a continuous function with continuous derivative satisfying the equations

$$\frac{d^2c}{dr^2} + \frac{2}{r}\frac{dc}{dr} = 0, \qquad \text{for } r \geq \bar{\varrho},$$

$$\frac{d^2c}{dr^2} + \frac{2}{r}\frac{dc}{dr} - \frac{\lambda c}{D_A + D_B} = 0, \qquad \text{for } r \leq \bar{\varrho}.$$

Solve this system. Use continuity of c at $r = 0$, continuity of c and its derivative at $r = \bar{\varrho}$, and the boundary condition at infinity ($c(r) \to c_\infty$ as $r \to \infty$) to determine unknown constants. Compute the total flux through the sphere boundary. It is equal to the rate constant of bimolecular reaction k multiplied by the concentration c_∞ of the chemical far from the reacting molecule. Dividing by c_∞, derive (6.38).

6.5 Derive the system of equations (6.41)–(6.43) for P_λ.

Hint: Let $c_i(r)$ be the concentration of molecules at distance r from the origin. Assuming that molecules only diffuse, their concentration at point r after the time interval Δt is given as

$$\int_0^\infty K(r, r'; \gamma)\, c_i(r')\, dr', \qquad (6.50)$$

where $K(r, r'; \gamma)$ is given by (6.42). Let us assume that the particles are removed, in the circle of radius $\bar{\varrho} = 1$ and centred at origin, with probability P_λ, and then diffuse for time Δt. Then (6.50) is modified to

$$c_{i+1}(r) = (1 - P_\lambda) \int_0^1 K(r, r'; \gamma)\, c_i(r')\, dr' + \int_1^\infty K(r, r'; \gamma)\, c_i(r')\, dr'.$$

Equation (6.41) is an equation for the fixed point of this iterative scheme. The rate of removal of particles (at steady state) during one time step is given by the right-hand side of equation (6.43). Comparing with κ, we obtain (6.43).

6.6 Compute the results in Figure 6.6(b) by implementing the compartment-based model of the reversible reactive boundary condition.

Hint: Generalize the compartment-based SSA from Section 4.5.

7

SSAs for Reaction–Diffusion–Advection Processes

Stochastic reaction–diffusion algorithms are suitable for modelling systems of interacting particles that are only transported by diffusion from place to place. In some applications, active transport (for example, by an electrical field, molecular and cellular motors, running, swimming or flying, all in response to external cues) also plays an important role. In this chapter, we show how active transport can be incorporated into the stochastic diffusion and reaction–diffusion algorithms we have introduced in previous chapters.

In Section 7.1 we add advection to both the SDE diffusion model (introduced in Section 4.1) and the compartment-based model of diffusion (introduced in Section 4.2). We show that the compartment-based modelling is described (in the limit of small compartments) by the same diffusion–advection PDE for the probability distribution as the SDE-based model. Stochastic reaction–diffusion–advection models generalizing both diffusion–advection approaches are discussed in Section 7.2.

In the following three sections, we discuss several applications. Using three examples, we illustrate that the modelling approaches discussed in this book can be used for studying a number of different systems consisting of many interacting "particles", where individual particles can range in size from small ions and molecules to individual cells and animals. In Section 7.3, we investigate velocity-jump processes and their applications to studying the collective behaviour of bacteria, using an example of an individual-based (agent-based) model. Another example of an agent-based model is presented in Section 7.4 where we discuss the collective behaviour of social insects.

Another application area of diffusion–advection processes – mathematical modelling of ions and ion channels – is discussed in Section 7.5. We conclude this chapter with the discussion of the Metropolis–Hastings algorithm (Markov chain Monte Carlo methods) which can be used to compute stationary (equilibrium) properties of complicated diffusion–advection problems.

7.1 SSAs for Diffusion–Advection Processes

In Section 4.1, we presented a diffusion model based on the system of uncoupled SDEs (4.1)–(4.3), which have drift coefficients equal to zero. A straightforward way to add advection (active transport) to this method is by modelling the trajectory $[X(t), Y(t), Z(t)]$ of a moving particle by the SDEs with non-zero drift coefficients, namely

$$X(t + dt) = X(t) + f_1(X(t), Y(t), Z(t), t)\, dt + \sqrt{2D}\, dW_x, \tag{7.1}$$

$$Y(t + dt) = Y(t) + f_2(X(t), Y(t), Z(t), t)\, dt + \sqrt{2D}\, dW_y, \tag{7.2}$$

$$Z(t + dt) = Z(t) + f_3(X(t), Y(t), Z(t), t)\, dt + \sqrt{2D}\, dW_z, \tag{7.3}$$

where $\mathbf{f} = [f_1, f_2, f_3]$ is a vector field describing the non-diffusive active transport, D is the (constant) diffusion coefficient and the subscripts in dW_x, dW_y, dW_z emphasize the fact that the noise terms are independent of each other. Let $p(x, y, z, t)\, dx\, dy\, dz$ be the probability that $X(t) \in [x, x + dx)$, $Y(t) \in [y, y + dy)$ and $Z(t) \in [z, z+dz)$ at time t. Then p evolves according to the (Fokker–Planck) equation (see Exercise 7.1)

$$\frac{\partial p}{\partial t} = D\left(\frac{\partial^2 p}{\partial x^2} + \frac{\partial^2 p}{\partial y^2} + \frac{\partial^2 p}{\partial z^2}\right) - \frac{\partial}{\partial x}(f_1\, p) - \frac{\partial}{\partial y}(f_2\, p) - \frac{\partial}{\partial z}(f_3\, p), \tag{7.4}$$

which can be equivalently rewritten as

$$\frac{\partial p}{\partial t} = \nabla \cdot (D\, \nabla p - p\, \mathbf{f}), \tag{7.5}$$

where ∇ is the gradient operator

$$\nabla p = \left(\frac{\partial p}{\partial x}, \frac{\partial p}{\partial y}, \frac{\partial p}{\partial z}\right). \tag{7.6}$$

In some applications, the vector field $\mathbf{f}$ is given as the gradient of a scalar (signal) field S, i.e. $\mathbf{f} = \nabla S(x)$. Then the SDEs (7.1)–(7.3) describe the behaviour of particles that prefer to move to areas with higher values of the signal $S(x)$. The signal field $S(x)$ can be, for example, an attracting chemical (e.g. food) if (7.1)–(7.3) models the movement of unicellular organisms (see Section 7.3 about chemotaxis), or an electrical potential if we are modelling ions (see Section 7.5). If $\mathbf{f} = \nabla S(x)$, then (7.5) reads as

$$\frac{\partial p}{\partial t} = \nabla \cdot (D\, \nabla p - p\, \nabla S), \tag{7.7}$$

which is sometimes called the chemotaxis equation. The techniques and results of Chapter 3 can all be used on the SDE-based diffusion–advection (or drift–diffusion) model (7.1)–(7.3).

The situation is slightly less straightforward when we consider compartment -based modelling. To avoid technicalities, we study an effectively one-dimensional problem. We assume that the vector field $\mathbf{f}$ is given as

$$\mathbf{f} = [\tilde{f}(x), 0, 0],$$

where $\tilde{f}$ is independent of y, z and t. This choice of $\mathbf{f}$ corresponds to active transport along the x-direction only. Then equation (7.1) reads as follows:

$$X(t + dt) = X(t) + \tilde{f}(X(t))\, dt + \sqrt{2D}\, dW_x. \tag{7.8}$$

The equations (7.2) and (7.3) are not coupled to (7.8). They simplify to (4.2) and (4.3). Since we only have diffusion in the y- and z-directions, we will focus on the behaviour along the x-axis. The Fokker–Planck equation for the SDE (7.8) is

$$\frac{\partial p}{\partial t}(x, t) = D\frac{\partial^2 p}{\partial x^2}(x, t) - \frac{\partial}{\partial x}(\tilde{f}(x)\, p(x, t)), \tag{7.9}$$

where $p(x, t)\, dx$ is the probability that $X(t) \in [x, x + dx)$. Our goal is to modify the compartment-based model (4.13) so that the limiting PDE is no longer the diffusion equation but the diffusion–advection equation (7.9).

In the model (4.13), we assumed that the rate of jumping between compartments is the same for all compartments, and is the same for a jump to the left as it is for a jump to the right. Advection can be added to this model by making the rate of jumping different for each compartment. Thus we modify (4.13) by writing

$$A_1 \underset{k_2^-}{\overset{k_1^+}{\rightleftarrows}} A_2 \underset{k_3^-}{\overset{k_2^+}{\rightleftarrows}} A_3 \underset{k_4^-}{\overset{k_3^+}{\rightleftarrows}} \cdots \underset{k_K^-}{\overset{k_{K-1}^+}{\rightleftarrows}} A_K, \tag{7.10}$$

where A_i is the number of molecules in the ith compartment $[(i - 1)h, ih]$ and k_i^+, k_i^- are positive rates of jumps from the ith compartment. As in (4.13), it is assumed that the one-dimensional computational domain $[0, L]$ is divided into K compartments of length $h = L/K$.

Let $p(\mathbf{n}, t)$ be the joint probability that $A_i(t) = n_i$, $i = 1, 2, \ldots, K$, and where as usual $\mathbf{n} = [n_1, n_2, \ldots, n_K]$. If we recall the operators $R_i, L_i \colon \mathbb{N}^K \to \mathbb{N}^K$, given by (4.20)–(4.21), then the chemical master equation corresponding to the system of chemical reactions given by (7.10) is

$$\frac{\partial p(\mathbf{n})}{\partial t} = \sum_{j=1}^{K-1} k_j^+ \{(n_j + 1)\, p(R_j\mathbf{n}) - n_j\, p(\mathbf{n})\} + \sum_{j=2}^{K} k_j^- \{(n_j + 1)\, p(L_j\mathbf{n}) - n_j\, p(\mathbf{n})\}.$$

The mean $\mathbf{M}(t) \equiv [M_1, M_2, \ldots, M_K]$, defined by (4.23), evolves according to the system of equations (see Exercise 7.2)

$$\frac{\partial M_i}{\partial t} = k_{i+1}^- M_{i+1} + k_{i-1}^+ M_{i-1} - (k_i^+ + k_i^-) M_i, \quad i = 2, \ldots, K-1, \qquad (7.11)$$

$$\frac{\partial M_1}{\partial t} = k_2^- M_2 - k_1^+ M_1, \qquad \frac{\partial M_K}{\partial t} = k_{K-1}^+ M_{K-1} - k_K^- M_K. \qquad (7.12)$$

We now choose the rates k_i^+ and k_i^- to be given by

$$k_i^+ = \frac{D}{h^2} + \frac{\tilde{f}(x_i)}{2h}, \qquad k_i^- = \frac{D}{h^2} - \frac{\tilde{f}(x_i)}{2h}, \qquad (7.13)$$

where x_i is the centre of the ith compartment $[(i-1)h, ih)$, namely $x_i = ih - h/2$. Since k_i^+ and k_i^- must be non-negative, we have implicitly assumed in (7.13) that

$$\left| \tilde{f}(x_i) \right| \leq \frac{2D}{h}, \quad \text{for } i = 1, 2, \ldots, K.$$

Substituting (7.13) into (7.11) gives

$$\frac{\partial M_i}{\partial t} = D \frac{M_{i+1} + M_{i-1} - 2M_i}{h^2} - \frac{\tilde{f}(x_{i+1}) M_{i+1} - \tilde{f}(x_{i-1}) M_{i-1}}{2h}. \qquad (7.14)$$

The classic deterministic description of diffusion advection processes is written in terms of a concentration $c(x, t)$, which can be approximated as $c(x_i, t) \approx M_i(t)/h$ where x_i is the centre of the ith compartment. Dividing (7.14) by h gives

$$\frac{\partial c}{\partial t}(x_i, t) \approx D \frac{c(x_i + h, t) + c(x_i - h, t) - 2c(x_i, t)}{h^2}$$
$$- \frac{\tilde{f}(x_i + h) c(x_i + h, t) - \tilde{f}(x_i - h) c(x_i - h, t)}{2h}.$$

Using a Taylor expansion on the right-hand side, we get

$$\frac{\partial c}{\partial t}(x_i, t) \approx D \frac{\partial^2 c}{\partial x^2}(x_i, t) - \frac{\partial}{\partial x}(\tilde{f}(x_i) c(x_i, t)),$$

which is the diffusion–advection equation (7.9). Thus we have confirmed that the appropriate advection term can be introduced to the compartment-based model of diffusion by changing the rates of diffusive jumps from D/h^2 to the rates given by (7.13). One can also show (Exercise 7.3) that (7.12) leads in the limit $h \to 0$ to the boundary conditions

$$D \frac{\partial c}{\partial x}(0) = \tilde{f}(0) c(0), \qquad D \frac{\partial c}{\partial x}(L) = \tilde{f}(L) c(L). \qquad (7.15)$$

In Figure 7.1, we present an illustrative example. We use $L = 1$ mm, $K = 40$, $h = L/K = 25 \, \mu m$ and $D = 10^{-4}$ mm^2 sec^{-1}. We consider $\tilde{f}(x) = S'(x)$, where

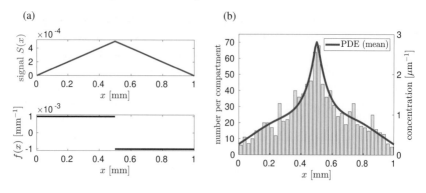

Figure 7.1 (a) *The signal $S(x)$ (red) and the drift coefficient $\tilde{f}(x)$ (blue) that are used in the illustrative example.* (b) *One realization of the Gillespie SSA (a5)–(d5) for the system of chemical reactions* (7.10). *The grey histogram shows numbers of molecules in each compartment at time t = 2 min. Solution of* (7.9) *with boundary conditions* (7.15) *is plotted as the red solid line.*

the signal $S(x)$ is given as a piecewise linear function with the maximum in the middle of the interval $[0, L]$; see Figure 7.1(a). We take

$$\tilde{f}(x) = S'(x) = \begin{cases} 10^{-3} \text{ mm}^{-1}, & \text{for } x \in [0, L/2], \\ -10^{-3} \text{ mm}^{-1}, & \text{for } x \in [L/2, L]. \end{cases} \qquad (7.16)$$

The drift coefficient $\tilde{f}(x)$ is plotted in Figure 7.1(a) as the blue line. We start with 25 molecules in each compartment, so that the initial density is uniform (with 1 molecule per μm). In Figure 7.1(b) we plot numbers of molecules in each compartment at time 2 min (one realization computed by the Gillespie SSA (a5)–(d5) applied to the system of chemical reactions (7.10)). We also plot the solution of the diffusion–advection equation (7.9) with boundary conditions (7.15) and the same (uniform) initial condition for comparison (red line). As discussed above, the red line can also be interpreted as the average of many realizations of the Gillespie SSA (a5)–(d5).

7.2 Reaction–Diffusion–Advection SSAs

In the previous section we presented a method for adding an additional advection term to the compartment-based model of diffusion. The same approach can be used to design a compartment-based reaction–diffusion–advection model. We explain the model by adding an additional drift term to the example presented in Section 6.1.

Our illustrative reaction–diffusion–advection model can be formulated as follows. As in Section 6.1, the molecules of chemical species A diffuse in the elongated domain $[0, L] \times [0, h] \times [0, h]$ (see Figure 6.1(a)) with the diffusion constant $D = 10^{-4}$ mm^2 sec^{-1}. They are also transported to the middle of the domain by the additional advective term (7.16). Initially, there are no molecules in the system. Molecules are produced in the part of the domain $[0, L/5] \times [0, h] \times [0, h]$ at a rate $k_2 = 2 \times 10^{-5}$ μm^{-3} sec^{-1} and are degraded at a rate $k_1 = 10^{-3}$ sec^{-1} in the whole domain. As in Sections 4.2 and 6.1, we divide the computational domain $[0, L] \times [0, h] \times [0, h]$ into $K = L/h = 40$ compartments of volume h^3 and we denote the number of molecules of chemical species A in the ith compartment $[(i - 1)h, ih) \times [0, h] \times [0, h]$ by A_i. Then our reaction–diffusion–advection process is described by the system of chemical reactions

$$A_1 \underset{k_2^-}{\overset{k_1^+}{\rightleftarrows}} A_2 \underset{k_3^-}{\overset{k_2^+}{\rightleftarrows}} A_3 \underset{k_4^-}{\overset{k_3^+}{\rightleftarrows}} \cdots \underset{k_K^-}{\overset{k_{K-1}^+}{\rightleftarrows}} A_K, \tag{7.17}$$

$$A_i \overset{k_1}{\longrightarrow} \emptyset, \qquad \text{for } i = 1, 2, \ldots, K, \tag{7.18}$$

$$\emptyset \overset{k_2}{\longrightarrow} A_i, \qquad \text{for } i = 1, 2, \ldots, K/5, \tag{7.19}$$

where k_i^+ and k_i^- are given by (7.13). This system is a generalization of the reaction–diffusion system (6.1)–(6.3). In fact, the production and degradation reactions (7.18)–(7.19) are identical to (6.2)–(6.3). To include advection, the chain of reactions (6.1) is modified to (7.17), which is the diffusion–advection model (7.10). The time evolution of the system (7.17)–(7.19) can be simulated using the Gillespie SSA (a5)–(d5). Starting with no molecules of A in the system, we present one realization of the Gillespie SSA (a5)–(d5) in Figure 7.2(a). We plot the numbers of molecules in each compartment at time $t = 15$ min as the grey histogram.

In Section 6.1, we formulated the reaction–diffusion master equation (6.4) for the system of reactions (6.1)–(6.3), and used it to derive the system of equations for the stochastic mean – equations (6.5)–(6.8). Dividing this system by h^3 and passing $h \to 0$, we obtained the PDE description (6.9) with zero-flux boundary conditions (6.10). In a similar way (Exercise 7.4), one can derive the PDE description corresponding to the reaction–diffusion–advection model (7.17)–(7.19). It describes the time evolution of the concentration, $a(x, t)$, of molecules of A at point x and time t and can be written as

$$\frac{\partial a}{\partial t} = D \frac{\partial^2 a}{\partial x^2} - \frac{\partial}{\partial x}(\tilde{f}a) + k_2 \chi_{[0, L/5]} - k_1 a, \tag{7.20}$$

(a)   (b)

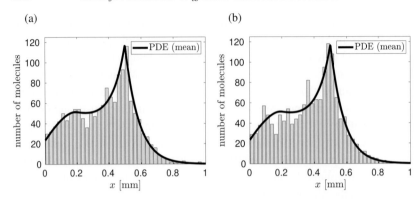

Figure 7.2 (a) *One realization of the Gillespie SSA (a5)–(d5) for the system of chemical reactions (7.17)–(7.19). The grey histogram shows numbers of molecules in each compartment at time t = 15 min. (b) Results of one realization of the SSA (a17)–(c17) which is modified so that (7.21) is replaced by the diffusion–advection model (7.22). Solution of (7.20) is plotted as the black solid line in both panels.*

where $\chi_{[0,L/5]}$ is the characteristic function of the interval $[0, L/5]$. The equation (7.20) provides a classic deterministic description of the reaction–diffusion–advection process. Using the initial condition $a(x,0) = 0$ and boundary conditions (7.15), we computed the solution of (7.20) numerically. It is plotted as the black solid line in Figure 7.2(a) for comparison. More precisely, we plot $a(x,t)h^3$, which expresses the results in terms of the number of molecules per compartment of volume h^3. As expected, the PDE (7.20) describes the time evolution of the stochastic mean well.

In Section 6.2, we presented the SSA (a17)–(c17), which is a molecular-based algorithm for simulating the reaction–diffusion example from Section 6.1. The diffusion trajectories of individual molecules were modelled using the SSA (a8)–(c8), i.e. the x-coordinate of a molecule at time $t + \Delta t$ was computed by

$$X(t + \Delta t) = X(t) + \sqrt{2D\,\Delta t}\,\xi \tag{7.21}$$

with the help of a normally distributed (with zero mean and unit variance) random number ξ. We can simulate the illustrative reaction–diffusion–advection model by the SSA (a17)–(c17) provided that (7.21) is modified to

$$X(t + \Delta t) = X(t) + \tilde{f}(X(t))\,\Delta t + \sqrt{2D\,\Delta t}\,\xi. \tag{7.22}$$

The results of one realization of the algorithm are shown in Figure 7.2(b). To visualize the results, we divide the elongated domain $[0, L] \times [0, h] \times [0, h]$ into

40 bins and we plot the numbers of molecules in bins at time 15 minutes as the grey histogram. Since we used the same number of bins as in the compartment-based model, the plot is directly comparable with Figure 7.2(a). We also plot the solution of (7.20) with boundary conditions (7.15) as the black solid line for comparison.

As in Chapter 6, we conclude that zeroth- and first-order chemical reactions can be easily introduced to both the compartment-based and the SDE-based diffusion–advection models. For such reactions the classic deterministic PDE (7.20) describes the average behaviour of both models over many stochastic realizations.

Higher-order chemical reactions do not present any special challenge in the compartment-based reaction-diffusion-advection approach. In the case of the SDE-based model, we have to be slightly careful if the reactants attract or repel each other. Such an interaction can also be captured by using (7.22) instead of (7.21) provided that the drift coefficient also depends on the positions of other molecules. However, the formulae for the reaction radius $\overline{\varrho}$ and the reaction probability P_λ (which were presented in Section 6.6) are no longer valid because they were derived under the assumption that all molecules are only transported by diffusion. We will return to the case of molecules that attract and/or repel each other in Section 7.5 where we discuss modelling ions and ion channels.

7.3 Bacterial Chemotaxis

In Section 4.3 we introduced velocity-jump processes. For such processes the position of a molecule $X(t)$ follows the deterministic equation (4.33) but $V(t)$, the velocity of the molecule, changes stochastically. Such stochastic processes can be used not only for the simulation of diffusing molecules but also as a description of the movement of unicellular organisms like bacteria. In this section, we present a generalization of the velocity-jump SSAs from Section 4.3 that is capable of describing bacterial behaviour.

The movement of individual biological agents often depends on external signal fields (for example, cells are attracted by nutrients and repelled by toxins). We start with a simple signal-dependent one-dimensional random walk which leads to the chemotaxis equation (7.7). As in Section 4.3, let us suppose that a particle moves along the x-axis at a constant speed s, but that at random instants of time it reverses direction. Now though, we bias the random walk by supposing that the particle is less likely to change direction when moving in a favourable direction, i.e. in the direction in which the signal function S is increasing. Specifically we suppose that the turning frequency is

$$\lambda = \lambda_0 \pm b \frac{\partial S}{\partial x}, \qquad (7.23)$$

where the sign depends on the direction of the particle movement: a plus sign is for the particles moving to the left and a minus sign is for the particles moving to the right. Let $p^+(x, t)$ be the density of the particles that are at (x, t) and are moving to the right, and let $p^-(x, t)$ be the density of particles that are at (x, t) and are moving to the left. Then $p^\pm(x, t)$ satisfy the equations (4.40)–(4.41). Using (7.23), we get

$$\frac{\partial p^+}{\partial t} + s \frac{\partial p^+}{\partial x} = -\left(\lambda_0 - b \frac{\partial S}{\partial x}\right) p^+ + \left(\lambda_0 + b \frac{\partial S}{\partial x}\right) p^-, \qquad (7.24)$$

$$\frac{\partial p^-}{\partial t} - s \frac{\partial p^-}{\partial x} = \left(\lambda_0 - b \frac{\partial S}{\partial x}\right) p^+ - \left(\lambda_0 + b \frac{\partial S}{\partial x}\right) p^-. \qquad (7.25)$$

The density of particles at (x, t) is given by the sum $n(x, t) = p^+(x, t) + p^-(x, t)$. It can be shown (see Exercise 7.5) that $n(x, t)$ satisfies the second-order PDE

$$\frac{1}{2\lambda_0} \frac{\partial^2 n}{\partial t^2} + \frac{\partial n}{\partial t} = \frac{s^2}{2\lambda_0} \frac{\partial^2 n}{\partial x^2} - \frac{bs}{\lambda_0} \frac{\partial}{\partial x}\left(n \frac{\partial S}{\partial x}\right). \qquad (7.26)$$

This is a hyperbolic version of the chemotaxis equation (7.7). As discussed in Section 4.3, the hyperbolic PDE (4.46) reduces for sufficiently large times to the parabolic diffusion equation (4.47). The same is true here (Hillen and Stevens, 2000). Thus we have shown that we can add advection to the velocity-jump model by changing the constant turning rate λ to the signal-dependent rate (7.23). In particular, it is straightforward to implement the diffusion–advection and reaction–diffusion–advection examples from Sections 7.1 and 7.2 as velocity-jump processes. We leave this to the reader as Exercises 7.6 and 7.7.

Velocity-jump processes can be used for modelling the motility of flagellated bacteria, such as *E. coli*, which is a widely used model organism in biological research. *E. coli* alternates between two modes of behaviour: a more or less straight motion with constant speed called "run" and a highly erratic motion called "tumble", which produces little translation but reorients the cell. The tumble time is much shorter than the average run time, i.e. a tumble can be viewed as an instantaneous change in velocity. Thus such a bacterium executes a velocity-jump process. When bacteria move in a favourable direction (that is, either in the direction of foodstuffs or away from noxious substances) the run times are increased, i.e. the turning frequency of the velocity-jump process decreases.

How does a bacterium know that it is moving in a favourable direction? Let $S(\mathbf{x}, t)$ be the concentration of an attracting chemical at point $\mathbf{x} \in \mathbb{R}^3$ and time t. Then the optimal direction to travel is the direction of the gradient of the attractant, ∇S. One way to estimate the gradient is to compare the value of concentration S at several points on the cell surface. If a bacterium was able to do this, then a three-dimensional analogue of (7.23) could be used to find places with high levels of the attractant. However, bacteria live in a noisy environment and they are too small to accurately estimate the concentration gradient by making "measurements" along their body lengths. Therefore, *E. coli* developed another strategy to find a favourable direction. Roughly speaking, it compares its current environment with the environment it visited a second ago. If the concentration of S is increasing along the cellular trajectory, then the bacterium is moving in a good direction and it is less likely to change it. On the other hand, if the concentration of S is decreasing along its trajectory, then the bacterium is more likely to turn.

To compare the quality of the current and previous environments, a bacterium must have a "memory": it must remember the quality of the previous environment. Of course, bacteria are very small and do not have a brain like humans; their "memory" is constructed in purely biochemical terms. *E. coli* detects the attractant concentration by receptors on its membrane. The information about the receptor occupancy is then carried by intracellular signalling molecules to flagellar motors. The signalling molecules are proteins which are subject to biochemical reactions, forming a so-called signal transduction network. Depending on the abundance of a particular intracellular protein, a bacterium changes its turning frequency. However, it is not the goal of this book to give the names of the proteins involved in the signal transduction. We refer the reader interested in detailed models to the biological literature (Barkai and Leibler, 1997). Here we present only a simplified toy model which can be easily implemented in the computer and which has the essential features of the signal transduction network (Erban and Othmer, 2004).

A bacterium will be modelled as a point that is described not only by its position $\mathbf{X}(t)$ and velocity $\mathbf{V}(t)$, but also by additional, internal, variables $\mathbf{Y}(t)$. In detailed models, the components of the internal vector $\mathbf{Y}(t)$ would stand for concentrations (or numbers) of molecules involved in the signal transduction and the dimension of $\mathbf{Y}(t)$ would be quite large. In our toy model, the vector $\mathbf{Y}(t)$ will have only two components, i.e. $\mathbf{Y}(t) = (Y_1(t), Y_2(t))$. To explain the time evolution of $\mathbf{Y}(t)$, let us denote the signal detected by a bacterium as $C(t)$. We have

$$C(t) = S(\mathbf{X}(t), t),$$

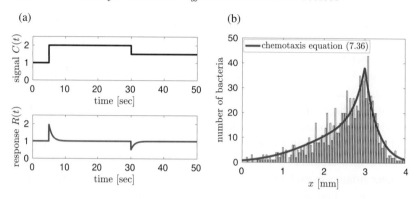

Figure 7.3 (a) *The signal C(t) (blue) and the corresponding response R(t) (red) computed by (7.27) and (7.28) for $t_e = 10^{-2}$ sec and $t_a = 1$ sec. (b) Simulation of 1000 bacteria using the SSA (a19)–(d19). The grey histogram shows numbers of bacteria in bins at time $t = 1$ hour. Solution of (7.36) is plotted as the red solid line.*

where $S(\mathbf{x}, t)$ is the concentration of extracellular signal. We postulate that $\mathbf{Y}(t)$ evolves according to the following system of ODEs (Othmer and Schaap, 1998):

$$\frac{dY_1}{dt} = \frac{C(t) - Y_1 - Y_2}{t_e}, \qquad \frac{dY_2}{dt} = \frac{C(t) - Y_2}{t_a}, \qquad (7.27)$$

where constants t_e and t_a are called the excitation time and the adaptation time, respectively. To mimic the behaviour of *E. coli*, the adaptation time must be chosen much larger (typically of the order of seconds) than the excitation time. In Figure 7.3(a), we present a typical response of the toy model (7.27) to step changes of the signal $C(t)$. For the purposes of this plot, we defined the response $R(t)$ of the signal transduction mechanism (7.27) as

$$R(t) = 1 + Y_1(t). \qquad (7.28)$$

We see that a step change in the incoming signal $C(t)$ is followed by a quick excitation (increase or decrease) of the response $R(t)$. Then $R(t)$ returns to its baseline level (adapts) on the time scale of seconds. This excitation–adaptation dynamics is common for many sensory systems. Therefore, it is important that our toy model can replicate it. In fact, any function of Y_1 will have dynamics qualitatively similar to $R(t)$. This can be seen by looking for the steady state of ODEs (7.27) under the condition that the signal detected by the cell is constant, i.e. $C(t) \equiv C_s$. Solving the corresponding steady-state equations

$$\frac{C_s - Y_1 - Y_2}{t_e} = 0, \qquad \frac{C_s - Y_2}{t_a} = 0,$$

we obtain $Y_1 = 0$ and $Y_2 = C_s$. In particular, Y_1 adapts to zero independently of the background signal C_s and any function of Y_1 will also possess the excitation–adaptation dynamics. It is worth noting that the second variable Y_2 does not adapt. This property of the toy model is also shared by more detailed models of signal transduction (Barkai and Leibler, 1997; Spiro et al., 1997) which are formulated as larger systems of ODEs. In order for a response to adapt, some other model variables must compensate for different levels of the background signal C_s.

We conclude this section by presenting a velocity-jump process with internal dynamics. This modelling framework is capable of describing the behaviour of *E. coli*. To simplify our presentation, we restrict our consideration to the toy model (7.27). Moreover, as in Section 4.3, we will only consider motion in one dimension. We assume that a bacterium moves along the *x*-axis at a constant speed *s*, but that at random instants of time it reverses its direction according to a Poisson process with the turning frequency

$$\lambda = \lambda_0 - b_1 Y_1, \tag{7.29}$$

where λ_0 and b_1 are positive constants. The turning frequency (which is a linear function of Y_1) possesses the excitation–adaptation dynamics discussed above. Moreover, if $C(t)$ increases along the cellular trajectory, then Y_1 will increase (see Figure 7.3(a)), and λ given by (7.29) will decrease, i.e. the cell is less likely to turn when moving in the direction of an increasing signal. If the cellular environment contains a spatially homogeneous signal, then $C(t)$ will be constant, Y_1 will adapt to zero and the turning frequency λ will be equal to the baseline turning frequency λ_0 (which is about 1 sec^{-1} for *E. coli*). A non-homogeneous environment can result in changes of Y_1 which translates to changes of λ as given by (7.29). Note that we are implicitly assuming that the change of the turning frequency $b_1 Y_1$ is less than or equal to λ_0. If this condition was not satisfied, then λ would be negative (which is obviously not allowed). Such restrictions could be removed by choosing some more complicated (nonlinear) function of Y_1 instead of (7.29).

Since $t_e \ll t_a$, we will further simplify the toy model (7.27) by considering the limiting case $t_e = 0$. Then the first equation in (7.27) implies that $Y_1 = C(t) - Y_2$. Substituting into (7.29), we obtain

$$\lambda = \lambda_0 - b_1 (C(t) - Y_2), \tag{7.30}$$

where Y_2 evolves according to the second equation in (7.27). Consequently, each bacterium will be described by its one-dimensional position $X(t)$, velocity $V(t)$ and internal state $Y_2(t)$. We will assume that the signal profile $S(x)$ is a fixed function (independent of time) and given by

$$S(x) = b_3 (b_2 - |x - b_2|),$$

where b_2 and b_3 are positive constants. In particular, $S(x)$ has a global maximum at the position $x = b_2$ and is linearly increasing (resp. decreasing) for $x \leq b_2$ (resp. $x \geq b_2$). Since $C(t) = S(X(t))$ and $t_e = 0$, the internal dynamics (7.27) reduces to the ODE

$$\frac{dY_2}{dt} = \frac{S(X(t)) - Y_2}{t_a} \tag{7.31}$$

and (7.30) reads as follows:

$$\lambda = \lambda_0 - b_1 S(X(t)) + b_1 Y_2(t). \tag{7.32}$$

To formulate the velocity-jump process with internal dynamics, we choose a small time step Δt. The position $X(t)$ is updated by (4.35) which is equal to

$$X(t + \Delta t) = X(t) + V(t) \Delta t. \tag{7.33}$$

Since the speed s is constant, there are only two possible values of the one-dimensional velocity $V(t)$, namely $\pm s$. In the algorithm, the velocity can only change its sign at the end of each time step. Consequently, the position of the cell at an (intermediate) time $t + \tau \in [t, t + \Delta t]$ can be, for $0 \leq \tau \leq \Delta t$, computed as $X(t + \tau) = X(t) + V(t) \tau$, where $X(t)$ and $V(t)$ are the values at the beginning of the time step. Substituting in (7.31) and using the fact that $S(x)$ is piecewise linear, we can solve (7.31) explicitly in the interval $[t, t + \Delta t]$ to get (see Exercise 7.8)

$$Y_2(t + \Delta t) = Y_2(t) \exp\left(-\frac{\Delta t}{t_a}\right) + S(X(t))\left(1 - \exp\left(-\frac{\Delta t}{t_a}\right)\right) \tag{7.34}$$

$$+ b_3 V(t)\left(\Delta t - t_a + t_a \exp\left(-\frac{\Delta t}{t_a}\right)\right) \times \begin{cases} 1, & \text{if } X(t) < b_2, \\ -1, & \text{if } X(t) > b_2. \end{cases}$$

The above formula is the exact solution of (7.31) provided that the cellular trajectory does not cross the point $x = b_2$ during the time step $[t, t + \Delta t]$. If this happens, the exact solution of (7.31) is more complicated. However, if the time step Δt is chosen small enough, it is not necessary to consider this technical detail. In fact, if $\Delta t \ll t_a$, then we can approximate

$$\exp\left(-\frac{\Delta t}{t_a}\right) \approx 1 - \frac{\Delta t}{t_a}$$

and (7.34) reads as follows:

$$Y_2(t + \Delta t) = Y_2(t) + \Delta t \frac{S(X(t)) - Y_2(t)}{t_a}, \tag{7.35}$$

which is the forward Euler scheme for solving the ODE (7.31). Either (7.34) or (7.35) can be used for updating the internal variable in the algorithm.

Thus the velocity-jump process with internal dynamics can be formulated as the following SSA. Starting with the given initial conditions for position $X(0)$, velocity $V(0) \in \{-s, s\}$ and internal state $Y_2(0)$, we perform four steps at time t:

(**a19**) Generate a random number r uniformly distributed in $(0, 1)$.
(**b19**) Compute the position of the molecule at time $t + \Delta t$ by (7.33).
(**c19**) Compute the internal variable Y_2 at time $t + \Delta t$ by (7.34).
(**d19**) Compute λ by (7.32). If $r < \lambda \Delta t$, then put $V(t + \Delta t) = -V(t)$.
Otherwise, $V(t + \Delta t) = V(t)$.
Then continue with step (a19) for time $t + \Delta t$.

In Figure 7.3(b), we present results computed by the SSA (a19)–(d19). We simulate the behaviour of 1000 bacteria for realistic choices of model parameters, namely $s = 20 \, \mu\text{m sec}^{-1}$, $t_e = 0$, $t_a = 1$ sec, $\lambda_0 = 1 \, \text{sec}^{-1}$, $b_1 = 1$, $b_2 = 3 \, \text{mm}$ and $b_3 = 4 \, \text{mm}^{-1}$. We use the time step $\Delta t = 10^{-2}$ sec which satisfies $\Delta t \ll t_a$. Consequently, (7.34) in the step (c19) could be replaced by the approximation (7.35) and we would obtain comparable results. With more complicated internal dynamics it is often impossible to solve the equations exactly (as we did in (7.34)) and numerical approximations of the form (7.35) are very helpful.

In Figure 7.3(b), we plot the spatial histogram of bacteria at time 1 hour, i.e. we divide the interval $[0, 4]$ mm into 100 bins and we plot numbers of bacteria in each bin. Initially, at time $t = 0$, all bacteria are positioned at $x = 1$ mm. We can clearly see the aggregation of bacteria around the maximum of the signal at $x = b_2 = 3$ mm.

From the toy model (7.27) it is possible to derive the evolution equation for cellular density $n(x, t)$ in the following form (Erban and Othmer, 2004):

$$\frac{\partial n}{\partial t} = \frac{s^2}{2\lambda_0} \frac{\partial^2 n}{\partial x^2} - \frac{b \, s^2 \, t_a}{\lambda_0 (1 + 2\lambda_0 t_a)(1 + 2\lambda_0 t_e)} \frac{\partial}{\partial x} \left(n \frac{\partial S}{\partial x} \right), \qquad (7.36)$$

which has the form of the chemotaxis equation (7.7). The solution of (7.36) is plotted in Figure 7.3(b) as the red solid line. If we simulated a larger group of bacteria, then the resulting spatial histogram would be less noisy and would (almost perfectly) match the red line.

In this section we have presented two velocity-jump processes in which the turning frequency depends on the extracellular signal S, either directly, as in (7.23), or indirectly through internal dynamics, as in (7.29). In both cases, the cellular population can (in the limit of many bacteria) be described by a chemotaxis equation of the form (7.7). However, it is worth noting that the chemotaxis equation (7.36) is valid only under the assumption that the signal gradients are sufficiently "shallow"; see Erban and Othmer (2004) for details. With increasing complexity of the model, the derivation of continuum PDEs becomes more challenging. If we consider two- and three-dimensional analogues of the models presented in this section, it is still possible to derive the chemotaxis equation (7.7), but the derivation is more technical (Hillen, 2001) (and again the simplifying assumptions of shallow signal gradients must be imposed (Erban and Othmer, 2005)).

With more complex models of intracellular dynamics the population-level behaviour might not reduce to the chemotaxis equation. Although this equation has been used in chemotaxis modelling since the pioneering work by Keller and Segel (1971*a,b*), it is often not suitable for the inclusion of intracellular processes. If a modeller is interested in the interplay between intracellular biochemistry and population-level behaviour, then the SSAs with internal dynamics provide an appropriate alternative to a deterministic PDE-based model.

More realistic models will also include interactions between individuals, which can be either direct through volume exclusion (only a finite number of cells can fit into a given volume) or indirect by modifying (releasing or consuming) signalling molecules in their environment. In the latter case, we also have to model the time evolution of extracellular signals. There have been a number of individual-based models of interacting cells developed in the literature (Dallon and Othmer, 1997). In the next section, we explain modelling interactions between individuals using a different application area.

7.4 Collective Behaviour of Locusts

The velocity-jump model with internal dynamics could be viewed as an example of an agent-based (individual-based) model. An agent (individual) uses a fixed set of rules based on communication with other agents and information about the environment in order to change its internal state and fulfil its design objective (Wooldridge, 2002; Franz and Erban, 2012). This type of description is useful for modelling collective animal behaviour (Sumpter, 2010). In this section, we will discuss experiments with locusts which can be described by an agent-based (individual-based) model written in one spatial dimension. This agent-based model can also be considered as an example of the so-called self-propelled particle model introduced by Vicsek and his co-authors (Vicsek et al., 1995).

Buhl et al. (2006) performed experiments with different numbers N of locusts (wingless juveniles) marching in a ring-shaped arena. To develop a simplified model of such an experimental set-up, we assume that locusts march along the unit circle

$$(\cos(\theta), \sin(\theta)) \qquad \text{for} \qquad \theta = [0, 2\pi). \qquad (7.37)$$

Let $\omega_i(t)$, $i = 1, 2, \ldots, N$, be the angular velocity of the ith locust. Let $\mathbf{X}_i(t) = (\cos(\theta_i(t)), \sin(\theta_i(t)))$ be the dimensionless position of the ith locust at time t. We assume that locusts can only sense other locusts if they are sufficiently close. For each locust, we define the set of its neighbours at time t by

$$\mathcal{J}_{i,R}(t) = \left\{ j \in \{1, 2, \ldots, N\} \,\middle|\, |\mathbf{X}_i - \mathbf{X}_j| \le R \right\}, \qquad (7.38)$$

where R is a dimensionless parameter and $|\mathbf{X}_i - \mathbf{X}_j|$ is the Euclidean distance between the ith and jth locust. Please note that we have $i \in \mathcal{J}_{i,R}(t)$ for simplicity. We define

$$U_i(t) = \frac{1}{|\mathcal{J}_{i,R}(t)|} \sum_{j \in \mathcal{J}_{i,R}(t)} \omega_j(t). \qquad (7.39)$$

Then the locust model is given by the following update equation:

$$\theta_i(t + \Delta t) = \theta_i(t) + \omega_i(t)\,\Delta t, \qquad (7.40)$$

$$\omega_i(t + \Delta t) = \omega_i(t) + (\text{sign}(U_i(t)) - \omega_i(t))\,\Delta t + \eta\,\sqrt{\Delta t}\,\xi_i, \qquad (7.41)$$

for $i = 1, 2, \ldots, N$, where η is a parameter, ξ_i is a normally distributed random number with zero mean and unit variance and "sign" is the signum function.

The model (7.40)–(7.41) is written as a coupled system of discretized ODEs and SDEs. It is not the same model as in Buhl et al. (2006), but it has similar properties and can also be used to fit their experimental data. Their individual-based model is written as a computer code that uses uniformly distributed random numbers and does not introduce time step Δt. Our SDE formulation will help us to analytically obtain some insight into the model behaviour.

Buhl et al. (2006) estimated an interaction radius between locusts to be 13.9 cm. This number is very small compared to the size of "marching bands" formed by locust nymphs (wingless juveniles) that can extend over many kilometres. The experimental ring-shaped arena built by Buhl et al. (2006) also has a circumference (2.5 metres) that is larger than the interaction radius. In particular, the parameter R in (7.39) should be chosen smaller than 2. We will discuss the biologically interesting case $R < 2$ at the end of this section. To get some insight into this problem, we will use $R = 2$ in what follows. If $R = 2$, then (7.38) implies

$$\mathcal{J}_{i,2}(t) = \{1, 2, \ldots, N\}, \qquad \text{for all } i = 1, 2, \ldots, N,$$

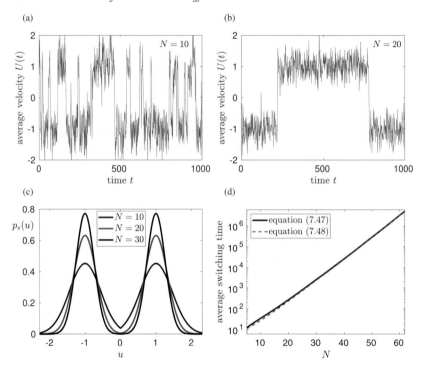

Figure 7.4 (a), (b) *Plot of the average velocity* (7.42) *as a function of time t computed by the model* (7.43) *for N* = 10 *(in panel (a)) and N* = 20 *(in panel (b)) locusts.* (c) *Stationary distribution* $p_s(u)$ *of the average velocity* (7.42) *given by* (7.46) *for N* = 10, *N* = 20 *and N* = 30 *locusts.* (d) *The average time taken for the group to change direction as a function of the number of locusts. We compare the results computed by* (7.47) *(blue line) with the results computed by* (7.48) *(red dashed line). There is a log scale on the vertical axis.*

and the local velocity average (7.39) is equal to the global velocity average

$$U(t) = \frac{1}{N} \sum_{j=1}^{N} \omega_j(t). \tag{7.42}$$

Then (7.41) reduces to

$$\omega_i(t + \Delta t) = \omega_i(t) + (\text{sign}(U(t)) - \omega_i(t)) \Delta t + \eta \sqrt{\Delta t} \, \xi_i. \tag{7.43}$$

In Figure 7.4(a) and (b), we present simulations of (7.43) for N = 10 and N = 20 locusts. We plot (7.42) as a function of time for parameters $\Delta t = 10^{-2}$ and $\eta = 2$ for N = 10 and N = 20.

In Figure 7.4(b), we observe that the average velocity of locusts switches between two favourable states, $+1$ and -1, which correspond to the whole locust group moving in an anticlockwise or clockwise direction, respectively. This is also the main observation in experiments and has been built into the advection term in models (7.41) or (7.43), which are chosen in such a way that each individual locust prefers to follow the direction of other locusts within its interaction radius. Such a behaviour is not restricted to locusts. For example, the ring-shaped experimental set-up was also used for groups of prawns by Mann et al. (2013) with similar conclusions.

Comparing Figure 7.4(a) and (b), we observe that the number of direction changes will decrease if the group size N increases. In the case of $R = 2$, we can further analyse this model. Adding equations (7.43) for $i = 1, 2, \ldots, N$, we obtain

$$\sum_{i=1}^{N} \omega_i(t + \Delta t) = \sum_{i=1}^{N} \omega_i(t) + \left(N \operatorname{sign}(U(t)) - \sum_{i=1}^{N} \omega_i(t) \right) \Delta t + \sum_{i=1}^{N} \eta \sqrt{\Delta t}\, \xi_i.$$

Dividing by N and using (7.42), we have

$$U(t + \Delta t) = U(t) + (\operatorname{sign}(U(t)) - U(t)) \Delta t + \frac{\eta}{N} \sqrt{\Delta t} \sum_{i=1}^{N} \xi_i.$$

Since the sum of N normally distributed random numbers ξ_i is a normally distributed random number (with zero mean and variance N), we can write

$$U(t + \Delta t) = U(t) + (\operatorname{sign}(U(t)) - U(t)) \Delta t + \frac{\eta}{\sqrt{N}} \sqrt{\Delta t}\, \xi,$$

where ξ is a normally distributed random number with zero mean and unit variance. Comparing with (3.5), we observe that $U(t)$ satisfies the SDE

$$U(t + dt) = U(t) + (\operatorname{sign}(U(t)) - U(t)) dt + \frac{\eta}{\sqrt{N}} dW. \tag{7.44}$$

Let $p(u, t)$ be the probability distribution of $U(t)$, i.e. $p(u, t)\, du$ is the probability that $U(t) \in [u, u + du)$, and let $p_s(u) = \lim_{t \to \infty} p(u, t)$ be the corresponding stationary distribution. Substituting into (3.36), we obtain

$$p_s(u) = C \exp\left[\frac{2N}{\eta^2} \int_0^u \left(\operatorname{sign}(y) - y \right) dy \right] = C \exp\left[\frac{N}{\eta^2} \left(2|u| - u^2 \right) \right], \tag{7.45}$$

where the normalization constant C is given by equation (3.37). Integrating equation (7.45), we can calculate C (see Exercise 7.9) and rewrite (7.45) as

$$p_s(u) = \frac{\sqrt{N}}{\eta \sqrt{\pi}} \left(1 + \operatorname{erf}\left(\frac{\sqrt{N}}{\eta} \right) \right)^{-1} \exp\left[\frac{N}{\eta^2} \left(-1 + 2|u| - u^2 \right) \right], \tag{7.46}$$

where the error function is defined by erf$(z) = (2/\sqrt{\pi}) \int_0^z \exp[-y^2]\,dy$. The stationary distribution $p_s(u)$ is plotted for $N = 10$, $N = 20$ and $N = 30$ locusts in Figure 7.4(c). We confirm our observation that the switches between the two macroscopic states of the system, corresponding to $U(t) = -1$ and $U(t) = +1$, will less likely occur for larger groups. We can also estimate the average time for the group to change its direction from $U(t) = -1$ to $U(t) = +1$ using equation (3.63) by

$$\frac{2N}{\eta^2} \int_{-1}^{1} \frac{1}{p_s(z)} \int_{-\infty}^{z} p_s(u)\,du\,dz = \pi\left(1 + \text{erf}\left(\frac{\sqrt{N}}{\eta}\right)\right)\text{erfi}\left(\frac{\sqrt{N}}{\eta}\right), \qquad (7.47)$$

where the imaginary error function is defined by erfi$(z) = (2/\sqrt{\pi}) \int_0^z \exp[y^2]\,dy$. The average switching time as a function of N for $\eta = 2$ is plotted in Figure 7.4(d) as the blue line. If N is large, then we can approximate erf and erfi in equation (7.47) to obtain

$$\frac{2\eta\sqrt{\pi}}{\sqrt{N}}\exp\left[\frac{N}{\eta^2}\right]. \qquad (7.48)$$

We plot this asymptotic result in Figure 7.4(d) as the red dashed line.

Formulae (7.47) and (7.48) were derived for $R = 2$, i.e. for the case where the local velocity average (7.39) can be substituted by the global velocity average (7.42) in equations (7.40)–(7.41). In Exercise 7.10, you are asked to investigate the behaviour of the model (7.40)–(7.41) for general values of R. In this case, both equations (7.40) and (7.41) have to be simulated. We are interested in the behaviour of the whole group (given by the global velocity average (7.42)), but we assume that individuals can only interact with their neighbours (where the neighbourhood size is given by R), which is a more realistic assumption on behaviour of individuals in large animal groups. Since such questions are difficult to investigate analytically, computer simulations are used in Exercise 7.10.

Another agent-based model formulated on the unit circle which leads to the formation of groups of animals moving clockwise and anticlockwise is investigated in Exercise 7.11. It is based on the velocity-jump process studied in Section 7.3. To simplify the analysis, we again assume that all individuals can communicate with each other. If this simplifying assumption was not used, the model behaviour could be investigated by deriving and analysing suitable PDEs (Erban and Haskovec, 2012). We will not study such a complication here, but instead turn our attention to another application area of diffusion–advection processes.

7.5 Ions and Ion Channels

Equations (7.1)–(7.3) can be used to model the trajectory of a charged particle (ion) in an external electrical field described by $\mathbf{f}$. These equations can be further modified to model a collection of ions, developing the so-called Brownian dynamics simulation approach. Using the heuristic notation of (4.96) we can write Newton's second law of motion for an ion of mass m in the form

$$m\frac{d\mathbf{V}}{dt}(t) = -\gamma\mathbf{V}(t) + \mathbf{F}_e + \gamma\sqrt{2D}\,\text{``}\frac{d\mathbf{W}}{dt}\text{''}, \tag{7.49}$$

where $\mathbf{V}(t) = [V_x(t), V_y t), V_z(t)]$ is the velocity of the ion. Now there are three forces on the right-hand side: (i) the frictional force $-\gamma\mathbf{V}(t)$, where γ is the frictional drag coefficient; (ii) the electrical force $\mathbf{F}_e$ that is exerted by neighbouring ions and/or by the external electrical field; and (iii) the random force describing the collisions with the surrounding molecules in the solution (e.g. the water molecules).

In Section 8.6, we will present a molecular dynamics approach to modelling ions, where we will explicitly describe the surrounding water molecules. Such simulations will use a very small time step equal to a femtosecond (10^{-15} sec). Brownian dynamics simulations in this section will use a thousand times larger time step which we can use to further simplify equation (7.49). If we multiply (7.49) by dt, we can interpret it as an SDE (compare with (4.51)). Dividing (7.49) by γ, we obtain

$$\frac{1}{\beta}\frac{d\mathbf{V}}{dt}(t) = -\mathbf{V}(t) + \frac{\mathbf{F}_e}{\gamma} + \sqrt{2D}\,\text{``}\frac{d\mathbf{W}}{dt}\text{''}, \tag{7.50}$$

where $\beta = \gamma/m$. The mass m of an ion is typically of order 10^{-25}–10^{-26} kg; for example, the mass of a sodium ion (Na^+) is 3.8×10^{-26} kg and the mass of a potassium ion (K^+) is 6.5×10^{-26} kg. The frictional drag coefficient γ is related to the diffusion coefficient D by the Einstein–Smoluchowski relation (4.95). The typical value of D for ions in aqueous solution at room temperature is $D \approx 10^{-3}$ mm^2 sec^{-1}. Consequently, (4.95) implies that $\beta = \gamma/m = k_B T/(Dm) \approx 10^{14}$ sec^{-1}. In stochastic simulations, we will use the time step $\Delta t = 10^{-12}$ sec, which is significantly larger than $1/\beta$. Consequently, we can assume that (7.50) is at equilibrium, that is,

$$\mathbf{V}(t) = \frac{\mathbf{F}_e}{\gamma} + \sqrt{2D}\,\text{``}\frac{d\mathbf{W}}{dt}\text{''}.$$

Substituting $\mathbf{V}(t) = d\mathbf{X}(t)/dt$, where $\mathbf{X}(t) = [X(t), Y(t), Z(t)]$ is the position of the ion, we obtain the SDE

$$\mathbf{X}(t + dt) = \mathbf{X}(t) + \gamma^{-1}\mathbf{F}_e\,dt + \sqrt{2D}\,d\mathbf{W}. \tag{7.51}$$

The main difference between (7.51) and equations (7.1)–(7.3) is that the force $\mathbf{F}_e$ depends not only on the position $\mathbf{X}(t)$ of the ion but also on the position of all the other ions in the system. This significantly complicates the analysis. Many results are only obtained by the computer simulation of a large system of coupled SDEs (7.51) where one (vector) equation (7.51) is written for each ion (particle) in the system. This approach is often called Brownian dynamics. In the rest of this section, we explore the consequences of interactions between ions using a simple model of an ion channel.

Ion channels and ion pores are proteins situated in cell membranes. They provide a path for rapid movement of ions and water, connecting the cell interior to its exterior through the hydrophobic barrier of the membrane. Channels and pores can be ion-selective, allowing only specific ions to permeate through them (for example, potassium, sodium, calcium or chloride ions), or non-specific, permeable to a variety of ions. In addition, many channels and pores can change their conformation between open and closed states spontaneously, or in response to stimuli such as a potential difference across the membrane or interaction with various molecules (neurotransmitters, second-messenger drugs). One remarkable feature of ion channels is that their narrowest part is often so small that only a single ion can get through, i.e. the passage of ions through the channel is in "single file". This means that individual-based models of ions (such as (7.51)) ought to be used inside the channel instead of deterministic PDE-based models for ion concentrations and charges (Corry et al., 2000). To simplify our presentation, we will suppose that the ion channel is a cylinder of length L and radius R (see Figure 7.5(a)), that is, we simulate the behaviour of ions in the three-dimensional domain

$$\Omega = \left\{ [x, y, z] \mid x \in [0, L] \text{ and } y^2 + z^2 \leq R^2 \right\}.$$

The boundary of the domain $\partial\Omega$ can be written as a union of three parts, namely $\partial\Omega = B_1 \cup B_2 \cup B_3$, where

$$B_1 = \left\{ [x, y, z] \mid x \in [0, L] \text{ and } y^2 + z^2 = R^2 \right\},$$
$$B_2 = \left\{ [x, y, z] \mid x = 0 \text{ and } y^2 + z^2 < R^2 \right\},$$
$$B_3 = \left\{ [x, y, z] \mid x = L \text{ and } y^2 + z^2 < R^2 \right\}.$$

We will assume reflective boundary conditions on the side B_1 of the cylinder Ω. In reality there are charged (and polarizable) molecules in the walls of the ion channel which attract or repel ions, so that the wall B_1 influences the behaviour of ions in a more complicated way. However, we ignore these interactions in what follows and treat B_1 as a reflective boundary.

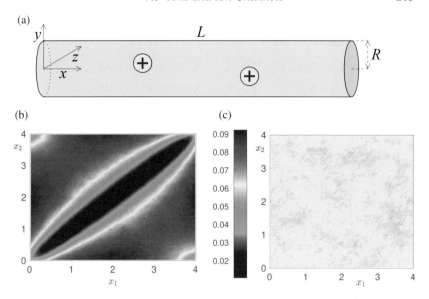

Figure 7.5 *Ion channel toy model* (7.56)–(7.57). (a) *The ion channel is modelled as a cylinder of length L and radius R.* (b) *The probability distribution function* $P_s(x_1, x_2)$ *defined by* (7.59) *estimated from the Brownian dynamics simulation. Parameters are given by* (7.58). (c) *The probability distribution function* $P_s(x_1, x_2)$ *estimated from the Brownian dynamics simulation in the case* $a_2 = 0$. *Other parameters are given by* (7.58).

The boundary conditions at boundaries B_2 and B_3 also need to be specified. In biological applications, the half space $\{x < 0\}$, which lies in front of the ion channel Ω, contains many ions which enter the channel from time to time through the disc B_2. To determine the correct flux of ions through B_2, we need to model the regions on both sides of the boundary B_2, either by simulating the individual ions by (7.51) or by PDE-based models (e.g. the so-called Poisson–Nernst–Planck PDEs) if the concentration of ions outside the channel is sufficiently high (Corry et al., 2000; Roux et al., 2004). The same is true for the boundary B_3. However, we will not cover this level of detail here. We will impose periodic boundary conditions at B_2 and B_3, i.e. an ion that leaves the domain through B_2 (resp. B_3) will return through B_3 (resp. B_2). This boundary condition is artificial. In particular, we will not get any useful information about the behaviour of individual ions in the parts of the channel that are close to B_2 and B_3, but the model will still be useful for understanding of the behaviour of ions inside the channel. One consequence of the periodic boundary condition is that the number of ions in the channel is conserved.

Assuming that there are only two (positive) ions in the channel, their dynamics are given as follows:

$$\mathbf{X}_1(t + dt) = \mathbf{X}_1(t) + \gamma^{-1}\mathbf{F}_{e;1}(\mathbf{X}_1(t), \mathbf{X}_2(t))\, dt + \sqrt{2D}\, d\mathbf{W}_1, \tag{7.52}$$

$$\mathbf{X}_2(t + dt) = \mathbf{X}_2(t) + \gamma^{-1}\mathbf{F}_{e;2}(\mathbf{X}_1(t), \mathbf{X}_2(t))\, dt + \sqrt{2D}\, d\mathbf{W}_2, \tag{7.53}$$

where $\mathbf{X}_1$ (resp. $\mathbf{X}_2$) is the position of the first (resp. second) ion. To simplify the model further, we assume that the force $\mathbf{F}_{e;1}(\mathbf{X}_1(t), \mathbf{X}_2(t))$ exerted on the first ion is only a sum of two forces: (1) a constant force pushing ions down the channel (which is due to the potential difference across the membrane), and (2) the electrical force exerted by the second ion. Using the Coulomb law, we have

$$\mathbf{F}_{e;1}(\mathbf{X}_1(t), \mathbf{X}_2(t)) = \mathbf{F}_0 + \frac{q_1 q_2}{4\pi\varepsilon_0\varepsilon_w} \frac{\mathbf{X}_1(t) - \mathbf{X}_2(t)}{|\mathbf{X}_1(t) - \mathbf{X}_2(t)|^3}, \tag{7.54}$$

where q_1 (resp. q_2) are charges of the first (resp. second) ion, ε_0 is the permittivity of free space, ε_w is the dielectric constant of water and $\mathbf{F}_0$ is the constant force down the channel, i.e.

$$\mathbf{F}_0 = [a_1\gamma, 0, 0]^\mathrm{T},$$

where a_1 is a constant. The equation for the force $\mathbf{F}_{e;2}(\mathbf{X}_1(t), \mathbf{X}_2(t))$ exerted on the second ion is similar to (7.54). The only difference is that the Coulomb force (second term) has the opposite sign. Let us denote

$$a_2 = \frac{q_1 q_2}{4\pi\varepsilon_0\varepsilon_w\gamma}. \tag{7.55}$$

Substituting (7.54) into (7.52)–(7.53), we obtain

$$\mathbf{X}_1(t + dt) = \mathbf{X}_1(t) + \left(a_1\, \mathbf{e}_1 + a_2 \frac{\mathbf{X}_1(t) - \mathbf{X}_2(t)}{|\mathbf{X}_1(t) - \mathbf{X}_2(t)|^3}\right) dt + \sqrt{2D}\, d\mathbf{W}_1, \tag{7.56}$$

$$\mathbf{X}_2(t + dt) = \mathbf{X}_2(t) + \left(a_1\, \mathbf{e}_1 + a_2 \frac{\mathbf{X}_2(t) - \mathbf{X}_1(t)}{|\mathbf{X}_1(t) - \mathbf{X}_2(t)|^3}\right) dt + \sqrt{2D}\, d\mathbf{W}_2, \tag{7.57}$$

where $\mathbf{e}_1 = (1, 0, 0)^\mathrm{T}$, and a_1, a_2 and D are parameters that need to be specified.

We choose parameter values that would be realistic for sodium (Na^+) or potassium (K^+) ions. Thus we take $D \approx 10^9 \text{ nm}^2 \text{ sec}^{-1}$. Since both ions carry an elementary charge, we have $q_1 = q_2 = 1.6 \times 10^{-19}\, \text{C}$. Substituting $\varepsilon_0 = 8.85 \times 10^{-12}\, \text{C}^2\, \text{N}^{-1}\, \text{m}^{-2}$, $\varepsilon_w = 80$ and $\gamma = k_B T/D = 4 \times 10^{-12} \text{ kg sec}^{-1}$ into formula (7.55), we obtain $a_2 = 7 \times 10^8 \text{ nm}^3 \text{ sec}^{-1}$. Considering that the ion channel is a cylinder with length $L = 4 \times 10^{-9}$ m and radius $R = 3 \times 10^{-10}$ m, we estimate the coefficient a_1 by $a_1 = q_1\Delta U/(L\gamma)$ where ΔU is the potential difference across the membrane. Using $\Delta U = 50$ mV, we get $a_1 = 5 \times 10^8 \text{ nm sec}^{-1}$.

As we saw in the previous paragraph, the parameters of Brownian dynamics simulations are either very large or very small, depending on the choice of units. Mathematicians usually solve this ambiguity by *non-dimensionalizing* the equations, that is, choosing units of length and time which are appropriate for the given problem. This is equivalent to choosing particular units of length and time in a computer simulation, so that the computed answer is not too far from 1 in these units (multiplying large numbers by small numbers on a computer introduces unnecessary rounding errors). In our case, we will choose a *nanometre* as the unit of length and a *nanosecond* as the unit of time. Then all parameters will be close to 1, namely

$$a_1 = 0.5, \quad a_2 = 0.7, \quad D = 1, \quad L = 4 \quad \text{and} \quad R = 0.3, \qquad (7.58)$$

where we have already dropped the units, i.e. all parameters can be considered dimensionless.

An illustrative result computed by the ion channel model described above is shown in Figure 7.5(b). We plot the probability distribution function $P_s(x_1, x_2)$ of finding the first ion at the x-coordinate x_1 and the second ion at the x-coordinate x_2. More precisely, we can define the time-dependent probability distribution $P(\mathbf{x}_1, \mathbf{x}_2, t)$ so that $P(\mathbf{x}_1, \mathbf{x}_2, t) \, d\mathbf{x}_1 \, d\mathbf{x}_2$ is the probability that $\mathbf{X}_1(t) \in [\mathbf{x}_1, \mathbf{x}_1 + d\mathbf{x}_1)$ and $\mathbf{X}_2(t) \in [\mathbf{x}_2, \mathbf{x}_2 + d\mathbf{x}_2)$. Then the stationary distribution plotted in Figure 7.5(b) is given by

$$P_s(x_1, x_2) = \lim_{t \to \infty} \int_{B_R} \int_{B_R} P(\mathbf{x}_1, \mathbf{x}_2, t) \, dy_1 \, dz_1 \, dy_2 \, dz_2, \qquad (7.59)$$

where $B_R = \{[y, z] \mid y^2 + z^2 < R^2\}$. To estimate $P_s(x_1, x_2)$ from the Brownian dynamics simulation, we run the simulation for a long time (10^5 nanoseconds) and record the position of both ions every time step ($\Delta t = 10^{-3}$ nanoseconds). The two-dimensional histogram of recorded positions is shown in Figure 7.5(b). We can clearly see that ions do not like to be at the same position, i.e. the probability is significantly lower along the diagonal $x_1 = x_2$. This is a consequence of the Coulomb interaction term in (7.56)–(7.57). Running the same simulation with $a_2 = 0$ and all other parameters given by (7.58), we can easily "switch off" the interaction term. The probability distribution function $P_s(x_1, x_2)$ for $a_2 = 0$ is plotted in Figure 7.5(c). As expected, it is more or less uniform. Thus we confirm that the correlation between the ion positions is solely due to the Coulomb term.

It is worth noting that $P_s(x_1, x_2)$ (for $a_2 \neq 0$) is also lower close to the corners $[0, L]$ and $[L, 0]$, but this is an unphysical consequence of our choice of the periodic boundary condition at boundaries B_2 and B_3. Better boundary behaviour could be obtained by coupling our model of an ion channel with a suitable description of ions in the exterior of the channel. Since there are many more ions outside the channel, there is a potential to use a coarser modelling approach there. We will discuss such multi-resolution approaches in Chapter 9, where we also present an illustrative example showing that Brownian dynamics can be coupled with a compartment-based stochastic reaction–diffusion approach to improve the efficiency of simulations. Such a multi-resolution approach is necessary for extending Brownian dynamics simulations of ions to larger parts of intracellular space; for applications to intracellular calcium dynamics, see Dobramysl et al. (2016).

The Brownian dynamics simulation presented above can be equivalently described in terms of a PDE for the probability density function $P(\mathbf{x}_1, \mathbf{x}_2, t)$ which is a function of seven variables. If the ion channel contained N ions, then the probability density function would be a function of $3N + 1$ variables. If we considered more realistic boundary conditions at boundaries B_2 and B_3, then the number of ions N would be itself a variable, which would further complicate the description of the ion channel in terms of probability distribution functions. Although it is possible to get a comparable plot to Figure 7.5(b) by solving a PDE for $P(\mathbf{x}_1, \mathbf{x}_2, t)$, this becomes very difficult for more complicated ion channel models (Chen et al., 2014).

Another challenging problem is to derive a simplified description of the system (e.g. a PDE) in terms of the concentration of ions $c(\mathbf{x}, t)$ which describes the number density of ions at position $\mathbf{x}$ and time t, and which is only a function of four variables (Bruna and Chapman, 2014). For discussion of different approaches to modelling ion channels, see for example Corry et al. (2000) and Roux et al. (2004). The description of an ion channel as a cylinder shown in Figure 7.5(a) is also highly simplified. In reality, the width of an ion channel varies along the membrane, and ions hop from one binding site to the other in the narrowest part of an ion channel (Berneche and Roux, 2003); this level of detail can be better captured by molecular dynamics approaches, which we will discuss in Section 8.6.

7.6 Metropolis–Hastings Algorithm

In the previous section, we presented a model of two interacting particles (ions). To compute the stationary distribution in Figure 7.5, we chose the finite time step $\Delta t = 10^{-12}$ sec and ran a long-time simulation of the model (7.56)–(7.57). In this section, we present a method which is useful for computing stationary distributions (equilibrium properties) of many interacting particles without the need of running long-time dynamical simulations. This is very useful in systems where the simulation time step must be chosen so small that Brownian dynamics (or molecular dynamics) simulations are computationally prohibitive. This method was first used by Metropolis et al. (1953) and his colleagues at Los Alamos to compute equilibrium properties of a system of rigid spheres, but the idea is quite general and it is now called the Metropolis or Metropolis–Hastings algorithm (Chib and Greenberg, 1995). We will explain it on a simple example.

Suppose that we want to generate many points $x^i \in [0, 1]$, $i = 1, 2, 3, \ldots,$ according to a given probability distribution function $Cp(x)$, where $p: [0, 1] \to [0, \infty)$ and

$$C = \left(\int_0^1 p(x) \, dx \right)^{-1} \tag{7.60}$$

is the normalization constant. We will generate these points in an iterative way. To that end, we will make use of an auxiliary function $q \equiv q(x, y): [0, 1] \times [0, 1] \to (0, \infty)$ which satisfies the symmetry assumption

$$q(x, y) = q(y, x), \qquad \text{for all } x \in [0, 1], \ y \in [0, 1], \tag{7.61}$$

and the normalization condition

$$\int_0^1 q(x, y) \, dy = 1, \qquad \text{for all } x \in [0, 1]. \tag{7.62}$$

The normalization condition guarantees that $q(x, \cdot)$ is, for every $x \in [0, 1]$, a probability distribution function. There are many functions q that satisfy the symmetry, normalization and positivity ($q > 0$) assumptions. For example, we could choose q to be constant and equal to 1; then $q(x, \cdot)$ would be the probability distribution function of uniformly distributed random numbers in $[0, 1]$. In particular, we assume that it is very easy to sample random numbers according to $q(x, \cdot)$.

The Metropolis algorithm computes the sequence x^i, $i = 1, 2, 3, \ldots,$ iteratively as follows. Given x^i, we first generate a candidate y for x^{i+1} according to the distribution $q(x^i, \cdot)$. Then we generate a random number r uniformly distributed in $(0, 1)$. If

$$r < \frac{p(y)}{p(x^i)}, \tag{7.63}$$

then[*] we put $x^{i+1} = y$. Otherwise, we put $x^{i+1} = x^i$. The resulting sequence x^1, x^2, x^3, x^4, … contains random numbers sampled according to the probability

---

[*] Of course, if $p(y)/p(x^i) \geq 1$ then there is no need to generate a random number, since the condition is always satisfied.

distribution $Cp(x)$, provided that x^1 is sampled from this distribution. This can be shown by mathematical induction as follows.

Let i be a positive integer and let us assume that x^i is distributed according to the probability distribution $Cp(x)$. Let $P[x^{i+1} \in (a,b)]$ be the probability that x^{i+1} is in the interval $(a,b) \subset [0,1]$ where $0 < a < b < 1$. There are two possible ways for x^{i+1} to be inside the interval (a,b): either the candidate y satisfied $y \in (a,b)$ and it was accepted by (7.63), or $x^i \in (a,b)$ and the candidate y was rejected by (7.63). Consequently, $P[x^{i+1} \in (a,b)]$ can be computed as follows:

$$P[x^{i+1} \in (a,b)] = \int_0^1 Cp(x^i) \int_a^b q(x^i,y) \min\left\{1, \frac{p(y)}{p(x^i)}\right\} dy\, dx^i$$
$$+ \int_a^b Cp(x^i) \int_0^1 q(x^i,y)\left(1 - \min\left\{1, \frac{p(y)}{p(x^i)}\right\}\right) dy\, dx^i.$$

Using (7.62), we get

$$P[x^{i+1} \in (a,b)] = \int_a^b Cp(x^i)\, dx^i$$
$$+ \int_0^1 Cp(x^i) \int_a^b q(x^i,y) \min\left\{1, \frac{p(y)}{p(x^i)}\right\} dy\, dx^i$$
$$- \int_a^b Cp(x^i) \int_0^1 q(x^i,y) \min\left\{1, \frac{p(y)}{p(x^i)}\right\} dy\, dx^i.$$

We write x^i instead of y and y instead of x^i in the second integral on the right-hand side and change the order of integration. We get

$$P[x^{i+1} \in (a,b)] = \int_a^b Cp(x^i)\, dx^i$$
$$+ \int_a^b \int_0^1 Cp(y)\, q(y,x^i) \min\left\{1, \frac{p(x^i)}{p(y)}\right\} dy\, dx^i$$
$$- \int_a^b \int_0^1 Cp(x^i)\, q(x^i,y) \min\left\{1, \frac{p(y)}{p(x^i)}\right\} dy\, dx^i.$$

We use (7.61) in the second term on the right-hand side to obtain

$$P[x^{i+1} \in (a,b)] = \int_a^b Cp(x^i)\, dx^i \tag{7.64}$$
$$+ \int_a^b \int_0^1 Cq(x^i,y)\left(p(y) \min\left\{1, \frac{p(x^i)}{p(y)}\right\}\right.$$
$$\left. - p(x^i) \min\left\{1, \frac{p(y)}{p(x^i)}\right\}\right) dy\, dx^i.$$

Considering the cases $p(x^i) > p(y)$ and $p(x^i) \leq p(y)$ separately, we deduce that

$$p(y) \min\left\{1, \frac{p(x^i)}{p(y)}\right\} - p(x^i) \min\left\{1, \frac{p(y)}{p(x^i)}\right\} = 0.$$

Consequently, the second term on the right-hand side of (7.64) vanishes and we get

$$P[x^{i+1} \in (a, b)] = \int_a^b C p(x^i)\,dx^i.$$

Thus, we deduce that x^{i+1} is distributed according to the probability distribution $Cp(x)$, provided that x^i was distributed according to this distribution. This concludes the proof.

The Metropolis algorithm gives us a sequence x^1, x^2, x^3, x^4, ..., which contains random numbers sampled according to the probability distribution $Cp(x)$, provided that x^1 is sampled from this distribution. In practice, if there is no easy way to determine x^1, we can choose it relatively arbitrarily, run the algorithm and disregard the beginning of the computed sequence. We will show how this algorithm works on the example presented in Section 7.1. In this case, the time evolution of the diffusion–advection process was described by the Fokker–Planck equation (7.9). After non-dimensionalization, the corresponding stationary distribution is given by (compare with (3.36))

$$p_s(x) = C \exp\left[\frac{S(x)}{D}\right], \tag{7.65}$$

where C is the normalization constant and the signal function $S(x)$ is a piecewise linear function which has its maximum in the middle of the interval $[0, L]$; see Figure 7.1(a). Since the particles are non-interacting in the example presented in Section 7.1, the computation of the stationary distribution is equivalent to sampling from the one-dimensional distribution (7.65). In particular, the Metropolis algorithm could be applied in the way we introduced above. However, the real strength of the Metropolis–Hastings algorithm is not in sampling points from the one-dimensional distribution $Cp(x)$, but in computing equilibrium properties of high-dimensional systems (many interacting particles). For example, considering a large collection of interacting particles (e.g. molecules), the probability distribution is often in the Boltzmann form $C \exp(-E/k_bT)$, where C is the normalization constant, k_B is the Boltzmann constant, T is the absolute temperature and E is the energy, which depends on the position of all the particles. Therefore we will present the algorithm in a way which can be used for the model that we studied in Section 7.1 with $N = 1000$ particles, even if we introduced interactions between them.

We consider a system of N particles. The Metropolis algorithm will iteratively compute a sequence of configurations of the system, i.e. it will compute vectors $\mathbf{x}^i = [x_1^i, x_2^i, \ldots, x_N^i]$ where x_k^i is the position of the kth particle, $k = 1, 2, \ldots, N$, at the ith iteration of the algorithm. In order to run this algorithm, we have to specify the initial configuration $\mathbf{x}^1$. Since we do not assume any prior knowledge about the system, we will distribute particles uniformly in $[0, L]$. At the ith iteration, we will create a proposed configuration $\mathbf{y}$ by randomly choosing one particle and displacing it by a small distance (controlled by parameter $h > 0$ below). We define the potential $\Phi \colon [0, L]^N \to \mathbb{R}$ by

$$\Phi(x_1, x_2, \ldots, x_N) = -\sum_{i=1}^{N} \frac{S(x_i)}{D}. \tag{7.66}$$

We extend Φ to $\mathbb{R}^N$ by postulating that $\Phi(\mathbf{x}) = \infty$ for $\mathbf{x} \notin [0, L]^N$. Then the following four steps will be performed at the ith iteration of the algorithm:

(a20) Choose randomly one particle, i.e. choose j uniformly distributed in the set of indices $\{1, 2, 3, \ldots, N\}$.

(b20) Generate a random number r_1 uniformly distributed in the interval $(0, 1)$ and compute the proposed system configuration $\mathbf{y}$ by
$y_j = x_j^i + h(r_1 - 0.5)$ and $y_k = x_k^i$ for $k \neq j$.

(c20) Compute $\Delta\Phi = \Phi(\mathbf{y}) - \Phi(\mathbf{x}^i)$.

(d20) Generate a random number r_2 uniformly distributed in the interval $(0, 1)$. If $r_2 < \exp(-\Delta\Phi)$, then put $\mathbf{x}^{i+1} = \mathbf{y}$.
Otherwise, put $\mathbf{x}^{i+1} = \mathbf{x}^i$.
Then continue with step (a20) to compute the next configuration.

To justify this algorithm, let us consider the case $N = 1$ for simplicity. Then the step (a20) can be omitted because we always choose the first particle. Moreover, the system configuration is no longer a vector. Using (7.66) and $N = 1$, we obtain that $\Phi \colon \mathbb{R} \to \mathbb{R}$ is given by

$$\Phi(x) = -\frac{S(x)}{D}, \quad \text{for } x \in [0, L], \tag{7.67}$$

and $\Phi(x) = \infty$ for $x \notin [0, L]$. In particular, all moves outside of $[0, L]$ will be rejected in the step (d20). The SSA (b20)–(d20) (for $N = 1$) will compute the sequence of numbers $x_1^1, x_1^2, x_1^3, \ldots$ in $[0, L]$. Using (7.65) and (7.67), we get $p_s(x) = C \exp[-\Phi(x)]$. In particular, steps (c20)–(d20) are equivalent to the acceptance criterion (7.63). Step (b20) defines the function $q(x, y)$ which satisfies both the symmetry and normalization conditions (7.61)–(7.62), where the domain $[0, 1]$ is replaced by $\mathbb{R}$. If q also satisfied the positivity condition

($q > 0$), then we would be sure that the algorithm converges to the stationary distribution. However, if h in step (b20) is small, then there exist configurations that cannot be proposed for the given value of x^i in step (b20). In particular, $q(x^i, y)$ is non-negative, but it is not strictly positive for all values of x^i and y. However, this will not cause any problem in this case and the algorithm will still converge, provided that h is non-zero. In fact, we have not explicitly used the positivity condition in the above derivation of the Metropolis algorithm. If x^i is sampled from the distribution $p_s(x) = C \exp[-\Phi(x)]$, then the symmetry and normalization conditions (7.61)–(7.62) guarantee that x^{i+1} will also be sampled from this distribution. In particular, the positivity of q is not necessary provided that every configuration y can be reached by the algorithm from the given value of x^i after a finite number of iterations.

In Figure 7.6(a), we present results computed by the algorithm (a20)–(d20) for different numbers of iterations I. In each case, we start with $N = 1000$ particles which are uniformly distributed in $[0, L]$. We run the algorithm for $I/10$ iterations. Then we run it for I iterations and we compute the average distribution of the resulting I configurations. We see that the results converge to the exact distribution given by (7.65). Results computed for a smaller number of iterations are biased towards the initial condition. In Figure 7.6(b), we compare results computed for two different potentials, potential (7.66) and

$$\Phi(x_1, x_2, \ldots, x_N) = -\sum_{i=1}^{N} \frac{S(x_i)}{D} + \sum_{i,j=1; i \neq j}^{N} \frac{a}{|x_i - x_j|}, \qquad (7.68)$$

(a)   (b)

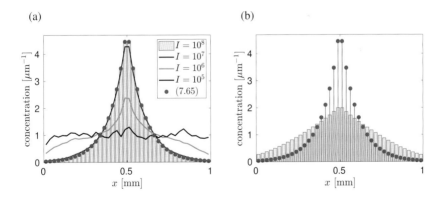

Figure 7.6 (a) *The results computed by the SSA (a20)–(d20) for different numbers of iterations I. Red dots (in both panels) correspond to the exact solution (7.65) which can be derived for the potential (7.66). (b) The results computed by the SSA (a20)–(d20) for potential (7.68) (grey histogram) compared with the exact result (7.65) for potential (7.66) (red dots).*

where $a > 0$. The second potential shows the strength of this algorithm, because it considers interactions between particles – particles are repelling each other. We use dimensionless parameters $D = a = 10^{-4}$, $L = 1$ and $h = 0.02$ in our simulations. The signal profile is given in Figure 7.1(a), which is non-dimensionalized by taking the unit of length equal to a millimetre. To enable direct comparison with Figure 7.1(b), we present Figure 7.6 using physical units.

Note that although the Metropolis–Hastings algorithm will generate a sequence of configurations which correctly samples the equilibrium distribution, this sequence is not related to the time evolution of the corresponding dynamical system. If we are interested in the dynamics, we need to use the methods discussed in Section 7.1. For example, Figure 7.1(b) presents the distribution of particles after time $t = 2$ min provided that the initial distribution is uniform. Comparing Figures 7.1(b) and 7.6(a), we observe that the maximum concentration in the middle of the domain is less than 3 μm^{-1} at time $t = 2$ min, but it is above 4 μm^{-1} at equilibrium. In particular, we would need to run our dynamical simulation in Section 7.1 for a longer time, if we wanted to calculate the equilibrium properties using Brownian dynamics or compartment-based SSAs.

The Metropolis algorithm can be further generalized. For example, the symmetry assumption (7.61) can be relaxed by considering a slightly more complicated acceptance criterion than the one given as (7.63). Finally, we note that the normalization constant (7.60) does not have to be known, because the algorithm does not make use of it, which is helpful in many applications. Although we discussed it here using a Brownian dynamics example, it can also provide an alternative approach to sample equilibrium properties of models written in terms of molecular dynamics, which we will discuss in the next chapter.

Exercises

7.1   Derive the equation (7.4) for $p(x, y, z, t)$.

7.2   Derive the equations (7.11)–(7.12) for the time evolution of mean vector $\mathbf{M}(t)$ of the diffusion–advection compartment-based model.

Hint: Use similar arguments as in the derivation of equations (4.24)–(4.25) and (1.15).

7.3   Derive the boundary conditions (7.15).

7.4 Derive the PDE (7.20) for the concentration $a(x, t)$ of the molecules of A.

7.5 Derive the PDE (7.26) for the density of bacteria.

Hint: Use similar arguments as in the derivation of (4.46) from (4.40)–(4.41).

7.6 Simulate the diffusion–advection example from Section 7.1 using the velocity-jump SSA (a10)–(c10) where the turning frequency λ is given by (7.23). Recompute the results of Figure 7.1.

7.7 Implement the reaction–diffusion–advection example from Section 7.2 as the velocity-jump process where the turning frequency λ is given by (7.23). Recompute the results from Figure 7.2.

7.8 Derive formula (7.34) by solving the ODE (7.31) in the interval $[t, t + \Delta t]$ with initial conditions $X(t)$, $V(t)$ and $Y_2(t)$.

7.9 Calculate the normalization constant C in the expression (7.45) for $p_s(u)$ and derive (7.46).

Hint: Integrate $p_s(u)$ in (7.45) in intervals $(-\infty, 0)$ and $(0, \infty)$ and add the results together to deduce

$$C = \left(\int_{\mathbb{R}} \exp\left[\frac{N}{\eta^2} \left(2|u| - u^2 \right) \right] du \right)^{-1} = \frac{\sqrt{N}}{\eta \sqrt{\pi}} \exp\left[-\frac{N}{\eta^2} \right] \left(1 + \mathrm{erf}\left(\frac{\sqrt{N}}{\eta} \right) \right)^{-1}.$$

7.10 Consider the model (7.40)–(7.41) from Section 7.4. Write a computer code which can simulate this model and investigate how the average time taken for the group to change its direction depends on the interaction radius $R \in (0, 2)$. Plot the average switching time as a function of R for $N = 10$ and $N = 20$. Plot the stationary distribution of the (total) average velocity given by (7.42) for $R = 0.5$ and $N = 20$. Compare with Figure 7.4(c) which was computed for $R = 2$.

7.11 Consider N animals following a velocity-jump process on the unit circle (7.37). Each animal moves either clockwise or anticlockwise with unit angular speed. Let $\omega_i(t) = \pm 1$, $i = 1, 2, \ldots, N$, be the angular velocity of the ith animal, where $+1$ is for the anticlockwise rotation and -1 is for the clockwise rotation. Let us define the average angular velocity of the whole group by (7.42), i.e. $U(t) \in [-1, 1]$. We will assume that the

ith animal changes its direction according to the Poisson process with the turning frequency

$$\lambda \equiv \lambda(\omega_i(t), U(t)) = 1 + 10\,(1 - \omega_i(t)\,U(t))^2.$$

(a) Let $N = 1$, i.e. we only consider one animal in this part of this exercise. Let $p^-(\theta, t)$ (resp. $p^+(\theta, t)$) be the probability that this animal moves clockwise (resp. anticlockwise). Write a system of partial differential equations which describe the time evolution of p^+ and p^-. Write a partial differential equation satisfied by $n = p^+ + p^-$.

(b) Let $N \geq 1$. Let $A(t)$ be the number of animals that move anticlockwise. Let $p(a, t)$ be the probability that $A(t) = a$. Write the master equation for $p(a, t)$.

(c) Let $N \gg 1$. Show that the system has two favourable states: either the group (most of the animals) moves clockwise or the group moves anticlockwise. What is the average time taken for the group to change its direction of movement?

7.12 Two particles move along the x-axis according to the velocity-jump process

$$X_1(t + \mathrm{d}t) = X_1(t) + V_1(t)\,\mathrm{d}t,$$
$$X_2(t + \mathrm{d}t) = X_2(t) + V_2(t)\,\mathrm{d}t,$$
$$V_1(t + \mathrm{d}t) = V_1(t) - \beta(V_1(t) - V_2(t))\,\mathrm{d}t + \beta\,\sqrt{2D}\,\mathrm{d}W_1,$$
$$V_2(t + \mathrm{d}t) = V_2(t) - \beta(V_2(t) - V_1(t))\,\mathrm{d}t + \beta\,\sqrt{2D}\,\mathrm{d}W_2,$$

where $X_i(t)$ and $V_i(t)$ are, respectively, the position and velocity of particle i, for $i = 1, 2$, and $\beta > 0$ and $D > 0$ are constants. As usual, $\mathrm{d}W_1$ and $\mathrm{d}W_2$ are independent noise terms. The initial conditions are $X_1(0) = 0$, $X_2(0) = 0$, $V_1(0) = v_0$ and $V_2(0) = -v_0$.

(a) Derive expressions for the average positions of the particles as a function of t and v_0, and the parameters D and β.

(b) Derive a closed system of seven ODEs that could be solved to give the average mean-square displacements of the particles, $\langle X_1^2(t) \rangle$ and $\langle X_2^2(t) \rangle$.

7.13 The Metropolis–Hastings algorithm was explained in Section 7.6 using a continuous probability distribution $Cp(x)$ defined for $x \in [0, 1]$. In this exercise, we will consider a discrete random variable which can only assume a countably infinite number of values. We will design a Metropolis–Hastings algorithm for the Poisson distribution which has naturally appeared in several places in this book (for example, in Section 1.2 as (1.24)). The Poisson distribution is characterized by the probability mass function (see Appendix)

$$Cp(k) = C \frac{\mu^k}{k!}, \qquad k = 0, 1, 2, \ldots,$$

with parameter $\mu > 0$ and the normalization constant $C = \exp[-\mu]$. Consider a discrete version of $q(x, y)$ given by

$$q_{ij} = \begin{cases} \frac{1}{2} & \text{if } j = i + 1 \text{ or } j = i - 1, \\ 0 & \text{otherwise,} \end{cases} \qquad i = 1, 2, \ldots,$$

and

$$q_{0j} = \begin{cases} \frac{1}{2} & \text{if } j = 1 \text{ or } j = 0, \\ 0 & \text{otherwise,} \end{cases}$$

for $j = 0, 1, 2, \ldots$ (note that q_{ij} satisfies discrete versions of conditions (7.61)–(7.62)).

Design a Metropolis–Hastings algorithm that will sample from $Cp(k)$ with the help of q_{ij}. Write a computer code that will use this algorithm to estimate the mean and variance of the Poisson distribution with $\mu = 2$. Compare your results with the known theoretical values of mean and variance.

8

Microscopic Models of Brownian Motion

In Section 3.2, we presented the SDE-based model for diffusion given by equations (3.15)–(3.17). They effectively postulate that the position of the diffusing molecule at time $t + \Delta t$ can be computed from its position at time t by adding a normally distributed random displacement. This model is further studied in Chapter 4, where it is shown to be a good limiting description of certain velocity-jump processes. In Section 4.6, velocity-jump process (4.90)–(4.91) is interpreted (in the formal limit $\Delta t \to 0$) as Newton's second law of motion (4.96) for the position $X(t)$ and velocity $V(t)$ of a particle with mass m_p as

$$\frac{dX}{dt}(t) = V(t), \tag{8.1}$$

$$m_p \frac{dV}{dt}(t) = F(t), \tag{8.2}$$

where the force on the particle, $F(t)$, is in equation (4.96) expressed as the sum of the frictional force $F_f(t) = -\gamma V(t)$ and a random force, namely as

$$F(t) = -\gamma V(t) + \gamma \sqrt{2D} \, \frac{\text{``}dW\text{''}}{dt}. \tag{8.3}$$

Here γ is the coefficient that relates the frictional drag force $F_f(t)$ to the particle velocity $V(t)$ and the random force is in quotes because derivatives of W are strictly speaking undefined (they are infinitely large). A precise mathematical meaning of equations (8.1)–(8.2) with force term (8.3) can only be given in terms of the corresponding SDE description (4.90)–(4.91). In the statistical physics literature, this description of the diffusing particle is often named after Langevin and we will call it the *Langevin equation* in this chapter. We have not used the Langevin terminology in previous chapters to avoid confusion with the chemical Langevin equation introduced earlier, which describes chemical reaction networks.

Force term (8.3) in the Langevin equation corresponds to collisions and interactions of the diffusing particle with surrounding (solvent) molecules. A natural question arises: When does equation (8.3) give a good description of interactions with the solvent? When it does not, can we get a more accurate description?

In this chapter, we provide answers to these questions by considering models that explicitly describe solvent molecules. We begin with theoretical models, where solvent molecules can also be called heat bath particles, terminology used in the literature to indicate that one heat bath particle does not have to be a molecule, but may represent a collection of real solvent molecules. In Section 8.1, we use a simple "one-particle" description of the solvent (heat bath) to motivate the generalized Langevin equation and the generalized fluctuation–dissipation theorem. This is followed by analysis of two theoretical (and relatively simple) models in Sections 8.3 and 8.4, which provide a lot of interesting analytical insights. We conclude this chapter with less analytically tractable, but more realistic, computational models, introducing molecular dynamics (molecular mechanics) in Section 8.6 and applying it to the Lennard-Jones fluid and to simulations of ions in aquatic solutions.

8.1 One-Particle Solvent Model

In this chapter, we study detailed models of Brownian motion, which explicitly describe the dynamics of solvent molecules. We are interested in the behaviour of a diffusing particle (molecule) of mass m_p that interacts with N solvent molecules. To get some initial insights into this problem, we restrict to a one-dimensional model where the position $X(t)$ and velocity $V(t)$ of the diffusing particle evolve according to (8.1)–(8.2). We denote the positions and velocities of the solvent molecules by the lower-case letters $x_i(t)$ and $v_i(t)$, respectively, where $i = 1, 2, \ldots, N$. In this section, we further simplify the model by assuming that $N = 1$. We drop the subscript $i = 1$ and denote the position and velocity of the single "solvent molecule" as $x(t) \equiv x_1(t)$ and $v(t) \equiv v_1(t)$, respectively. Force term $F(t)$ is no longer given by (8.3) but by interaction with the "solvent molecule". We assume that this interaction is given as a harmonic spring with spring constant k, i.e. $F(t) = k\,(x(t) - X(t))$. Then equations (8.1)–(8.2) read as follows:

$$\frac{dX}{dt}(t) = V(t), \tag{8.4}$$

$$m_p \frac{dV}{dt}(t) = k\,(x(t) - X(t)). \tag{8.5}$$

Equations (8.4)–(8.5) are coupled with the evolution equations for the "solvent molecule" of mass m_s, which will again be given through Newton's second law of motion (compare with (4.96) and (8.3)) by

$$\frac{dx}{dt}(t) = v(t), \tag{8.6}$$

$$m_s \frac{dv}{dt}(t) = k\left(X(t) - x(t)\right) - \gamma_s v(t) + \gamma_s \sqrt{2D_s} \; \frac{\text{"}dW\text{"}}{dt}, \tag{8.7}$$

where the three terms on the right-hand side of (8.7) represent three contributions to the force on the "solvent molecule": a spring force with spring constant k, a friction force with friction coefficient γ_s, and a random force, written in the same formal way as in (8.3). In what follows, we refer to model (8.4)–(8.7) as the *one-particle solvent model*.

The one-particle solvent model (8.4)–(8.7) is formally a system of four equations for four variables $X(t)$, $V(t)$, $x(t)$ and $v(t)$. While the first three equations (8.4)–(8.6) have clear physical interpretation and are mathematically well defined, we have to be more careful with the last equation (8.7). From the physical point of view, the friction and random terms in (8.7) implicitly take into account that there are many other solvent molecules in the system which interact with our one explicitly described "solvent molecule". We have already seen that the diffusion coeficient D_s in (8.7) is related to temperature through the Einstein–Smoluchowski relation (4.95). Consequently, the friction and random terms in (8.7) can be viewed as a "stochastic thermostat", keeping the simulated system at the right temperature. We will discuss the temperature control in more detail in Section 8.6 when we introduce molecular dynamics.

One could argue that other solvent molecules that implicitly interact with the "solvent molecule" at position $x(t)$ would also interact with the diffusing particle at position $X(t)$, i.e., the friction and noise terms should not only be in equation (8.7), but also on the right-hand side of equation (8.5) for the diffusing particle. However, the goal of this chapter is to understand what form the interaction between the diffusing molecule and surrounding solvent can take. We therefore only include physical forces exerted by explicitly modelled solvent molecules on the right-hand side of equation (8.2). Thus, in the case of the one-particle solvent model, we only include the single spring term on the right-hand side of (8.5).

The position and velocity of the diffusing particle, $X(t)$ and $V(t)$, are the important variables of the model, while the solvent position and velocity, $x(t)$ and $v(t)$, are two additional auxiliary variables. From the mathematical point of view, we do not have to write the two auxiliary variables as $x(t)$ and $v(t)$. In Exercise 8.1, you are asked to transform the one-particle solvent model (8.4)–(8.7) into an equivalent model written in terms of four variables $X(t)$, $V(t)$, $U(t)$ and $Z(t)$, where

$$U(t) = \frac{k\,(x(t) - X(t))}{m_p} \quad \text{and} \quad Z(t) = \frac{k\,v(t)}{m_p}. \tag{8.8}$$

To be more precise, we interpret equation (8.7) as an SDE, using our computational definition. Then the one-particle solvent model (8.4)–(8.7) is given by

$$X(t + dt) = X(t) + V(t)\,dt, \tag{8.9}$$

$$V(t + dt) = V(t) + U(t)\,dt, \tag{8.10}$$

$$U(t + dt) = U(t) + (Z(t) - \alpha_1\,V(t))\,dt, \tag{8.11}$$

$$Z(t + dt) = Z(t) - \alpha_4\,(Z(t) + \alpha_2\,U(t))\,dt + \alpha_3\alpha_4\,dW, \tag{8.12}$$

where U is the acceleration of the diffusing particle, and α_j, $j = 1, 2, 3, 4$, are positive parameters given by

$$\alpha_1 = \frac{k}{m_p}, \quad \alpha_2 = \frac{k}{\gamma_s}, \quad \alpha_3 = \frac{k\,\sqrt{2D_s}}{m_p} \quad \text{and} \quad \alpha_4 = \frac{\gamma_s}{m_s}. \tag{8.13}$$

To analyse the one-particle solvent model (8.9)–(8.12), we first consider its behaviour in the limit $\alpha_4 \to \infty$. Dividing equation (8.12) by α_4 and passing to the limit $\alpha_4 \to \infty$, we get

$$Z(t)\,dt \approx -\alpha_2\,U(t)\,dt + \alpha_3\,dW.$$

Substituting for Z in equation (8.11), we reduce the one-particle solvent model (8.9)–(8.12) into a system of three equations for the position, velocity and acceleration of the diffusing particle:

$$X(t + dt) = X(t) + V(t)\,dt, \tag{8.14}$$

$$V(t + dt) = V(t) + U(t)\,dt, \tag{8.15}$$

$$U(t + dt) = U(t) - (\alpha_1\,V(t) + \alpha_2\,U(t))\,dt + \alpha_3\,dW. \tag{8.16}$$

Our next goal is to derive a system of equations for the important variables X and V. This will be given in the form of the so-called *generalized Langevin equation*. We first use the computational definition (3.5) to interpret equation (8.16) as

$$U(t + \Delta t) = (1 - \alpha_2\,\Delta t)\,U(t) - \alpha_1\,V(t)\,\Delta t + \alpha_3\,\sqrt{\Delta t}\,\xi. \tag{8.17}$$

In a computer simulation, we use equation (8.17) iteratively at times $t = 0, \Delta t, 2\Delta t, \ldots$, namely we use the following k equations to compute the value of U at time $k\Delta t$, for $k = 1, 2, 3, \ldots$:

$$U(\Delta t) = (1 - \alpha_2 \, \Delta t) \, U(0) - \alpha_1 \, V(0) \, \Delta t + \alpha_3 \, \sqrt{\Delta t} \, \xi_0, \tag{8.18}$$

$$U(2\Delta t) = (1 - \alpha_2 \, \Delta t) \, U(\Delta t) - \alpha_1 \, V(\Delta t) \, \Delta t + \alpha_3 \, \sqrt{\Delta t} \, \xi_1, \tag{8.19}$$

$$U(3\Delta t) = (1 - \alpha_2 \, \Delta t) \, U(2\Delta t) - \alpha_1 \, V(2\Delta t) \, \Delta t + \alpha_3 \, \sqrt{\Delta t} \, \xi_2, \tag{8.20}$$

$$\vdots$$

$$U(k\Delta t) = (1 - \alpha_2 \, \Delta t) \, U((k-1)\Delta t) - \alpha_1 \, V((k-1)\Delta t) \, \Delta t + \alpha_3 \, \sqrt{\Delta t} \, \xi_{k-1}, \tag{8.21}$$

where $\xi_1, \xi_2, \xi_3, \ldots, \xi_{k-1}$ are random numbers which are independently sampled from the normal distribution with zero mean and unit variance. Substituting equation (8.18) for $U(\Delta t)$ in equation (8.19), then equation (8.19) for $U(2\Delta t)$ in equation (8.20), and continuing until (8.21), we can solve these equations for $U(k\Delta t)$, $k = 1, 2, 3, \ldots$, in the following form (see Exercise 8.2):

$$U(k\Delta t) = (1 - \alpha_2 \, \Delta t)^k \, U(0) - \alpha_1 \, \Delta t \sum_{i=1}^{k} (1 - \alpha_2 \, \Delta t)^{i-1} \, V((k-i)\Delta t)$$

$$+ \alpha_3 \, \sqrt{\Delta t} \sum_{i=1}^{k} (1 - \alpha_2 \, \Delta t)^{i-1} \, \xi_{k-i}. \tag{8.22}$$

Equation (8.22) depends on the initial acceleration, $U(0)$. Since $(1 - \alpha_2 \, \Delta t) < 1$, the first term on the right-hand side will become smaller and smaller as we increase k. Put another way, as time progresses, the initial value of the acceleration will be forgotten by the system. Therefore, considering sufficiently long times $t = k\Delta t$, we neglect the first term on the right-hand side of (8.22). The last term on the right-hand side is the sum of independent normally distributed random variables, i.e. it is also normally distributed with variance equal to the sum of the individual variances, which can be approximated by

$$\alpha_3^2 \, \Delta t \sum_{i=1}^{k} (1 - \alpha_2 \, \Delta t)^{2i-2} = \alpha_3^2 \, \Delta t \, \frac{(1 - \alpha_2 \, \Delta t)^{2k} - 1}{(1 - \alpha_2 \, \Delta t)^2 - 1} \approx \frac{\alpha_3^2 \, \Delta t}{1 - (1 - \alpha_2 \, \Delta t)^2} \approx \frac{\alpha_3^2}{2\alpha_2},$$

where, in the first approximation, we have again used the fact that k is sufficiently large so that we can neglect $(1 - \alpha_2 \, \Delta t)^{2k}$, while in the second approximation, we used the fact that Δt is small (so that $\alpha_2 \, \Delta t \ll 1$). Consequently, equation (8.22) can be simplified as

$$U(t) = -\alpha_1 \, \Delta t \sum_{i=1}^{k} (1 - \alpha_2 \, \Delta t)^{i-1} \, V((k-i)\Delta t) + \frac{\alpha_3}{\sqrt{2\alpha_2}} \, \xi(t), \tag{8.23}$$

where $t = k\Delta t$ is (sufficiently large) time and $\xi(t)$ is a random number which is sampled from the normal distribution with zero mean and unit variance.

The summation on the right-hand side can be further approximated by an integral. Using approximation $(1 - \alpha_2 \Delta t)^i \approx \exp(-i\alpha_2 \Delta t)$, which is valid for sufficiently small Δt (so that $(i\alpha_2 \Delta t)^2 \ll 1$), we obtain

$$\Delta t \sum_{i=1}^{k} (1 - \alpha_2 \Delta t)^{i-1} V((k-i)\Delta t) \approx \sum_{i=1}^{k} \exp(-\alpha_2 (i-1)\Delta t) V((k-i)\Delta t) \Delta t$$

$$\approx \int_0^t \exp(-\alpha_2 \tau) V(t - \tau) \, d\tau, \qquad (8.24)$$

where the last approximation is valid in the limit $\Delta t \to 0$. Substituting into (8.23), we obtain

$$U(t) = -\alpha_1 \int_0^t \exp(-\alpha_2 \tau) V(t - \tau) \, d\tau + \frac{\alpha_3}{\sqrt{2\alpha_2}} \xi(t). \qquad (8.25)$$

We define the friction kernel $\kappa(\tau)$ for $\tau \geq 0$ and noise term $R(t)$ for $t \geq 0$ by

$$\kappa(\tau) = \alpha_1 \exp(-\alpha_2 \tau), \qquad (8.26)$$

$$R(t) = \frac{\alpha_3}{\sqrt{2\alpha_2}} \xi(t). \qquad (8.27)$$

Then, substituting (8.25) into (8.14)–(8.15), we obtain the generalized Langevin equation in the form

$$\frac{dX}{dt} = V(t), \qquad (8.28)$$

$$\frac{dV}{dt} = -\int_0^t \kappa(\tau) V(t - \tau) \, d\tau + R(t), \qquad (8.29)$$

which is written in the form of Newton's second law of motion; see equations (8.1)–(8.2). However, unlike in the case of the standard Langevin equation with force term (8.3), we have no mathematical issues in writing the generalized Langevin equation as a system of differential equations, because the noise term $R(t)$ is a continuous function. It depends on the noise terms at previous times, which we can formally quantify by computing its autocorrelation function $\langle R(t) R(t - \tau) \rangle = \mathrm{E}[R(t) R(t - \tau)]$, for $\tau > 0$, where the average is taken over all sufficiently large values of t. To evaluate the autocorrelation function, we use the last term in equation (8.22) as the discretized description of the noise term $R(t)$. Considering $t = k\Delta t$ and $\tau = j\Delta t$, we have $\langle R(t) R(t - \tau) \rangle = \langle R(k\Delta t) R((k - j)\Delta t) \rangle$, which implies

$$\langle R(t) R(t - \tau) \rangle = \alpha_3^2 \, \Delta t \left\langle \sum_{i=1}^{k} (1 - \alpha_2 \Delta t)^{i-1} \xi_{k-i} \sum_{\ell=1}^{k-j} (1 - \alpha_2 \Delta t)^{\ell-1} \xi_{k-j-\ell} \right\rangle.$$

Using $E[\xi_i \xi_j] = 0$ for $i \neq j$ and $E[\xi_i^2] = 1$, we have

$$\langle R(t) R(t - \tau) \rangle = \alpha_3^2 \, \Delta t \sum_{\ell=1}^{k-j} (1 - \alpha_2 \, \Delta t)^{j+2\ell-2}.$$

Considering that k can be as large as we want (while j is fixed), we can approximate

$$\langle R(t) R(t - \tau) \rangle \approx \alpha_3^2 \, \Delta t \, (1 - \alpha_2 \, \Delta t)^j \sum_{\ell=0}^{\infty} (1 - \alpha_2 \, \Delta t)^{2\ell} = \frac{\alpha_3^2 \, \Delta t \, (1 - \alpha_2 \, \Delta t)^j}{1 - (1 - \alpha_2 \, \Delta t)^2}.$$

Finally, using that $\alpha_2 \Delta t \ll 1$, we have

$$\langle R(t) R(t - \tau) \rangle \approx \frac{\alpha_3^2 \, \exp(-\alpha_2 \, \tau)}{2 \alpha_2},$$

which can be, using definition (8.26), rewritten as the *generalized fluctuation–dissipation theorem*

$$\langle R(t) R(t - \tau) \rangle = \frac{\alpha_3^2}{2 \alpha_1 \alpha_2} \kappa(\tau). \tag{8.30}$$

Although we have derived the generalized Langevin equation (8.28)–(8.29) and the generalized fluctuation–dissipation theorem (8.30) in a simplified case of the three-equation model (8.14)–(8.16), it is, in fact, a very general description of more complicated solvent models, as we will see in Section 8.3. We leave it as Exercise 8.3 to generalize the results of this section to the full one-particle solvent model expressed in terms of four variables as (8.9)–(8.12). In this case, the generalized Langevin equation (8.28)–(8.29) holds with the following definition of the friction kernel:

$$\kappa(\tau) = \alpha_1 \, \exp\left(-\frac{\alpha_4 \, \tau}{2}\right) \left(\frac{\alpha_4}{\alpha_5} \sinh\left(\frac{\alpha_5 \, \tau}{2}\right) + \cosh\left(\frac{\alpha_5 \, \tau}{2}\right)\right), \tag{8.31}$$

where $\alpha_5 = \sqrt{\alpha_4(\alpha_4 - 4\alpha_2)}$ and noise term $R(t)$ is again Gaussian with zero mean and with an equilibrium autocorrelation function satisfying the generalized fluctuation–dissipation theorem given by (8.30) with $\kappa(\tau)$ given by (8.31). We note that the auxiliary coefficient α_5 is a square root of a real negative number for $\alpha_4 < 4\alpha_2$. However, formula (8.31) is still valid in this case – for $\alpha_4 < 4\alpha_2$ it can be rewritten in terms of sine and cosine functions, taking into account that α_5 is purely imaginary and using identities given in the hint of Exercise 8.3. The friction kernel $\kappa(\tau)$, given by equation (8.31), is visualized in Figure 8.1(a) for different values of parameter α_4. We confirm that friction kernel (8.31) approaches friction kernel (8.26) in the limit $\alpha_4 \to \infty$ (see also Exercise 8.4).

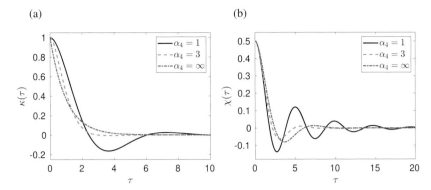

Figure 8.1 (a) *Friction kernels given by* (8.31) *for* $\alpha_4 = 1$ *(black solid line) and* $\alpha_4 = 3$ *(green dashed line). The limiting kernel for* $\alpha_4 = \infty$*, given by* (8.26), *is shown using the red dot-dashed line. Other parameter values are* $\alpha_1 = \alpha_2 = 1$. (b) *Velocity autocorrelation functions* $\chi(\tau)$ *corresponding to each of the three kernels. Results obtained by* (8.56) *or by application of the SSA* (a21)–(c21).

8.2 Generalized Langevin Equation

In the previous Section 8.1 and in the next Section 8.3, the generalized Langevin equation is derived for specific models. In either case, it can be stated as a system of equations (8.28)–(8.29), where the friction kernel is a function $\kappa: [0, \infty) \to \mathbb{R}$ satisfying $\kappa(\tau) \to 0$ as $\tau \to \infty$, and the noise term satisfies the generalized fluctuation–dissipation theorem (compare with (8.30))

$$\langle R(t)\, R(t - \tau) \rangle = \frac{\sigma^2}{\kappa(0)}\, \kappa(\tau), \qquad \text{for} \quad t \geq \tau \geq 0, \tag{8.32}$$

where σ^2 is the variance of the noise term, i.e. $\sigma^2 = \langle R(t)\, R(t) \rangle$. Friction kernels $\kappa(\tau)$ for one-particle solvent models are plotted in Figure 8.1(a). Before we derive friction kernels $\kappa(\tau)$ for more complicated solvent models, we discuss the following two natural questions:

(1) Can we derive Brownian motion from the generalized Langevin equation?
(2) How can we solve the generalized Langevin equation on a computer?

To answer the first question, let us first look at the simple three-equation model (8.14)–(8.16) and choose

$$\alpha_1 = \alpha_2 \beta \qquad \text{and} \qquad \alpha_3 = \alpha_2 \beta \sqrt{2D}, \tag{8.33}$$

where β and D are positive constants. Then equation (8.16) reads as follows:

$$U(t + dt) = U(t) - \alpha_2 \left(\beta\, V(t) + U(t) \right) dt + \alpha_2 \beta \sqrt{2D}\, dW. \tag{8.34}$$

Next, we consider its behaviour in the limit $\alpha_2 \to \infty$. Dividing equation (8.34) by α_2 and passing to the limit $\alpha_2 \to \infty$, we get

$$U(t)\,dt \approx -\beta\,V(t)\,dt + \beta\,\sqrt{2D}\,dW.$$

Substituting for U in equation (8.15), we reduce the three-equation model (8.14)–(8.16) into a system of two equations for the position and velocity of the diffusing particle, namely

$$X(t+dt) = X(t) + V(t)\,dt, \tag{8.35}$$
$$V(t+dt) = V(t) - \beta\,V(t)\,dt + \beta\,\sqrt{2D}\,dW, \tag{8.36}$$

which is the Langevin description (4.49)–(4.50). In Section 4.3, we have already shown that this Langevin description leads to Brownian motion for $\beta \to \infty$. If we substitute the above scaling (8.33) into the corresponding friction kernel (8.26), we can rewrite it as

$$\kappa(\tau) = \beta\,\alpha_2 \exp(-\alpha_2\,\tau). \tag{8.37}$$

In particular, we have

$$\int_0^\infty \kappa(\tau)\,d\tau = \beta \tag{8.38}$$

for any $\alpha_2 > 0$. The limiting friction kernel can then be formally derived by passing to the limit $\alpha_2 \to \infty$ in equation (8.37). Denoting

$$\kappa_\infty(\tau) = \lim_{\alpha_2 \to \infty} \kappa(\tau), \tag{8.39}$$

we observe that the limiting friction kernel satisfies $\kappa_\infty(\tau) = 0$ for $\tau > 0$, with $\kappa_\infty(0)$ equal to infinity, and

$$\int_0^\infty \kappa_\infty(\tau)\,d\tau = \beta. \tag{8.40}$$

Thus the limiting friction kernel is not a function in the usual sense, but what is known as a generalized function or distribution. A basic example of such generalized functions is the Dirac delta function, which is zero everywhere except one point (the origin) where it is equal to infinity, and this infinity is such that the function integrates to one. We would have to introduce the theory of distributions if we wanted to make this definition mathematically precise. Then we would also see that κ_∞ is the Dirac delta function multiplied by 2β, taking into account the integral property (8.40) (the factor of 2 arises because the integral is over $[0, \infty)$ and not $(-\infty, \infty)$). However, as in other places in this book, we avoid technical mathematical details by looking at a discretized problem, which is also relevant to our second question: How can we solve the generalized Langevin equation on a computer?

Considering time step Δt, we rewrite the generalized Langevin equation (8.28)–(8.29) by discretizing the time derivatives as

$$X(t + \Delta t) = X(t) + V(t)\,\Delta t, \tag{8.41}$$

$$V(t + \Delta t) = V(t) - \Delta t \int_0^t \kappa(\tau)\,V(t - \tau)\,d\tau + R(t)\,\Delta t. \tag{8.42}$$

Next, we set $t = k\Delta t$, where k is an integer, and we also discretize the integral on the right-hand side of (8.42). There are many ways to do this. We have already seen one possible discretization of this integral, given by equation (8.24), during the derivation of the generalized Langevin equation for a specific choice of kernel κ. In general, friction kernel $\kappa \colon [0, \infty) \to \mathbb{R}$ satisfies $\kappa(\tau) \to 0$ as $\tau \to \infty$. We therefore assume that there exists an integer value k_m such that $\kappa(\tau)$ is negligible for $\tau > k_m \Delta t$. Assuming that time $t = k\Delta t$ is sufficiently large, so that $k \geq k_m$, we discretize the integral on the right-hand side of (8.42) as follows:

$$\int_0^t \kappa(\tau)\,V(t - \tau)\,d\tau \approx \sum_{i=0}^{k_m} \kappa(i\Delta t)\,V((k - i)\Delta t)\,\Delta t,$$

i.e. we use $k_m + 1$ values of the friction kernel, evaluated at $\kappa(0)$, $\kappa(\Delta t)$, $\kappa(2\Delta t)$, ..., $\kappa(k_m\Delta t)$, to approximate the integral. Substituting into (8.42), we obtain for $t = k\Delta t$, $k = 0, 1, 2, \ldots$,

$$X((k + 1)\Delta t) = X(k\Delta t) + V(k\Delta t)\,\Delta t, \tag{8.43}$$

$$V((k + 1)\Delta t) = V(k\Delta t) - (\Delta t)^2 \sum_{i=0}^{\min\{k_m, k\}} \kappa(i\Delta t)\,V((k - i)\Delta t) + R(k\Delta t)\,\Delta t, \tag{8.44}$$

where the minimum, $\min\{k_m, k\}$, indicates that we use fewer terms to approximate the integral for small values of time $t = k\Delta t < k_m\Delta t$. In order to use equation (8.44) to compute the velocity at time $t + \Delta t$, we also need a method for generating suitable random numbers. In what follows, we assume that the noise is normally distributed with known variance $\sigma^2 = \langle R(t)\,R(t) \rangle$. We want to generate the noise term satisfying the generalized fluctuation–dissipation theorem (8.32). Since we only calculate the state of the system at integer multiples of Δt, the generalized fluctuation–dissipation theorem (8.32) can be rewritten by using $t = k\Delta t$ and $\tau = j\Delta t$ as follows:

$$\langle R(k\Delta t)\,R((k - j)\Delta t) \rangle = \frac{\sigma^2}{\kappa(0)}\,\kappa(j\Delta t), \qquad \text{for} \quad \min\{k_m, k\} \geq j \geq 0. \tag{8.45}$$

In our algorithm, denoted as the SSA (a21)–(c21) below, we generate one normally distributed random number with zero mean and unit variance at each time step. We denote by ξ_k such a number generated in the kth time step. This is similar to the computational definition of the SDE, equation (3.5), used in the SSA (a6)–(b6), where we also generate one normally distributed random number per time step. However, to satisfy (8.45), the noise term $R(t)$ must not only depend on ξ_k, but also on the normally distributed random numbers that are generated in previous time steps, namely on $\xi_{k-1}, \xi_{k-2}, \ldots, \xi_{k-\min\{k_m, k\}}$. We therefore write the noise term at time $t = k\Delta t$ as a linear combination

$$R(k\Delta t) = c_0\, \xi_k + \sum_{i=1}^{\min\{k_m, k\}} c_i\, R((k - i)\Delta t) \tag{8.46}$$

with coefficients c_i yet to be determined. Substituting this form of $R(t)$ into equation (8.45), we deduce (see Exercise 8.5) that coefficients $c_1, c_2, \ldots, c_{\min\{k_m, k\}}$ can be, for each k, obtained as a solution of a suitable system of linear equations. Namely, we define $\ell \times \ell$ matrices A_ℓ, for $\ell = 1, 2, \ldots, k_m$, by

$$A_\ell = \begin{pmatrix} \kappa(0) & \kappa(\Delta t) & \kappa(2\Delta t) & \cdots & \kappa((\ell - 1)\Delta t) \\ \kappa(\Delta t) & \kappa(0) & \kappa(\Delta t) & \cdots & \kappa((\ell - 2)\Delta t) \\ \kappa(2\Delta t) & \kappa(\Delta t) & \kappa(0) & \cdots & \kappa((\ell - 3)\Delta t) \\ \vdots & \vdots & \vdots & \ddots & \vdots \\ \kappa((\ell - 1)\Delta t) & \kappa((\ell - 2)\Delta t) & \kappa((\ell - 3)\Delta t) & \cdots & \kappa(0) \end{pmatrix}. \tag{8.47}$$

Then coefficients $c_1, c_2, \ldots, c_{\min\{k_m, k\}}$ can be obtained for each k as the solution of the following system of linear equations:

$$A_{\min\{k_m, k\}} \begin{pmatrix} c_1 \\ c_2 \\ c_3 \\ \vdots \\ c_{\min\{k_m, k\}} \end{pmatrix} = \begin{pmatrix} \kappa(\Delta t) \\ \kappa(2\Delta t) \\ \kappa(3\Delta t) \\ \vdots \\ \kappa(\min\{k_m, k\}\, \Delta t) \end{pmatrix}. \tag{8.48}$$

In particular, we see that we have to calculate coefficients c_i only during the first k_m time steps. For all subsequent times $t = k\Delta t \geq k_m\Delta t$, we use the same coefficients $c_1, c_2, \ldots, c_{k_m}$, which can be precomputed as solutions of

$$A_{k_m} \begin{pmatrix} c_1 \\ c_2 \\ c_3 \\ \vdots \\ c_{k_m} \end{pmatrix} = \begin{pmatrix} \kappa(\Delta t) \\ \kappa(2\Delta t) \\ \kappa(3\Delta t) \\ \vdots \\ \kappa(k_m\Delta t) \end{pmatrix}. \tag{8.49}$$

Finally, generalized fluctuation–dissipation theorem (8.32) also implies (see Exercise 8.6) that the coefficient c_0 should be chosen as

$$c_0 = \sigma \sqrt{1 - \sum_{j=1}^{\min\{k_m, k\}} \frac{c_j \, \kappa(j\Delta t)}{\kappa(0)}}, \tag{8.50}$$

which can again be precomputed as one single number for $k \geq k_m$. Given initial conditions $X(0)$ and $V(0)$ and setting $k = 0$, the following three steps are performed at the kth iteration of our algorithm (i.e. at time $t = k\Delta t$) for solving the generalized Langevin equation (8.28)–(8.29):

(a21) Generate a normally distributed (with zero mean and unit variance) random number ξ_k.

(b21) Calculate the noise term $R(k\Delta t)$ by (8.46) where coefficients c_i, $i = 1, 2, \ldots, \min\{k_m, k\}$, are given by solving linear system (8.48) and coefficient c_0 is computed by (8.50).

(c21) Compute $X((k + 1)\Delta t)$ and $V((k + 1)\Delta t)$ from the values of position and velocity in previous time steps by (8.43)–(8.44).

Then increase k by 1 and continue with step (a21).

In Figure 8.2(a), we present ten illustrative realizations computed by the SSA (a21)–(c21). We use friction kernel $\kappa(\tau)$ given by equation (8.31) and parameters $\alpha_1 = \alpha_2 = \alpha_3 = \alpha_4 = 1$, $k_m = 1200$ and $\Delta t = 10^{-2}$. This friction kernel is visualized in Figure 8.1(a) as the black solid line and has been previously derived for the original one-particle solvent model (8.9)–(8.12). However, even if we start with the same initial condition $X(0) = V(0) = 0$, we cannot expect that the SSA (a21)–(c21) would produce trajectories that would be comparable to the initial time evolution of the one-particle solvent model (8.9)–(8.12). While the SSA (a21)–(c21) does exactly solve the generalized Langevin equation (8.28)–(8.29), this equation was only derived from the one-particle solvent model for sufficiently large times t. We have neglected terms that decay to zero as $t \to \infty$, but these terms do influence transient dynamics. Thus, in Figure 8.2(a), we plot the time evolution of $X(t) - X(100)$ for $t \geq 100$, which is comparable with the long-term dynamics of the original one-particle solvent model (8.9)–(8.12).

In general, the SSA (a21)–(c21) can be used for arbitrary friction kernel $\kappa(\tau)$. Starting with initial conditions $X(0)$ and $V(0)$ prescribed at a single time point, $t = 0$, initial dynamics of the SSA (a21)–(c21) can be simply viewed as an approach to initialize the history of velocity V, while the subsequent simulation (for times $t \gg k_m\Delta t$) has real physical meaning and is comparable to dynamics of detailed microscopic models, from which the friction kernel can be estimated.

(a) (b)

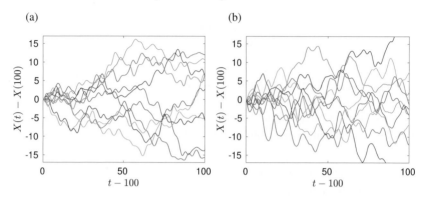

Figure 8.2 (a) *Ten realizations of the SSA (a21)–(c21) for friction kernel* $\kappa(\tau)$ *given by (8.31) for* $\alpha_4 = 1$ *starting with initial conditions* $X(0) = V(0) = 0$. *We plot* $X(t) - X(100)$ *as a function of* $t - 100$, *ignoring initial dynamics. Other parameter values are* $\alpha_1 = \alpha_2 = \alpha_3 = 1$, $k_m = 1200$ *and* $\Delta t = 10^{-2}$. (b) *Ten realizations of the SSA (a21)–(c21) for friction kernel* $\kappa_e(\tau)$ *given by (8.75) for* $\beta = \omega_m = k_B T = m_p = 1$, $k_m = 120$ *and* $\Delta t = 10^{-1}$. *Starting with initial conditions* $X(0) = V(0) = 0$, *we again plot* $X(t) - X(100)$ *as a function of* $t - 100$.

Having presented a method for solving the generalized Langevin equation on a computer, we now return to our first question: Can we derive Brownian motion from the generalized Langevin equation? In Figure 8.2(a), we observe that trajectories $X(t)$ randomly spread from their starting position $X(100)$. In what follows, we identify $X(100)$ with the origin, as we did in Figure 8.2, shifting time by 100. Using (4.94), we can then use the mean-square displacement $\langle X^2 \rangle(t)$ to define the corresponding diffusion coefficient by

$$D = \frac{1}{2} \lim_{t \to \infty} \frac{d\langle X^2 \rangle}{dt}(t), \qquad (8.51)$$

where the limit $t \to \infty$ indicates that we have derived equation (4.94) for sufficiently large times. Formula (8.51) gives us a relatively straightforward way to estimate the diffusion constant from simulations. We calculate $\langle X^2 \rangle(t)$ by averaging over multiple realizations and then estimate the derivative of $\langle X^2 \rangle(t)$ for large times. In this calculation, we implicitly assume that the limit in (8.51) exists, i.e. $\langle X^2 \rangle(t)$ is a linear function of time for large times. If it was not the case, then the process would not describe diffusion. Averaging over 10^6 realizations of the SSA (a21)–(c21), formula (8.51) gives diffusion coefficient $D \approx 0.5$ for the dimensionless parameters used in Figure 8.2(a).

In Figure 8.1(b), we plot the equilibrium velocity autocorrelation function, which is defined as

$$\chi(\tau) = \lim_{t \to \infty} \langle V(t) \, V(t - \tau) \rangle, \tag{8.52}$$

for $\tau \in [0, \infty)$. It can again be estimated from long-time evolution of the SSA (a21)–(c21) by computing how correlated is the current velocity (at time t) with velocity in previous times. In Exercise 8.7, we also show that it can be related to the variance of the noise term, σ^2, and friction kernel $\kappa(\tau)$ through the formula

$$\chi(\tau) = \frac{\sigma^2}{\kappa(0)} \, \mathscr{L}^{-1} \left(\frac{1}{s + \mathscr{L}[\kappa](s)} \right), \tag{8.53}$$

where $\mathscr{L}^{-1}$ denotes Laplace inversion and $\mathscr{L}[\kappa](s)$ denotes the Laplace transform of $\kappa(\tau)$ defined by

$$\mathscr{L}[\kappa](s) = \int_0^\infty \kappa(t) \exp(-st) \, dt. \tag{8.54}$$

For example, substituting friction kernel (8.31) into formula (8.53), we obtain

$$\chi(\tau) = \frac{\sigma^2}{\kappa(0)} \, \mathscr{L}^{-1} \left(\frac{s^2 + \alpha_4 s + \alpha_2 \alpha_4}{s^3 + \alpha_4 s^2 + (\alpha_1 + \alpha_2 \alpha_4)s + \alpha_1 \alpha_4} \right), \tag{8.55}$$

which can be Laplace inverted as follows. The polynomial in the denominator, $p(s) = s^3 + \alpha_4 s^2 + (\alpha_1 + \alpha_2 \alpha_4)s + \alpha_1 \alpha_4$, has positive coefficients. Since $p(-\alpha_4) < 0 < p(0)$, it has one negative real root in interval $(-\alpha_4, 0)$, which we denote by b_1. The other two roots (b_2 and b_3 say) may be real or complex, but if they are complex they will be complex conjugates since $p(s)$ has real coefficients. Assuming that the real part of each root is negative, we first find the partial fraction decomposition of the rational function in (8.55) as

$$\frac{s^2 + \alpha_4 s + \alpha_2 \alpha_4}{s^3 + \alpha_4 s^2 + (\alpha_1 + \alpha_2 \alpha_4)s + \alpha_1 \alpha_4} = \frac{a_1}{s - b_1} + \frac{a_2}{s - b_2} + \frac{a_3}{s - b_3},$$

where $a_i \in \mathbb{C}$ are constants (which depend on α_1, α_2 and α_4). Then we can rewrite (8.55) as

$$\chi(\tau) = \frac{\sigma^2}{\kappa(0)} \left(a_1 \exp(b_1 \tau) + a_2 \exp(b_2 \tau) + a_3 \exp(b_3 \tau) \right). \tag{8.56}$$

The results computed by (8.56) are shown in Figure 8.1(b). The same results can also be obtained using the SSA (a21)–(c21). Equation (8.56) provides a "reasonably closed" formula for the velocity autocorrelation function. We note that although it may include complex exponentials, the resulting $\chi(\tau)$ is always real. Passing to the limit $\alpha_4 \to \infty$ in (8.55), we can obtain perhaps an even "more closed" analytical result. It is left as Exercise 8.8 to show that the limit $\alpha_4 \to \infty$ in (8.55) leads to

$$\chi(\tau) = \frac{\sigma^2}{\kappa(0)} \mathcal{L}^{-1} \left(\frac{s + \alpha_2}{s^2 + \alpha_2 s + \alpha_1} \right)$$

(8.57)

$$= \frac{\sigma^2}{\kappa(0)} \exp\left(-\frac{\alpha_2 \tau}{2}\right) \left(\cosh\left(\sqrt{\alpha_2^2 - 4\alpha_1}\, \frac{\tau}{2} \right) + \frac{\alpha_2 \sinh\left(\sqrt{\alpha_2^2 - 4\alpha_1}\, \frac{\tau}{2} \right)}{\sqrt{\alpha_2^2 - 4\alpha_1}} \right),$$

which is the velocity autocorrelation function of the three-equation model (8.14)–(8.16). It is plotted in Figure 8.1(b) using the red dot-dashed line, where we note that the square root of $(\alpha_2^2 - 4\alpha_1)$ is a purely imaginary number for the parameters, $\alpha_1 = \alpha_2 = 1$, used in Figure 8.1(b). In this case (8.57) can also be rewritten in terms of sine and cosine functions (instead of sinh and cosh).

The above analysis shows that, for the one-particle solvent model, we can derive closed formulae for statistics obtained by the SSA (a21)–(c21). However, for more complicated models and friction kernels, computer simulations will often be our (only) method of choice.

It is worth noting that the velocity autocorrelation function, $\chi(\tau)$, can provide an estimate of the diffusion constant. Such a result is independent of whether we know $\chi(\tau)$ analytically or whether it was estimated by using the SSA (a21)–(c21). To derive it, we first rewrite the derivative of the mean-square displacement in definition (8.51) as follows:

$$\frac{\mathrm{d}\langle X^2 \rangle}{\mathrm{d}t}(t) = \frac{\mathrm{d}}{\mathrm{d}t} \left\langle \int_0^t V(\tau)\, \mathrm{d}\tau \int_0^t V(s)\, \mathrm{d}s \right\rangle = \frac{\mathrm{d}}{\mathrm{d}t} \int_0^t \int_0^t \langle V(s)\, V(\tau) \rangle\, \mathrm{d}\tau\, \mathrm{d}s.$$

(8.58)

Next, we want to differentiate the integral on the right-hand side. To do this, we use the symmetry $\langle V(\tau)\, V(s) \rangle = \langle V(s)\, V(\tau) \rangle$, which implies

$$\int_0^t \int_0^t \langle V(s)\, V(\tau) \rangle\, \mathrm{d}\tau\, \mathrm{d}s = 2 \int_0^t \int_0^s \langle V(s)\, V(\tau) \rangle\, \mathrm{d}\tau\, \mathrm{d}s.$$

Substituting this into (8.58), we can differentiate the integral. Substituting into equation (8.51), we then obtain

$$D = \lim_{t \to \infty} \int_0^\infty \langle V(t)\, V(t - \tau) \rangle\, \mathrm{d}\tau = \int_0^\infty \chi(\tau)\, \mathrm{d}\tau.$$

(8.59)

Thus, a (numerical) integration of velocity autocorrelation function $\chi(\tau)$ gives us another way to estimate diffusion coefficient D. For the simulation presented in Figure 8.2(a), formula (8.59) gives $D \approx 0.5$, which is the same estimate as we have obtained by (8.51). In the case of the one-particle solvent model (8.9)–(8.12), we can also use (8.59) to get an analytical result. Since our formula for the friction kernel $\kappa(\tau)$ defined in (8.31) is perhaps slightly simpler to integrate than our result (8.56) for $\chi(\tau)$, we first observe that equation (8.59) is effectively the Laplace transform of χ evaluated at $s = 0$ (see definition (8.54)). Using (8.53), we can rewrite (8.59) as follows:

$$D = \int_0^\infty \chi(\tau)\,d\tau = \frac{\sigma^2}{\kappa(0)\int_0^\infty \kappa(\tau)\,d\tau}. \tag{8.60}$$

Integrating the friction kernel $\kappa(\tau)$ defined in (8.31), we obtain

$$\int_0^\infty \kappa(\tau)\,d\tau = \frac{\alpha_1}{\alpha_2},$$

and using (8.30), we have

$$\frac{\sigma^2}{\kappa(0)} = \frac{\alpha_3^2}{2\alpha_1\alpha_2}.$$

Substituting into (8.60), we obtain

$$D = \frac{\alpha_3^2}{2\alpha_1^2}, \tag{8.61}$$

which is indeed equal to 1/2 for our parameters, $\alpha_1 = \alpha_3 = 1$, used in Figure 8.2(a). Interestingly this result does not depend on α_4, i.e. formula (8.61) holds for the one-particle solvent model (8.9)–(8.12) for any α_4. It is also true in the limit $\alpha_4 \to \infty$, which leads to the three-equation model (8.14)–(8.16). At the beginning of this section, we used the scaling (8.33) and studied the limit $\alpha_2 \to \infty$ in equation (8.34) to derive the Langevin description (8.35)–(8.36). Substituting the scaling (8.33) into (8.61), we obtain that D is indeed equal to the diffusion constant used in the Langevin description. Thus, the limit $\alpha_2 \to \infty$ together with scaling (8.33) corresponds to a limiting process that removes velocity correlations, while keeping the same diffusion constant D.

Convergence to the Langevin equation can also be deduced from the discretized generalized Langevin equation, given by equations (8.43)–(8.44). If we choose a discretized friction kernel satisfying $\kappa(0) = \beta/\Delta t$ and $\kappa(j\Delta t) = 0$, for $j \geq 1$, and substitute it into (8.43)–(8.44), we directly obtain the Langevin description (4.49)–(4.50). This specific choice of friction kernel $\kappa(\tau)$ shares some properties with the Dirac delta function and satisfies equation (8.40). In Section 4.3, we have already shown that the Langevin description leads to Brownian motion for $\beta \to \infty$. Thus, we conclude that the generalized Langevin equation can indeed describe Brownian motion. In general, it describes diffusion of particles with correlated noise and velocities, but it can also describe uncorrelated diffusion in certain limits.

8.3 Solvent as Harmonic Oscillators

In Section 8.1, we modelled the solvent by one particle that is coupled through a harmonic spring with the diffusing particle of interest. Such a huge simplification has helped us to get some initial insights into this model. In this section, we generalize equations (8.4)–(8.5) by assuming that the diffusing particle of interest is coupled with N solvent molecules. Then the position $X(t)$ and velocity $V(t)$ of the diffusing particle evolve according to

$$\frac{dX}{dt}(t) = V(t), \tag{8.62}$$

$$m_p \frac{dV}{dt}(t) = \sum_{i=1}^{N} k_i \left(x_i(t) - X(t) \right), \tag{8.63}$$

where $x_i(t)$ is the position of the ith solvent molecule, which interacts with the diffusing particle through a harmonic spring with spring constant k_i, $i = 1, 2, \ldots, N$. Equations (8.62)–(8.63) are coupled with the evolution equations for solvent molecules. We assume that the ith solvent molecule has mass $m_{s,i}$. Using Newton's second law of motion, we get

$$\frac{dx_i}{dt}(t) = v_i(t), \tag{8.64}$$

$$m_{s,i} \frac{dv_i}{dt}(t) = k_i \left(X(t) - x_i(t) \right). \tag{8.65}$$

Comparing (8.7) and (8.65), we observe that equation (8.7) includes additional friction and random forces. They are necessary in the one-particle solvent model to implicitly model all other solvent molecules in the system. We do not include such forces in (8.65), because we assume that we explicitly model all solvent molecules (N is considered to be "large"). In practice, it is impossible to include all solvent molecules in simulations (because N would have to be "very large") and friction and noise terms are still included to control the temperature of the simulated system. We will return to this in Section 8.6, when we discuss molecular dynamics. Thus model (8.62)–(8.65) is still a theoretical simplified picture, which helps us to get some additional insights into microscopic models of Brownian motion.

For each i, the solvent equations (8.64)–(8.65) describe a forced harmonic oscillator: each of the solvent molecules has a natural frequency of oscillation $\omega_i = \sqrt{k_i/m_{s,i}}$. These equations can be solved (Exercise 8.9) to get

$$x_i(t) = x_i(0) \cos \left(\omega_i t \right) + \frac{v_i(0)}{\omega_i} \sin \left(\omega_i t \right) + \omega_i \int_0^t \sin \left(\omega_i (t - \tau) \right) X(\tau) \, d\tau, \tag{8.66}$$

where $x_i(0)$ is the initial position of the ith molecule and $v_i(0)$ is its initial velocity. Using integration by parts and equation (8.62), we have

$$\omega_i \int_0^t \sin\left(\omega_i(t-\tau)\right) X(\tau)\,d\tau = X(t) - \cos\left(\omega_i t\right) X(0) - \int_0^t \cos\left(\omega_i(t-\tau)\right) V(\tau)\,d\tau.$$

Substituting into (8.66) and assuming without loss of generality that the diffusing particle started at the origin, i.e. $X(0) = 0$, we obtain

$$x_i(t) - X(t) = x_i(0)\cos\left(\omega_i t\right) + \frac{v_i(0)}{\omega_i}\sin\left(\omega_i t\right) - \int_0^t \cos\left(\omega_i(t-\tau)\right) V(\tau)\,d\tau.$$

Substituting into (8.63), we have

$$\frac{dV}{dt}(t) = \sum_{i=1}^{N} \frac{k_i}{m_p}\left(x_i(0)\cos\left(\omega_i t\right) + \frac{v_i(0)}{\omega_i}\sin\left(\omega_i t\right) - \int_0^t \cos\left(\omega_i \tau\right) V(t-\tau)\,d\tau\right).$$

Thus the evolution of the diffusing particle can again be formally written in the form of the generalized Langevin equation (8.28)–(8.29) provided that the friction kernel $\kappa(\tau)$ and noise term $R(t)$ are given by

$$\kappa(\tau) = \frac{1}{m_p}\sum_{i=1}^{N} m_{s,i}\,\omega_i^2 \cos\left(\omega_i \tau\right), \tag{8.67}$$

$$R(t) = \frac{1}{m_p}\sum_{i=1}^{N} x_i(0)\,m_{s,i}\,\omega_i^2 \cos\left(\omega_i t\right) + v_i(0)\,m_{s,i}\,\omega_i \sin\left(\omega_i t\right), \tag{8.68}$$

where masses $m_{s,i}$ and frequencies $\omega_i = \sqrt{k_i/m_{s,i}}$ are yet to be specified. The dynamics of the diffusing particle will also depend on our choice of the initial positions and velocities of solvent oscillators. Since we do not have much information about them, it is natural to assume that $x_i(0)$ and $v_i(0)$ are both independently sampled according to some probability distributions. If we sample them according to their equilibrium distributions, then the noise autocorrelation function can be expressed as

$$\langle R(t)\,R(t-\tau)\rangle = \langle R(\tau)\,R(0)\rangle = \frac{1}{m_p^2}\sum_{i=1}^{N}\langle x_i^2(0)\rangle\,m_{s,i}^2\,\omega_i^4 \cos\left(\omega_i \tau\right), \tag{8.69}$$

where we have used the fact that $x_i(0)$ and $v_i(0)$ are independent and have zero mean, i.e. $\langle x_i(0)\rangle = \langle v_i(0)\rangle = 0$. As in Sections 4.3 and 4.6, we associate the kinetic energy of a particle with the absolute temperature T through the relation $m_{s,i}\langle v_i^2(0)\rangle/2 = k_B T/2$. The average potential energy of a harmonic oscillator, $k_i\langle x_i^2(0)\rangle/2$, is also equal to the same value $k_B T/2$. Thus, we have

$$\langle x_i^2(0)\rangle = \frac{k_B T}{m_{s,i}\,\omega_i^2} \qquad \text{and} \qquad \langle v_i^2(0)\rangle = \frac{k_B T}{m_{s,i}}. \tag{8.70}$$

Substituting into (8.69) and using (8.67), we obtain the generalized fluctuation–dissipation theorem in the form

$$\langle R(t)\,R(t-\tau)\rangle = \frac{k_B T}{m_p^2}\sum_{i=1}^{N} m_{s,i}\,\omega_i^2\cos(\omega_i\tau) = \frac{k_B T}{m_p}\,\kappa(\tau). \qquad (8.71)$$

Therefore, we have two options to simulate a solvent as a system of harmonic oscillators. In both cases, we first need to specify the mass $m_{s,i}$ and frequency $\omega_i = \sqrt{k_i/m_{s,i}}$ of each oscillator. Then, a computationally intensive option to simulate the system would be to sample initial positions and velocities, $x_i(0)$ and $v_i(0)$, from normal distributions with variances given by (8.70) and evolve the system by solving ODEs (8.62)–(8.65). Alternatively, if we are only interested in the time evolution of position $X(t)$ and velocity $V(t)$ of the diffusing particle, we can use the SSA (a21)–(c21) with friction kernel given by (8.67). The results of such a calculation are shown in Figure 8.2(b), where ten realizations computed by the SSA (a21)–(c21) are visualized.

In order to use the SSA (a21)–(c21), or properties of the generalized Langevin equation derived in Section 8.2, we have to choose the masses $m_{s,i}$ and frequencies $\omega_i = \sqrt{k_i/m_{s,i}}$ in such a way that the friction kernel (8.67) satisfies $\kappa(\tau) \to 0$ as $\tau \to \infty$. Clearly, if we choose all the ω_i equal to the same constant, i.e. $\omega_i = \omega_m$, then the friction kernel (8.67) is just a multiple of $\cos(\omega_m\tau)$, which does not satisfy $\kappa(\tau) \to 0$ as $\tau \to \infty$. Thus we instead assume that ω_i are sampled according to a suitable probability distribution $p(\omega)$ with given mean ω_m. To simplify our discussion we assume that all spring constants k_i, $i = 1, 2, \ldots, N$, are equal to the same value k, i.e.

$$k_i = m_{s,i}\,\omega_i^2 = k, \qquad \text{where} \qquad k = \frac{2\beta\,m_p\,\omega_m}{N\,\pi}. \qquad (8.72)$$

Here, β is a new parameter, which is simply equal to k multiplied by a suitable combination of the already defined parameters m_p, ω_m and N. The relation (8.72) between k and β is chosen in such a way that the final friction kernel, derived as κ_e below, satisfies the integral property (8.40). Using assumption (8.72), the friction kernel (8.67) simplifies to

$$\kappa(\tau) = \frac{2\beta\omega_m}{\pi}\,\frac{1}{N}\sum_{i=1}^{N}\cos(\omega_i\tau).$$

Letting $N \to \infty$, we can replace this sum by an integral,

$$\kappa(\tau) = \frac{2\beta\omega_m}{\pi}\int_0^\infty \cos(\omega\tau)\,p(\omega)\,d\omega, \qquad (8.73)$$

where $p(\omega)$ is the probability distribution from which we sample the frequencies ω_i. A simple choice is to sample ω_i uniformly in interval $[0, 2\omega_m]$. Then ω_m is indeed the mean frequency and equation (8.73) implies

$$\kappa_u(\tau) = \frac{\beta}{\pi} \int_0^{2\omega_m} \cos(\omega\tau)\, d\omega = \frac{\beta \sin(2\omega_m\tau)}{\pi \tau}. \tag{8.74}$$

In Figure 8.3(a), we plot friction kernel $\kappa_u(\tau)$ for three different values of ω_m. We observe that $\kappa_u(0) \to \infty$ as $\omega_m \to \infty$, and we also have

$$\int_0^\infty \kappa_u(\tau)\, d\tau = \frac{\beta}{2},$$

which is the integral property (8.40) up to a factor of $1/2$. Of course, we could get the integral property (8.40) without the additional factor $1/2$, if we included this factor in our definition of k in equation (8.72). We did not do this, because we had in mind when writing the relation (8.72) our second example of the frequency distribution. In the second example, we sample ω_i from exponential distribution with mean ω_m. Then equation (8.73) implies

$$\kappa_e(\tau) = \frac{2\beta}{\pi} \int_0^\infty \cos(\omega\tau) \exp\left(-\frac{\omega}{\omega_m}\right) d\omega = \frac{2\beta}{\pi} \frac{\omega_m}{\omega_m^2 \tau^2 + 1}. \tag{8.75}$$

Then

$$\int_0^\infty \kappa_e(\tau)\, d\tau = \beta,$$

i.e. we obtain exactly the integral property (8.40). Moreover, we can define the limiting friction kernel

$$\kappa_\infty(\tau) = \lim_{\omega_m \to \infty} \kappa_e(\tau), \tag{8.76}$$

which satisfies $\kappa_\infty(\tau) = 0$ for $\tau > 0$ and $\kappa_\infty(0) = \infty$. Thus the limiting kernel is again a generalized function which has the same properties as the limiting kernel studied in equation (8.39). In particular, we can use the results of Section 8.2 to show that the harmonic oscillator solvent model leads to Brownian motion for the friction kernel κ_e. We note that we cannot use the limit $\omega_m \to \infty$ to define a limiting kernel of the friction kernel obtained for the uniform distribution, κ_u in (8.74), because this limit does not exist for any fixed $\tau > 0$.

In Figure 8.3(b), we plot friction kernel $\kappa_e(\tau)$ for three different values of ω_m. The results illustrate that the limiting kernel (8.76) is indeed a multiple of the Dirac delta function. Ten realizations of the SSA (a21)–(c21) for friction kernel $\kappa_e(\tau)$ with dimensionless parameters $\beta = \omega_m = k_B T = m_p = 1$ are plotted in Figure 8.2(b). We plot $X(t) - X(100)$ as a function of $t - 100$ to show long-term dynamics, the regime for which the generalized Langevin equation is derived from the solvent described as harmonic oscillators. Using (8.51), we

(a)   (b)

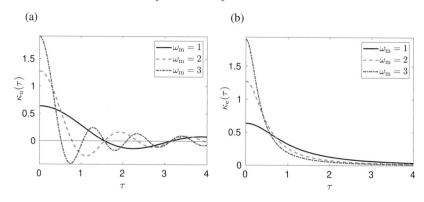

Figure 8.3 (a) *Friction kernel $\kappa_u(\tau)$ given by (8.74) for $\omega_m = 1$ (black solid line),*
$\omega_m = 2$ (green dashed line) and $\omega_m = 3$ (red dot-dashed line). (b) Friction kernel
$\kappa_e(\tau)$ given by (8.75) for $\omega_m = 1$ (black solid line), $\omega_m = 2$ (green dashed line)
and $\omega_m = 3$ (red dot-dashed line). We use $\beta = 1$ in both panels.

estimate the diffusion constant from 10^6 realizations of the SSA (a21)–(c21)
as $D \approx 1$. We can also use results (8.60) and (8.40) to obtain

$$D = \frac{\sigma^2}{\kappa_e(0) \displaystyle\int_0^\infty \kappa_e(\tau)\, d\tau} = \frac{\sigma^2}{\kappa_e(0)\beta}. \tag{8.77}$$

Using (8.71), we get

$$\sigma^2 = \frac{k_B T}{m_p} \kappa_e(0).$$

Substituting into (8.77), we obtain

$$D = \frac{k_B T}{\beta m_p},$$

which is the Einstein–Smoluchowski relation derived in Section 4.6 as equation (4.95).

8.4 Solvent as Points Colliding with the Diffusing Particle

In Section 8.3, we described the solvent as a system of harmonic oscillators. The advantage of this simplified description is that we can make a lot
of analytic progress, but this model also includes some unrealistic artefacts.
For example, in this model the force between two solvent particles increases
with their separation, while in real molecular systems, forces are negligible

for particles that are far apart. Real force terms are therefore not given in terms of harmonic springs, and we discuss more complicated force terms in Section 8.5. Before then we introduce another analytically tractable solvent model which lies at the other theoretical extreme: it only includes interactions between nearby particles (in fact, they only interact when they are in contact) and does not have any long-range forces at all. We consider a (heavy) diffusing particle with mass m_p, described as a ball with radius R, with the coordinates of its centre denoted by $\mathbf{X} = [X_1, X_2, X_3]$ and its velocity by $\mathbf{V} = [V_1, V_2, V_3]$. The diffusing particle collides with a large number of (light) point solvent particles with masses $m_s \ll m_p$, positions $\mathbf{x}^i$ and velocities $\mathbf{v}^i$, $i = 1, 2, 3, \ldots$. The collisions of particles are without friction, which means that post-collision velocities can be computed using the conservation of momentum and energy. Assuming that $\mathbf{v}^j$ is the velocity of the heat bath molecule that collided with the heavy molecule, we obtain (see Exercise 8.10)

$$\widetilde{\mathbf{V}} = [\mathbf{V}]^\| + \frac{\mu - 1}{\mu + 1} \, [\mathbf{V}]^\perp + \frac{2}{\mu + 1} \, [\mathbf{v}^j]^\perp, \tag{8.78}$$

$$\widetilde{\mathbf{v}}^j = [\mathbf{v}^j]^\| + \frac{1 - \mu}{\mu + 1} \, [\mathbf{v}^j]^\perp + \frac{2\mu}{\mu + 1} \, [\mathbf{V}]^\perp, \tag{8.79}$$

where tildes denote post-collision velocities, superscripts $\perp$ denote projections of velocities on the line through the centre of the molecule and the collision point on its surface, superscripts $\|$ denote tangential components of velocity, and we define μ as the dimensionless ratio of masses,

$$\mu = \frac{m_p}{m_s}. \tag{8.80}$$

To show that this solvent model can describe Brownian motion, we consider the limit $\mu \to \infty$. This limit can be achieved in a number of different ways. For example, we can keep m_s fixed and pass $m_p \to \infty$, or we can keep m_p fixed and pass $m_s \to 0$. In both cases, the effect of a single collision on the velocity of the heavy molecule becomes negligible. To get a non-trivial limiting behaviour, we need to simultaneously increase the number and velocities of solvent molecules as we increase μ. It is then possible to derive the Langevin description (8.1)–(8.2) with force term (8.3) for the position $[X_1, X_2, X_3]$ and velocity $[V_1, V_2, V_3]$ of the diffusing molecule, which can be written in the form

$$X_i(t + dt) = X_i(t) + V_i(t) \, dt, \tag{8.81}$$

$$V_i(t + dt) = V_i(t) - \beta \, V_i(t) \, dt + \beta \, \sqrt{2D} \, dW_i, \quad i = 1, 2, 3, \tag{8.82}$$

where D is the diffusion coefficient and $\beta = \gamma/m_p$ is the friction coefficient. This description can be further reduced to the SDE model of diffusion, given by equations (4.1)–(4.3), in the overdamped limit $\beta \to \infty$. To obtain the Langevin description (8.81)–(8.82) we need to increase the density of solvent particles around the heavy molecule as we pass to the limit $\mu \to \infty$. In this section, we work in infinite three-dimensional space $\mathbb{R}^3$ and leave the technicality of boundary conditions and finite domains to Section 9.2 in the next chapter. We assume that the density of solvent particles in any finite subdomain $\Omega \in \mathbb{R}^3$ is given by $\lambda_\mu|\Omega|$, where $|\Omega|$ is the volume of Ω and the constant λ_μ is given by

$$\lambda_\mu = \frac{3}{8R^2} \sqrt{\frac{(\mu + 1)\beta}{2\pi D}}. \tag{8.83}$$

More precisely, we assume that the number of solvent particles in a set $\Omega \in \mathbb{R}^3$ is distributed according to the Poisson distribution with mean $\lambda_\mu|\Omega|$, which is sometimes shortened to saying that heat bath particles are distributed according to the (spatial) Poisson distribution (or process) with density λ_μ.

In addition to positions $\mathbf{x}^i$ of heat bath particles, we also have to initialize their velocities $\mathbf{v}^i$, $i = 1, 2, 3, \ldots$. Again, we assume that their velocities are taken from a distribution with a suitably scaled mean. We assume that each initial velocity $\mathbf{v}^i = [v_1^i, v_2^i, v_3^i]$ is sampled according to the distribution with distribution function

$$f_\mu(\mathbf{v}^i) = \frac{1}{\sigma_\mu^3 (2\pi)^{3/2}} \exp\left[-\frac{(v_1^i)^2 + (v_2^i)^2 + (v_3^i)^2}{2\sigma_\mu^2} \right], \tag{8.84}$$

where

$$\sigma_\mu = \sqrt{(\mu + 1) D \beta}. \tag{8.85}$$

This is again a Maxwell distribution – compare to (4.60). Using equations (8.83) and (8.85), we see that both $\lambda_\mu \to \infty$ and $\sigma_\mu \to \infty$ in the limit $\mu \to \infty$, i.e. we have more and faster heat bath particles in this limit, but equation (8.78) also implies that the effect of each individual collision is getting smaller (as $\mu \to \infty$). The scalings of λ_μ and σ_μ, equations (8.83) and (8.85), are chosen in such a way that the position and velocity of the heavy molecule, $\mathbf{X}$ and $\mathbf{V}$, converge to the solution of (8.81)–(8.82) in the limit $\mu \to \infty$.

Our goal for the remainder of this section is to verify that (8.82) is indeed the right equation for the velocity of the heavy particle. We want to replace the explicit simulation of the solvent particles and their collisions with the heavy particle with an effective equation for the position and velocity of the heavy particle alone. In particular we want to motivate the choices (8.83) and (8.85) for the parameters λ_μ and σ_μ. For a rigorous proof of the convergence to (8.82) the reader is referred to Dürr et al. (1981).

We simplify the discussion by assuming that in the limit of large μ the velocity of the heavy particle evolves according to the discretized SDE

$$V_i(t + \Delta t) = V_i(t) + f_i(t) \Delta t + g_i(t) \sqrt{\Delta t}\, \xi_i, \tag{8.86}$$

where Δt is a small time step, and $\boldsymbol{\xi} = [\xi_1, \xi_2, \xi_3]$ is a vector of three normally distributed random numbers each with zero mean and unit variance, but the drift and diffusion coefficients $f_i(t)$ and $g_i(t)$ are unknown. The normally distributed noise is a result of the central limit theorem applied to many small velocity jumps caused by collisions with solvent molecules. To determine $f_i(t)$ and $g_i(t)$ we will match the mean and variance of the velocity jump in the microscopic solvent model with those given by (8.86).

Let $\mathbf{y} = [y_1, y_2, y_3]$ be a given point on the surface of the heavy molecule at time t, i.e.

$$\left| \mathbf{y} - \mathbf{X}(t) \right| = \sqrt{(y_1 - X_1(t))^2 + (y_2 - X_2(t))^2 + (y_3 - X_3(t))^2} = R.$$

First, we find a formula for an auxiliary distribution that describes the average change of the velocity of the heavy molecule due to collisions near the surface point $\mathbf{y}$. We denote this by $\psi_i(\mathbf{y})$, so that $\psi_i(\mathbf{y})\, dA$ is the average change of the ith component of the velocity of the heavy molecule caused by collisions with heat bath particles in the time interval $(t, t + \Delta t)$ on the surface area element dA centred at $\mathbf{y}$.

To estimate $\psi_i(\mathbf{y})\, dA$, consider the situation that is schematically shown in Figure 8.4. Here, a heat bath particle, which was at point $\mathbf{x}$ at time t, collides with the heavy molecule at time $t + \tau \in (t, t + \Delta t)$ at the surface point, which had coordinate $\mathbf{y}$ at time t. Since Δt is small, and the velocity jump in the

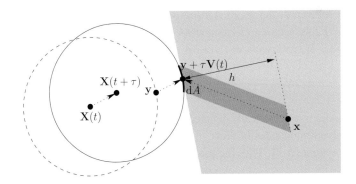

Figure 8.4 *A schematic of the collision of a heat bath particle, which was at point* $\mathbf{x}$ *at time t, with the heavy molecule at its surface point, which had coordinate* $\mathbf{y}$ *at time t. The collision occurs at time* $t + \tau$ *and the collision point is at* $\mathbf{y} + \tau \mathbf{V}(t)$.

heavy molecule due to collisions with heat bath particles is small, to a first approximation we may take $\mathbf{V}(t)$ to be constant in $(t, t + \Delta t)$ when calculating the collisions (equation (8.86) implies that the change in $\mathbf{V}$ is $O(\sqrt{\Delta t})$). Thus the coordinate of the surface point at the collision time $t + \tau$ is equal to $\mathbf{y} + \tau \mathbf{V}(t)$. Since the heat bath molecule moved from $\mathbf{x}$ to the collision point $\mathbf{y} + \tau \mathbf{V}(t)$, its pre-collision velocity was

$$\mathbf{v} = \frac{\mathbf{y} + \tau \mathbf{V}(t) - \mathbf{x}}{\tau} = \mathbf{V}(t) + \frac{(\mathbf{y} - \mathbf{x})}{\tau}.$$

Using equation (8.78), we can write the change of the velocity of the heavy molecule due to the collision as

$$\widetilde{\mathbf{V}} - \mathbf{V} = \frac{2}{\mu + 1} [\mathbf{v} - \mathbf{V}(t)]^{\perp} = \frac{2}{\mu + 1} \left((\mathbf{v} - \mathbf{V}(t)) \cdot \frac{(\mathbf{y} - \mathbf{X}(t))}{R} \right) \frac{(\mathbf{y} - \mathbf{X}(t))}{R}.$$
(8.87)

The position $\mathbf{x}$ of the heat bath particle must be in the half space that lies above the plane tangent to the heavy molecule at the collision point $\mathbf{y} + \tau \mathbf{V}(t)$. This half space is schematically denoted by light shading in Figure 8.4. This means that the component of the velocity $\mathbf{v}$ in the direction $\mathbf{y} - \mathbf{X}(t)$ must be negative. Thus the allowable velocities may be parametrized by

$$\mathbf{v} = -c_1 \frac{(\mathbf{y} - \mathbf{X}(t))}{R} + c_2 \boldsymbol{\eta}_2 + c_3 \boldsymbol{\eta}_3,$$

where $c_1 > 0$, $c_2 \in \mathbb{R}$ and $c_3 \in \mathbb{R}$, and $(\mathbf{y} - \mathbf{X}(t))/R$, $\boldsymbol{\eta}_2$, $\boldsymbol{\eta}_3$ is an orthornormal basis in $\mathbb{R}^3$. Then (8.87) reads as follows:

$$\widetilde{\mathbf{V}} - \mathbf{V} = -\frac{2}{(\mu + 1)R} \left(c_1 + \mathbf{V}(t) \cdot \frac{(\mathbf{y} - \mathbf{X}(t))}{R} \right) (\mathbf{y} - \mathbf{X}(t)).$$
(8.88)

Now, for a given $\mathbf{v}$, the set of starting points $\mathbf{x}$ for solvent particles that will collide with the surface area element dA centred on $\mathbf{y}$ during the time $(t, t + \Delta t)$ is a cylinder of cross-sectional area dA and perpendicular height

$$h = \Delta t \, (\mathbf{V}(t) - \mathbf{v}) \cdot \frac{(\mathbf{y} - \mathbf{X}(t))}{R} = \Delta t \left(c_1 + \mathbf{V}(t) \cdot \frac{(\mathbf{y} - \mathbf{X}(t))}{R} \right).$$

This cylinder is schematically denoted by dark shading in Figure 8.4. The number of solvent molecules in this cylinder is Poisson distributed with mean λ_μ times its volume $h \, dA$. Thus the probability of collision of a solvent molecule with velocity in $(\mathbf{v}, \mathbf{v} + d\mathbf{v})$ with the surface area element dA during the time $(t, t + \Delta t)$ is

$$\lambda_\mu \left(c_1 + \mathbf{V}(t) \cdot \frac{(\mathbf{y} - \mathbf{X}(t))}{R} \right) \Delta t \, dA \, f_\mu \left(-c_1 \frac{\mathbf{y} - \mathbf{X}(t)}{R} + c_2 \boldsymbol{\eta}_2 + c_3 \boldsymbol{\eta}_3 \right) d\mathbf{v}. \quad (8.89)$$

To find the average change in velocity of the heavy molecule due to collisions with dA during the time $(t, t + \Delta t)$ we multiply (8.88) by (8.89) and integrate over all possible velocities $\mathbf{v}$ to obtain

$$
\psi_j(\mathbf{y})\, dA = -\Delta t\, dA \frac{2\,\lambda_\mu\,(y_j - X_j(t))}{(\mu + 1)\,R} \int_0^\infty \int_{-\infty}^\infty \int_{-\infty}^\infty \left(c_1 + \mathbf{V}(t) \cdot \frac{(\mathbf{y} - \mathbf{X}(t))}{R} \right)^2
$$
$$
\times f_\mu \left(-c_1 \frac{\mathbf{y} - \mathbf{X}(t)}{R} + c_2\,\boldsymbol{\eta}_2 + c_3\,\boldsymbol{\eta}_3 \right) dc_3\, dc_2\, dc_1. \tag{8.90}
$$

Substituting (8.84) for f_μ and integrating over c_2 and c_3, we have

$$
\psi_j(\mathbf{y}) = -\frac{\lambda_\mu\,(y_j - X_j(t))\,\Delta t\,\sqrt{2}}{(\mu + 1)\,R\,\sigma_\mu\,\sqrt{\pi}} \int_0^\infty \left(c_1 + \mathbf{V}(t) \cdot \frac{(\mathbf{y} - \mathbf{X}(t))}{R} \right)^2 \exp\left[-\frac{c_1^2}{2\sigma_\mu^2} \right] dc_1.
$$

Integrating over c_1, we deduce

$$
\psi_i(\mathbf{y}) = -\frac{\lambda_\mu\,\sigma_\mu^2\,(y_i - X_i(t))\,\Delta t}{(\mu + 1)\,R} - \frac{4\,\lambda_\mu\,\sigma_\mu\,(y_i - X_i(t))\,\Delta t}{(\mu + 1)\,R^2\,\sqrt{2\pi}}\,\mathbf{V}(t) \cdot (\mathbf{y} - \mathbf{X}(t))
$$
$$
- \frac{\lambda_\mu (y_i - X_i(t))\,\Delta t}{(1 + \mu) R^3}\,(\mathbf{V}(t) \cdot (\mathbf{y} - \mathbf{X}(t)))^2. \tag{8.91}
$$

Using this result, we can compute the drift coefficient $f_i(t)$ in equation (8.86) by integrating over the surface of the heavy molecule

$$
S = \left\{ \mathbf{y} \in \mathbb{R}^3 \,\middle|\, |\mathbf{y} - \mathbf{X}(t)| = R \right\}
$$

and equating to the average velocity jump in (8.86), to give

$$
f_i(t)\,\Delta t = \int_S \psi_i(\mathbf{y})\, dA. \tag{8.92}
$$

Substituting (8.91) into (8.92) and noting that $\int_S (y_i - X_i(t))^k\, dA = 0$ for odd k, we find

$$
f_i(t) = -\frac{4\,\lambda_\mu\,\sigma_\mu\,V_i(t)}{(\mu + 1)\,R^2\,\sqrt{2\pi}} \int_S (y_i - X_i(t))^2\, dA.
$$

Using (8.83) and (8.85) and evaluating the surface integral, we obtain

$$
f_i(t) = -\beta\,V_i(t), \qquad \text{for } i = 1, 2, 3, \tag{8.93}
$$

i.e. we have indeed derived the drift term in equation (8.82).

To estimate the noise term, we calculate the variance in the velocity jump in the ith direction from time t to time $t + \Delta t$ and equate this to $g_i^2(t)\,\Delta t$. We first note that, since the mean velocity jump is $O(\Delta t)$, to leading order the variance is simply the second moment of the velocity jump. We therefore square (8.88), multiply by (8.89) and integrate over all possible $\mathbf{v}$ to give the variance in the *i*th component of the velocity jump of the heavy molecule due to collisions with dA during the time $(t, t + \Delta t)$ as

$$\phi_i(\mathbf{y}) \, dA = \Delta t \, dA \, \frac{4\lambda_\mu(y_i - X_i(t))^2}{(\mu + 1)^2 R^2} \int_0^\infty \int_{-\infty}^\infty \int_{-\infty}^\infty \left(c_1 + \mathbf{V}(t) \cdot \frac{(\mathbf{y} - \mathbf{X}(t))}{R} \right)^3$$

$$\times f_\mu \left(-c_1 \frac{\mathbf{y} - \mathbf{X}(t)}{R} + c_2 \, \boldsymbol{\eta}_2 + c_3 \, \boldsymbol{\eta}_3 \right) dc_3 \, dc_2 \, dc_1.$$

Integrating as before over the surface of the heavy molecule, and substituting (8.84)–(8.85) for f_μ and (8.83) for λ_μ, gives (Exercise 8.11)

$$g_i^2 \, \Delta t = \int_S \phi_i(\mathbf{y}) \, dA = 2\beta^2 D \Delta t + O(\mu^{-1} \Delta t) \tag{8.94}$$

for $i = 1, 2, 3$. Therefore, in the limit $\mu \to \infty$,

$$g_i = \beta \, \sqrt{2D}, \qquad \text{for } i = 1, 2, 3, \tag{8.95}$$

as required. Thus we have shown that our choice of λ_μ and f_μ, given in equations (8.83)–(8.85), indeed implies that the heavy particle satisfies the Langevin equation (8.82) in the limit $\mu \to \infty$. Equation (8.82) is also the Langevin description (4.49)–(4.50) in three spatial dimensions. In Section 4.3, we have already shown that this Langevin description leads to Brownian motion for $\beta \to \infty$. Thus we have presented another theoretical model that leads to Brownian motion in a certain limit. Unlike the harmonic oscillator heat bath studied in Section 8.3, the model presented here only includes short-range (contact, uncorrelated) interactions and does not require the generalized Langevin description to describe correlations. In the more realistic models discussed in the rest of this chapter, forces between molecules include both short-range and long-range interactions, so we will see a combination of the effects discussed in Sections 8.3 and 8.4.

8.5 Forces Between Atoms and Molecules

In physics textbooks, we often learn that there are only four fundamental forces. Two of them (called strong and weak interactions) are relevant in the subatomic world and govern interactions between elementary particles (such as neutrons, protons or electrons). They are not important in the applications studied in this book. This leaves us with the two other fundamental forces – gravitational and electromagnetic interactions – with the latter one being the source of all intermolecular interactions. This might, on the face of it, be a surprising observation, when we compare it with chemistry textbooks, which include many different names for possible forces between atoms and molecules, including ionic bonds, hydrogen bonding or hydrophobic interactions (Rowlinson, 2002; Israelachvili, 2011). Such a classification is important to help us understand how different molecular structures interact, but we will not go into this level of detail in our book. Instead, we discuss only two types of interatomic interactions, which will be used in our illustrative molecular dynamics simulations in Section 8.6.

Consider a particle (an atom or a molecule) of mass m_p with its centre of mass at position $\mathbf{X} = [X_1, X_2, X_3]$ with velocity $\mathbf{V} = [V_1, V_2, V_3]$ evolving according to Newton's second law of motion,

$$\frac{d\mathbf{X}}{dt} = \mathbf{V}, \tag{8.96}$$

$$m_p \frac{d\mathbf{V}}{dt} = -\nabla\Phi, \tag{8.97}$$

where $\Phi \equiv \Phi(\mathbf{X}) \colon \mathbb{R}^3 \to \mathbb{R}$ is a scalar potential function, which is such that the force on the particle is given by the negative of the gradient of the potential:

$$\mathbf{F} = -\nabla\Phi \equiv -\left[\frac{\partial\Phi}{\partial X_1}, \frac{\partial\Phi}{\partial X_2}, \frac{\partial\Phi}{\partial X_3} \right]. \tag{8.98}$$

If we suppose that the particle at position $\mathbf{X}$ only interacts with one other particle at position $\mathbf{x} = [x_1, x_2, x_3]$, the potential of the force between them usually depends on the distance between the two particles, i.e. $\Phi \equiv \Phi(r)$, where

$$r = |\mathbf{X} - \mathbf{x}| = \sqrt{\sum_{i=1}^{3}(X_i - x_i)^2}.$$

Then equation (8.97) can be rewritten as

$$m_p \frac{d\mathbf{V}}{dt} = -\Phi'(r)\,\frac{\mathbf{r}}{r}, \tag{8.99}$$

where $\mathbf{r} = \mathbf{X} - \mathbf{x}$ is the vector connecting the two particles and $\Phi'(r)$ is the derivative of the potential. When we considered harmonic spring interactions in Sections 8.1 and 8.3, Newton's second law of motion was given, in the one-dimensional case, as equations (8.4)–(8.5) for the one-particle solvent model. To write these in the form (8.96)–(8.97), we can define the harmonic potential

$$\Phi_{\text{ha}}(r) = \frac{k}{2} r^2, \tag{8.100}$$

where k is the spring constant. Then indeed $\mathbf{F} = -\nabla\Phi_{\text{ha}} = -k(\mathbf{X} - \mathbf{x}) = -k\mathbf{r}$. One criticism of harmonic springs is that they are extremely long-ranged: the force between two particles increases with their distance. In real molecular systems, forces are negligible for two particles that are far apart and their potential decreases with distance as r^{-a} for $r \to \infty$, for some parameter $a > 0$. Two commonly used examples include the Coulomb potential (with $a = 1$) and the Lennard-Jones potential (with $a = 6$).

In Exercise 8.12, you are asked to show that interaction potentials with $a < 3$ can still be considered long-ranged, while interaction potentials with $a > 3$ are short-ranged. Roughly speaking, we do not get a full picture by considering intermolecular and interatomic forces for two interacting particles in isolation. We have to consider that the particle at position $\mathbf{X}$ interacts with many particles at positions $\mathbf{x}^i = [x_1^i, x_2^i, x_3^i]$, $i = 1, 2, \ldots$. If these particles are uniformly distributed with density ϱ, there are $4\pi r^2 \, dr \times \varrho$ particles in a region of space that have their distance (from the particle at position $\mathbf{X}$) between r and $r + dr$. Since this number is increasing with r, the potential of the force has to decay sufficiently quickly to zero as $r \to \infty$ (with $a > 3$) to neglect the contribution of forces exerted by molecules that are far away from the particle at position $\mathbf{X}$. In particular, potentials describing Coulomb or gravitational interactions are long-ranged because they are both proportional to r^{-1} (i.e. $a = 1$). We have already discussed Coulomb's law in equation (7.54). The corresponding Coulomb potential (in vacuum) is given by

$$\Phi_c(r) = \frac{k_e Q q}{r}, \tag{8.101}$$

where Q and q are the charges of the two interacting particles (one at position $\mathbf{X}$ and one at position $\mathbf{x}$) and $k_e = 8.99 \times 10^9 \, \text{N m}^2 \, \text{C}^{-2}$ is the Coulomb constant.

In Section 8.4, we presented an example of a theoretical model based on extremely short-ranged interactions: two particles interact when they are in contact. It is possible to write this model in the potential form (8.96)–(8.97), if we introduce the corresponding hard-sphere potential

$$\Phi_{hs}(r) = \begin{cases} \infty, & \text{for } r \le R, \\ 0, & \text{for } r > R, \end{cases} \tag{8.102}$$

where R is the radius of the particle (ball). Assuming that heat bath particles do not interact with each other, we have been able to show in Section 8.4 that this model directly leads to the Langevin description (without any need to introduce the generalized Langevin model to describe correlations). However, the infinities in the potential Φ_{hs} lead to computational issues when we attempt to solve equations (8.96)–(8.97) using standard numerical methods for ODEs. Thus, in practice, steric (volume exclusion) effects are often modelled not by (8.102) but by continuous potentials.

In the 1920s and 1930s, Lennard-Jones (1924, 1931) studied potentials of the form $\lambda r^{-a_1} - \mu r^{-a_2}$, where λ and μ are positive constants and a_1 and a_2 are positive integers. The case $a_1 = 12$ and $a_2 = 6$ now bears his name. In applications, it is used to describe interactions between neutral atoms and molecules. We write it as

$$\Phi_{\mathrm{LJ}}(r) = 4\varepsilon\left[\left(\frac{\sigma}{r}\right)^{12} - \left(\frac{\sigma}{r}\right)^{6}\right], \tag{8.103}$$

where ε and σ are positive parameters. The graphs of the Lennard-Jones potential (8.103) and the negative of its derivative as a function of distance r are given in Figure 8.5(a) and (b), respectively. The parameters ε and σ in (8.103) have physical units of energy and length, respectively. We use them to annotate the axis labels in Figure 8.5. The derivative of the Lennard-Jones potential,

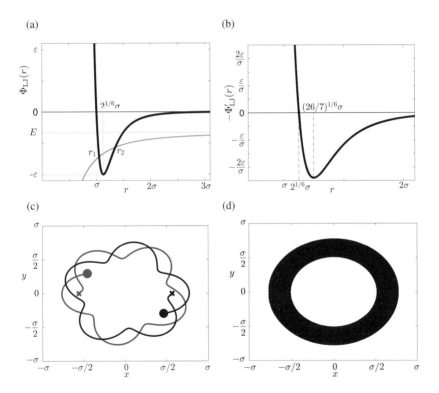

Figure 8.5 (a) *The Lennard-Jones potential,* $\Phi_{\mathrm{LJ}}(r)$, *given by equation (8.103) (blue line) together with the right-hand side of equation (8.104) (green line).* (b) *The derivative of the Lennard-Jones potential,* $\Phi'_{\mathrm{LJ}}(r)$, *as a function of r.* (c) *Initial trajectories of two isolated particles that interact through the Lennard-Jones force term.* (d) *Long-time trajectory of the blue particle.*

$\Phi'_{LJ}(r)$, has physical units of force (expressed in ε/σ). We observe that there is a unique minimum at $r_{min} = 2^{1/6}\sigma$ of the Lennard-Jones potential, where the right-hand side in equation (8.99) is zero, i.e. $\Phi'_{LJ}(r_{min}) = 0$. If we separate two particles by a distance r_{min} and they are at rest (no initial velocity), then equation (8.99) predicts that they will not move – they will stay at this separation forever.

Now let us give these particles a non-zero initial velocity. Consider an isolated system of two particles of identical mass m, which are initially at positions $\mathbf{X}(0) = [-r_{min}/2, 0, 0]$ and $\mathbf{x}(0) = [r_{min}/2, 0, 0]$ with velocities $\mathbf{V}(0) = [v_0, v_0, 0]$ and $\mathbf{v}(0) = [-v_0, -v_0, 0]$, respectively, where $v_0 > 0$ is small (we will shortly discuss what this means). Then the motion of the particles will be restricted to the x–y plane and the centre of mass of the system will be at fixed position at the origin. In Figure 8.5(c), we plot the initial trajectories of both particles, each particle starting from the corresponding cross and finishing at the corresponding dot.

In Figure 8.5(c), we observe that the distance between particles oscillates around r_{min}, which is what one would expect for small initial velocities. If the distance r between particles is greater than r_{min}, then equation (8.99) implies that they will be attracted together, while they will be repelled at separations less than r_{min}. Our initial condition for velocity implies that the sum of kinetic and potential energies of the system is $E = 2mv_0^2 - \varepsilon$. The initial velocity v_0 should be small enough so that $2mv_0^2 < \varepsilon$, i.e. total energy $E < 0$. Then, using conservation of energy, we can calculate the minimum and maximum possible distance between particles as the two solutions, r_1 and r_2, of (see Exercise 8.13)

$$\Phi_{LJ}(r) = E - \frac{(E + \varepsilon)}{2} \frac{r_{min}^2}{r^2}, \tag{8.104}$$

as is shown, for $E = -\varepsilon/3 < 0$, in Figure 8.5(a). If we increase E, the difference between r_2 and r_1 increases. The trajectories in Figure 8.5(c) have been computed for $E = -\varepsilon/50$, where $r_1 \approx 1.03\,\sigma$ and $r_2 \approx 1.54\,\sigma$. For energies $E > 0$ there is only one solution r to equation (8.104), so that there is no bound on the separation of particles from above, and they escape from the influence of each other.

In Figure 8.5(c), we also observe some clockwise rotation. Here, one could find some similarities to Kepler's laws of planetary motion. Indeed, the same equation (8.99) would describe the behaviour of two planets, provided that we use the gravitational potential (proportional to r^{-1}). However, unlike in the case of planets, the orbits are not closed. In fact, if we simulate the trajectories for a very long time, then each particle visits every single point between two circles with radii r_1 and r_2 as is shown (for the blue particle) in Figure 8.5(d), by plotting a very long blue trajectory which is filling the annulus. Indeed, the French mathematician Joseph Bertrand proved in the nineteenth century that there are only two types of potentials that give closed orbits. One of them is the harmonic potential (8.100) and the other type includes potentials proportional to r^{-1}, which is the form of both gravitational and Coulomb potentials, given by (8.101).

The above example with $N = 2$ particles was useful for providing some initial insights into the properties of the Lennard-Jones potential (8.103), and to discuss its similarities and differences from the Coulomb potential (8.101). Considering $N = 3$ particles, it was already known to Henri Poincaré that a three-particle (three-planet) gravitational system can produce chaotic dynamics, a dynamical behaviour that is very sensitive to initial conditions. We can expect the same with the case of $N = 3$ particles that interact through the Lennard-Jones potential or for much larger values of N, which are discussed in the next section.

8.6 Molecular Dynamics

There exist a number of dedicated textbooks that provide an introduction to molecular dynamics (MD), including books by Frenkel and Smit (2002) and Alen and Tildesley (2017), covering many technical details of how molecular simulations can be performed in practice. In this section, we do not attempt to replicate this literature. Our goal is to provide a quick introduction to MD methods which will help us to put them into context with other parts of our book.

In biological applications, MD approaches are used to study how complex biomolecules move, deform and interact with other molecules over time. In mathematical terms, MD can be described by systems of ODEs (or SDEs) for the positions and velocities of atoms or beads (representing collections of atoms), which can also be subject to algebraic constraints. The resulting equations are then solved on a computer using a suitable numerical approach. Leimkuhler and Matthews (2015) present a detailed discussion of the numerical methods developed for MD simulations. Here, we explain MD using relatively simple examples, leaving many technical details to the literature.

MD methods discussed in this section are based on classical mechanics and are sometimes called molecular mechanics approaches in the computational chemistry literature (Lewars, 2016). Let us first reformulate our previous examples using the notation commonly used to describe MD models. Considering a molecular system involving N particles moving in three-dimensional physical space, we can describe its state at time t as a point in $6N$-dimensional phase space, i.e. $[\mathbf{q}(t), \mathbf{p}(t)] \in \mathbb{R}^{6N}$, where $\mathbf{q}(t) \in \mathbb{R}^{3N}$ and $\mathbf{p}(t) \in \mathbb{R}^{3N}$ are (column) vectors of positions and momenta, respectively. Using this notation, the case $N = 2$ studied in the previous Section 8.5 can be written in terms of $\mathbf{q}(t) = [\mathbf{X}(t), \mathbf{x}(t)] \in \mathbb{R}^6$ and $\mathbf{p}(t) = [m_p \mathbf{V}(t), m_s \mathbf{v}(t)] \in \mathbb{R}^6$, where m_p and m_s are the masses (now possibly different) of the two particles.

The dynamics of the state variable $[\mathbf{q}(t), \mathbf{p}(t)]$ can be written in terms of Hamiltonian dynamics as a system of ODEs,

$$\frac{d\mathbf{q}}{dt} = \mathbf{M}^{-1}\mathbf{p}, \tag{8.105}$$

$$\frac{d\mathbf{p}}{dt} = = -\nabla U(\mathbf{q}), \tag{8.106}$$

where $U(\mathbf{q})$ is the potential energy function and $\mathbf{M}$ is a diagonal $3N \times 3N$ matrix with the masses of the corresponding particles on the diagonal. In the case of $N = 2$ studied in Section 8.5, the potential energy function $U(\mathbf{q})$ is equal to $\Phi(|\mathbf{X} - \mathbf{x}|)$ and M is a diagonal 6×6 matrix which has the vector $[m_p, m_p, m_p, m_s, m_s, m_s]$ on its diagonal. Then equations (8.105)–(8.106) can be rewritten in the form (8.96)–(8.97) describing the behaviour of the particle at position $\mathbf{X}$. In a similar way (see Exercise 8.14), we can rewrite the harmonic oscillator MD model (8.62)–(8.65) in the Hamiltonian form (8.105)–(8.106).

One of the main limitations of all-atom MD simulations is that their direct application to the modelling of intracellular behaviour is restricted to modelling processes in relatively small domains over relatively short time intervals. Although all-atom MD simulations of systems consisting of a million atoms have been reported in the literature (Tarasova et al., 2017), it is beyond the reach of state-of-the-art computers to simulate intracellular processes that include transport of molecules between different parts of a cell, because this would require simulations of trillions of atoms. MD methods are therefore often used to simulate a small representative part of the intracellular space. We will illustrate MD here using two examples. Later, in Chapter 9, we discuss some modern multi-resolution methods that connect MD approaches with coarser stochastic reaction–diffusion methods and enable simulations with MD-level resolution in larger computational domains.

Our first illustrative MD example is a simulation of a noble gas, say argon. In this case, the simulated particles are neutral atoms, which only interact with each other through the Lennard-Jones force terms, given as derivatives of the Lennard-Jones potential (8.103). The resulting system is sometimes called a Lennard-Jones fluid. Here, we present a result computed by simulation of $N = 512$ atoms interacting through the Lennard Jones potential. More precisely, atoms are simulated in a cubic box (of fixed volume V), and we keep the temperature T constant, i.e. a computational chemist would say that we simulate a Lennard-Jones fluid with constant N, V and T.

To keep the number of particles, N, and the volume, V, fixed, we implement periodic boundary conditions, as shown schematically (in two dimensions) in Figure 8.6(a). Whenever a simulated atom leaves the cubic domain, it re-enters through the opposite side. Therefore, to calculate the total force on an atom at position $\mathbf{X}$ (denoted green in Figure 8.6(a)), we not only have to take into account the forces exerted by the atoms in the same cubic box (denoted red in Figure 8.6(a)), but also by their copies in neighbouring boxes (denoted blue). In this way, we would also have to take into account the forces exerted by the copies of the green atom itself (denoted light blue). In practice, we make use of the fact that the Lennard-Jones potential is short-ranged, i.e. we can neglect forces between atoms that are sufficiently far apart, say, further apart than a half of the side of the simulated cube. Then each red particle corresponds to only one force term on the right-hand side of the equation corresponding to the particle at position $\mathbf{X}$ in system (8.106) (or in equation (8.97)).

Equations (8.105)–(8.106) on their own do not correspond to constant temperature simulations, but to constant energy simulations. To see this, we define the Hamiltonian (total energy) corresponding to ODEs (8.105)–(8.106) which is the sum of kinetic and potential energies

$$H(\mathbf{q}, \mathbf{p}) = \frac{1}{2}\mathbf{p}^{\mathrm{T}}\mathbf{M}^{-1}\mathbf{p} + U(\mathbf{q}), \tag{8.107}$$

where we note that ^{T} in $\mathbf{p}^{\mathrm{T}}$ denotes the transpose of vector $\mathbf{p}$, while we have used (italic font) T to denote the temperature. In Exercise 8.15, we show that Hamiltonian (8.107) is constant for all solutions of ODEs (8.105)–(8.106), i.e. equations (8.105)–(8.106) with periodic boundary conditions can be used to model systems that keep constant N, V and energy (this is called the microcanonical ensemble in statistical mechanics). In biological applications, it is more common to use MD simulations with constant temperature (the so-called canonical ensemble). To keep the temperature T constant, we have a number of approaches to choose from. In our examples, we will use the extended variable method developed by Nosé (1984) and Hoover (1985), which introduces one additional auxiliary variable $\xi \equiv \xi(t)$ and modifies the right-hand side of equation (8.106) as follows:

$$\frac{d\mathbf{p}}{dt} = -\nabla U(\mathbf{q}) - \xi\mathbf{p}. \tag{8.108}$$

This is coupled with ODEs (8.105) and with the following evolution equation for ξ:

(a)

(b)

(c)

(d)

(e)

(f)

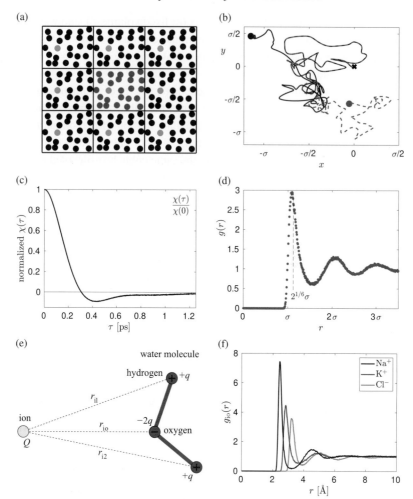

Figure 8.6 (a) *A schematic two-dimensional illustration of periodic boundary conditions.* (b) *Trajectory of one atom computed by short-time MD simulation of the Lennard-Jones fluid with N = 512 atoms (blue line). The same simulation with a very small change of initial conditions (red dashed line).* (c) *Normalized velocity autocorrelation function $\chi(\tau)$ estimated from long-time MD simulation of the Lennard-Jones fluid.* (d) *Radial distribution function $g(r)$ estimated, using equation (8.111), from long-time MD simulation of the Lennard-Jones fluid.* (e) *A schematic of an SPC/E water molecule together with an ion.* (f) *Radial distribution functions, g_{io}, of the oxygen atom distribution around each ion, estimated from long-time MD simulations of ions in the SPC/E water.*

$$\mu \frac{\mathrm{d}\xi}{\mathrm{d}t} = \mathbf{p}^{\mathrm{T}}\mathbf{M}^{-1}\mathbf{p} - N_{\mathrm{d}}k_B T, \tag{8.109}$$

where $\mu > 0$ is a parameter, T is the temperature, k_B is the Boltzmann constant and $N_{\mathrm{d}} = 3N$ is the number of degrees of freedom. The rationale behind the Nosé–Hoover approach for controlling temperature is similar to our discussion of temperature in Sections 4.3 and 4.6, or in equation (8.70), where we have associated the kinetic energy of one particle with absolute temperature T as $3k_B T/2$. The kinetic energy of the system of N particles is $\mathbf{p}^{\mathrm{T}}\mathbf{M}^{-1}\mathbf{p}/2$. Equations (8.108)–(8.109) make sure that it fluctuates around $3Nk_B T/2 = N_{\mathrm{d}}k_B T/2$, i.e. that we have

$$T = \frac{1}{N_{\mathrm{d}}k_B} \left\langle \mathbf{p}^{\mathrm{T}}\mathbf{M}^{-1}\mathbf{p} \right\rangle. \tag{8.110}$$

For example, if "measured temperature" $\mathbf{p}^{\mathrm{T}}\mathbf{M}^{-1}\mathbf{p}/(N_{\mathrm{d}}k_B)$ is temporarily higher than the desired value T, equation (8.109) implies that the auxiliary variable ξ is increasing. Equation (8.108) then slows down particles, thereby decreasing the "measured temperature" of the system.

Denoting $\mathbf{z}(t) = [\mathbf{q}(t), \mathbf{p}(t), \xi(t)]$, we can rewrite the system of $(6N+1)$ ODEs (8.105), (8.108) and (8.109) as a general system of ODEs, $\dot{\mathbf{z}} = \mathbf{f}(\mathbf{z})$, where $\mathbf{f}: \mathbb{R}^{6N+1} \to \mathbb{R}^{6N+1}$ is the right-hand side corresponding to (8.105), (8.108) and (8.109). To solve these ODEs, we need to prescribe the initial positions and velocities of all atoms and choose a suitable numerical method. Some ODE solvers work better than others with MD; see, for example, the book by Leimkuhler and Matthews (2015). Here, we implement the commonly used velocity Verlet algorithm to solve ODEs (8.105), (8.108) and (8.109). Since $N = 512 = 8^3$, we can easily prepare an artificial initial condition by arranging atoms into an $8 \times 8 \times 8$ regular lattice and giving them random initial velocities. We can then prepare a more realistic initial condition by starting with the artificial initial condition and solving ODEs (8.105), (8.108) and (8.109) for "some time". Importantly, quantities of interest do not depend on our specific choice of initial conditions, as we illustrate in an example.

In Figure 8.6(b), we plot a relatively short trajectory of one of the $N = 512$ simulated atoms (blue line) starting from the cross and finishing with the blue dot. Strictly speaking, the real trajectory is three-dimensional and we are here plotting its projection onto the x–y plane. Mathematically, we have just been solving a large system of $(6N + 1) = 3073$ ODEs (8.105), (8.108) and (8.109), which, given the initial positions and velocities of all atoms, uniquely predict their trajectories. Next, we very slightly change the initial position of one atom in our system. Its x-coordinate is changed by only $10^{-6}\%$, while the initial positions and velocities of all other atoms (including the atom whose trajectory was plotted in Figure 8.6(b)) are kept the same. We recompute the trajectory and plot it as the red dashed line in Figure 8.6(b). Both trajectories start from the cross and they are initially not distinguishable, but we see that they soon become completely different. Due to the chaotic nature of MD simulations, we cannot predict a single trajectory. We are more interested in computing quantities that contain some meaningful (predictable, average) information – quantities that are not so sensitive to small changes in initial conditions. Two examples are plotted in Figure 8.6(c) and (d).

In Figure 8.6(c), we plot the velocity autocorrelation function, $\chi(\tau)$, which is defined in (8.52). More precisely, we normalize it by $\chi(0)$ and plot the ratio $\chi(\tau)/\chi(0)$. It can be easily estimated from long-time MD simulations by computing how correlated is the current velocity (at time t) with the velocity at previous times. In a similar way, we can define and estimate other correlation functions, like the autocorrelation function for forces or the correlation function between forces and velocities. This information can then be used to estimate the kernel, $\kappa(\tau)$, of the corresponding generalized Langevin equation (8.29) by a number of methods developed in the literature (Shin et al., 2010; Gottwald et al., 2015; Jung et al., 2017). Indeed, if we compare Figures 8.1(b) and 8.6(c), we observe that the velocity autocorrelation function of our MD model has a similar behaviour to the velocity autocorrelation function of simple models, so one would also expect that the model can be described by the generalized Langevin equation with a numerically estimated kernel, $\kappa(\tau)$. This can be used for developing a coarse-grained stochastic description.

Coarse-grained models are useful for decreasing the computational intensity of simulations. They can be designed to match key macroscopic properties of MD models, like the diffusion constant, but they will obviously lack some details which can only be obtained from MD simulations. An example is plotted in Figure 8.6(d), where we plot the so-called radial distribution function. It is a density of (neighbouring) atoms as a function of the distance r from a given atom (here, the density of 1 corresponds to the density in the bulk), i.e. the radial distribution function tells us how far are the neighbouring atoms, on average, from a given atom. More precisely, if $n(r)$ particles are situated at a distance between r and $r + \Delta r$ from a given particle, we can estimate the radial distribution function $g(r)$ by averaging $n(r)$ over a long-time MD simulation:

$$g(r) \approx \frac{V}{N} \frac{\langle n(r) \rangle}{4\pi r^2 \Delta r}. \tag{8.111}$$

This quantity is plotted using red dots in Figure 8.6(d). The radial distribution function, $g(r)$, could then be formally defined by (8.111) in the limit $\Delta r \to 0$. We see that there are no atoms in close proximity to a given atom (for $r \ll \sigma$)

and the first peak of $g(r)$ is around $r_{\min} = 2^{1/6}\sigma$, which is consistent with the properties of the Lennard-Jones potential (8.103) discussed in Section 8.5 (compare with Figure 8.5(a) and (b)).

The above illustrative MD simulation has been based on solving ODEs (8.105), (8.108) and (8.109). Other MD methods can be written as systems of SDEs with additive noise, i.e. as $d\mathbf{z} = \mathbf{f}(\mathbf{z})\,dt + \mathbf{B}\,d\mathbf{W}$, where $\mathbf{z}(t)$ is a suitable state variable (for example, $\mathbf{z}(t) = [\mathbf{q}(t), \mathbf{p}(t)]$), $\mathbf{B}$ is a constant matrix and $\mathbf{W}$ is a vector of independent standard Wiener processes. An example is Langevin dynamics (Bussi and Parrinello, 2007; Leimkuhler and Matthews, 2013) which replaces equation (8.106) with a system of SDEs:

$$d\mathbf{p} = -\nabla U(\mathbf{q})\,dt - \gamma_{MD}\,\mathbf{p}\,dt + \sqrt{2\,\gamma_{MD}\,k_B T}\,\mathbf{M}^{1/2}\,d\mathbf{W}, \qquad (8.112)$$

where $\gamma_{MD} > 0$ is the parameter of the thermostat. This approach adds additional noise and friction terms to equation (8.106) in such a way that results obtained by SDEs (8.105) and (8.112) correspond to MD simulations at temperature T. Adding a similar stochastic Langevin-type thermostat to the Nosé–Hoover dynamics (8.108)–(8.109), we can also obtain the so-called Nosé–Hoover–Langevin method; see Leimkuhler and Matthews (2015) for the discussion of advantages and disadvantages of each MD approach. We conclude this section with our second illustrative MD example – simulations of ions in water.

There have been several MD models of liquid water developed in the literature. The simplest models include three sites in total – two hydrogen atoms and an oxygen atom – to describe one water molecule, while more complicated water models include four, five or six sites (Mark and Nilsson, 2001; Huggins, 2012). In this book, we use the three-site extended simple point charge (SPC/E) model of water (Berendsen et al., 1987) which was previously used for MD simulations of ions in aquatic solutions (Lee and Rasaiah, 1996; Koneshan et al., 1998). We investigate aquatic solutions of three ions (K^+, Na^+ and Cl^-) at 25°C using MD parameters presented in Lee and Rasaiah (1996).

An SPC/E water molecule is schematically shown, together with an ion, in Figure 8.6(e), where we denote by r_{io} (resp., r_{i1} and r_{i2}) the distance between the ion and the oxygen site (resp., the first and second hydrogen sites). In the SPC/E model, the charges ($q = 0.4238\,e$) on hydrogen sites are a distance $1\,\text{Å} = 10^{-10}\,\text{m}$ from the Lennard-Jones centre at the oxygen site, which has negative charge $-2q = -0.8476\,e$. The HOH angle is $109.47°$. The pair potential between the water molecule and the ion is then given by

$$\Phi_{LJ}(r_{io}) + \frac{k_e Q q_o}{r_{io}} + \frac{k_e Q q_h}{r_{i1}} + \frac{k_e Q q_h}{r_{i2}}, \qquad (8.113)$$

where Q is the charge of the ion (which is $+e$ for Na^+ and K^+, and $-e$ for Cl^-), k_e is the Coulomb constant and Φ_{LJ} is the Lennard-Jones potential defined by (8.103). The parameters of the Lennard-Jones potential are chosen to be $\varepsilon = 0.5216$ kJ mol^{-1} (for all three ions considered) and σ is equal to 2.876 Å (for Na^+), 3.250 Å (for K^+) and 3.785 Å (for Cl^-). The masses of each atom are well-known numbers which can be found in the periodic table.

The interactions between two SPC/E water molecules have a similar form to (8.113). They include all Coulomb interactions between charges on each site of each water molecule. They also include the Lennard-Jones term (8.103) between the oxygen sites with $\varepsilon = 0.6502$ kJ mol^{-1} and $\sigma = 3.169$ Å. Finally, we also have to make sure that we satisfy the constraints of fixed length of OH bonds and fixed HOH angle. We use the RATTLE algorithm (Andersen, 1983) to impose constraints between atoms of the same water molecule.

We consider a cube of side $L = 24.83$ Å containing 511 water molecules and one ion, i.e. we have $8^3 = 512$ molecules in our simulation box. To control temperature we again use the Nosé–Hoover thermostat (8.108)–(8.109) and the number of particles is kept constant by implementing periodic boundary conditions as shown schematically in Figure 8.6(a). In particular, we assume that our simulation box is surrounded by periodic copies of itself. Then the long-range (Coulombic) interactions have to be handled with care. Several different approaches have been developed in the literature, including Ewald summation and the reaction field method (Perera et al., 1995; Nymand and Linse, 2000). We use a cutoff sphere of radius $L/2$ and the reaction field correction as implemented in Koneshan et al. (1998). This approach is more suitable for some multi-resolution methods (discussed later in Chapter 9) than the Ewald summation technique. The MD time step is chosen as $\Delta t = 10^{-3}$ ps = 1 fs, which is a typical MD time step used in applications.

Once we prepare our simulation, we can estimate statistics of interest from long-time MD simulations. An example is shown in Figure 8.6(f), where we plot a radial distribution function. As discussed in (8.111), it gives a density of (neighbouring) atoms as a function of the distance r from a given atom. In Figure 8.6(f), we plot $g_{io}(r)$, which we define as the density of oxygen sites (of each water molecule) around the simulated ion. Since the oxygen site is negatively charged, it is not surprising that the first peak is closer to the ion for positively charged cations (Na^+ and K^+) than for a negatively charged anion (Cl^-). Alternatively, we could define a radial distribution function for positively charged hydrogen sites and we would obtain that the first peak is closer for the simulated anion (Cl^-) than for cations. We could also define radial distribution functions between water molecules or investigate relative orientations of water molecules around ions.

Detailed MD simulations can be used for studying single-ion solvation characteristics, and there are still a number of unsolved problems in this research area (Hunenberger and Reif, 2011). MD simulations allow us to consider different ions, or even hypothetical ions where we modify some parameters (for example, the ionic charge) that are difficult to change in experiments. Such MD simulations provide valuable insights into many questions in this field. We could also use other water models and other parameter sets for ions, which could to some extent change our computed results. The results in Figure 8.6(f) have been obtained from three independent long-time MD simulations (one for each ion), where we used one million time steps of length 1 fs, i.e. we have simulated each system over a nanosecond – a relatively very short time compared to times used in other chapters of this book. This motivates the development of coarse-grained stochastic models and multi-resolution methods studied in the next chapter.

Exercises

8.1 Show that the one-particle solvent model (8.4)–(8.7) can be equivalently written as equations (8.9)–(8.12) for four variables $X(t)$, $V(t)$, $U(t)$ and $Z(t)$, where $U(t)$ and $Z(t)$ are given by (8.8).

8.2 Solve equations (8.18)–(8.21), i.e. derive solution (8.22) for $U(k\Delta t)$.

8.3 Consider the full one-particle solvent model expressed in terms of four variables as (8.9)–(8.12). Show that the corresponding friction kernel $\kappa(\tau)$ can be written in the form (8.31). Prove the generalized fluctuation–dissipation theorem (8.30) for this $\kappa(\tau)$.

Hint: If $\alpha_4 < 4\alpha_2$, then $\alpha_5 = i\,|\alpha_5|$ is a purely imaginary number and formula (8.31) can be rewritten in terms of sine and cosine functions using $\sinh(i\,|\alpha_5|) = i\,\sin(|\alpha_5|)$ and $\cosh(i\,|\alpha_5|) = \cos(|\alpha_5|)$.

8.4 Consider limit $\alpha_4 \to \infty$ in (8.31). Show that $\kappa(\tau)$ in (8.31) converges to $\kappa(\tau)$ that we obtained for the three-variable model, namely to equation (8.26), in the limit $\alpha_4 \to \infty$.

8.5 Assume that the noise term $R(t)$ is given as a linear combination of normally distributed random numbers in equation (8.46). Find coefficients c_0, c_1, c_2, ..., c_{k_m} so that $R(t)$ satisfies the generalized fluctuation–dissipation theorem given in equation (8.45).

Hint: Substitute (8.46) into equation (8.45) to derive linear system (8.48).

8.6 Derive equation (8.50) for computation of coefficient c_0.

8.7 Derive equation (8.53) relating velocity autocorrelation function $\chi(\tau)$ and friction kernel $\kappa(\tau)$.

Hint: Application of the Laplace transform on equations of the form (8.29) turns the integral term (a convolution) into a product and differentiation into multiplication by the transform variable s. For example, applying the Laplace transform to (8.29) gives $s\mathscr{L}[V](s) - V(0) = -\mathscr{L}[\kappa](s)\mathscr{L}[V](s) + \mathscr{L}[R](s)$. You can also multiply (8.29) by $V(0)$ and average to get a similar equation for $\chi(\tau)$, which can then be Laplace transformed to derive (8.53).

8.8 Derive equation (8.57) for velocity autocorrelation function $\chi(\tau)$ of the three-equation model (8.14)–(8.16) by passing to the limit $\alpha_4 \to \infty$ in (8.53) and inverting the Laplace transform in (8.57).

8.9 Solve equations (8.64)–(8.65) and derive formula (8.66) for positions of heat bath particles of the solvent modelled as a system of harmonic oscillators.

8.10 Use the conservation of momentum and energy to derive formulae for post-collision velocities of the colliding diffusing (heavy) and solvent (light) particles, given in equations (8.78)–(8.79).

8.11 Derive formula (8.94) for the effective noise term g_i.

8.12 Consider a three-dimensional domain, a ball of radius $\overline{R}$, which contains a particle, a ball of radius $R < \overline{R}$, at its centre. Let r denote the distance from the common centre of both balls and assume that we have other particles uniformly distributed in the spherical shell $R \leq r \leq \overline{R}$, which all exert force (8.98) on the particle at the origin. Use the potential $\Phi(r) = r^{-a}$, where $a > 0$ is a constant, and denote the density of particles by ϱ. Calculate the total energy of their interactions, i.e. evaluate

$$\int_R^{\overline{R}} \Phi(r)\,\varrho\,4\pi r^2 \, dr,$$

and show that it can be considered independent of $\overline{R}$, provided that $\overline{R}$ is sufficiently large and $a > 3$. What happens for $a < 3$?

8.13 Use conservation of energy to derive equation (8.104).

Hint: The second term on the right-hand side of equation (8.104) arises from the angular velocity, which is non-zero even at the turning points r_1 and r_2.

8.14 Consider the solvent described as a system of harmonic oscillators studied in Section 8.3 and described by system of ODEs (8.62)–(8.65). Define the potential energy function $U(\mathbf{q})$ and state variable $[\mathbf{q}(t), \mathbf{p}(t)]$ so that equations (8.62)–(8.65) can be equivalently rewritten in the Hamiltonian form (8.105)–(8.106).

8.15 Differentiate the Hamiltonian (8.107) with respect to time and use ODEs (8.105)–(8.106) to show that the Hamiltonian (8.107) is constant for all solutions of ODEs (8.105)–(8.106).

9

Multiscale and Multi-Resolution Methods

In Chapters 6 and 8, we have discussed a number of approaches to stochastic reaction–diffusion modelling, which can be schematically summarized in the following diagram, where we have highlighted three main classes of methods. Going from left to right, we are decreasing both the level of model detail and its computational cost:

> molecular dynamics $\longrightarrow$ Brownian dynamics $\longrightarrow$ compartment-based SSA.

On the right, we have the compartment-based reaction–diffusion SSA, introduced in Section 6.1. It is the coarsest approach, which is based on dividing the computational domain into (well-mixed) compartments and describing the state of the system by the numbers of molecules in each compartment. Diffusion is modelled as jumps between neighbouring compartments, and the compartment-based SSA postulates that chemical reactions can only occur if the reacting molecules are in the same compartment.

Going from right to left in our diagram (against the arrows), we arrive at Brownian dynamics. An example has been introduced in Section 6.2 as a reaction–diffusion SSA based on the SDE diffusion model. Brownian dynamics simulations compute trajectories of individual biomolecules like we did in the SSA (a17)–(c17) by using steps (a8)–(c8) of the SDE diffusion model. This Brownian dynamics model effectively postulates that simulated molecules are points. Bimolecular (second-order) reactions occur whenever reacting molecules are "sufficiently close", but molecules do not have any volume. They can pass through each other. The term "Brownian dynamics" can also mean models with more detailed descriptions of diffusing molecules. For example, they can be described as hard balls (which cannot pass through each other), or even as particles with more complicated internal structure. A molecule could be represented as several beads (balls), representing parts of the simulated biomolecule, connected by springs. Such models can be obtained from all-atom molecular dynamics (MD) by an approach called ultra-coarse-graining. For example, Dama et al. (2013) ultra-coarse-grained a protein called actin to a model described by four beads connected by three springs. Adding more details to the description of molecules, we have moved (against the arrows in our diagram) towards the last class of approaches, collectively described here as MD.

All-atom MD has been the most detailed method mentioned in this book in Section 8.6, where we have illustrated MD on two examples. What is called MD in this book could also be called molecular mechanics in the computational chemistry literature, where it could be considered as one of the coarser models, if it is compared with more accurate quantum mechanics calculations. In particular, the arrow from MD to Brownian dynamics in our diagram really represents a multitude of different approaches, including the generalized Langevin description (8.28)–(8.29) in the middle.

Starting from all-atom MD, we can coarse-grain it by replacing a group of atoms with one coarse-grained particle. For example, each group of five atoms, say, could be replaced by a single particle. Such a coarse-grained MD approach can decrease the computational intensity of MD simulations and enable quicker simulations of quantities of interest. From coarse-grained MD we can then continue along the arrows in our diagram all the way to the compartment-based SSA, losing details and decreasing the computational cost of simulations. While all-atom MD is typically restricted to simulations of small systems containing a few biomolecules, the compartment-based SSA enables stochastic reaction–diffusion simulations of whole cells. The choice of a suitable method will depend on questions of interest. If a modeller is interested in too many details in a large domain, then the simulations can become computationally prohibitive.

In some applications a detailed modelling approach is important only in certain parts of the domain, whilst in the remainder of the domain a coarser, less detailed, method could be used. This is the case, for example, in the modelling of ion channels which we have introduced in Section 7.5 using a Brownian dynamics approach. Ions pass through a channel in single file and an individual-based model has to be used to accurately compute the discrete, stochastic, current in the channel. On the other hand the positions of individual ions are less important away from the channel where copy numbers may be very large. Another example is the stochastic reaction–diffusion modelling of filopodia, which are dynamic finger-like protrusions used by eukaryotic motile cells to probe their environment and help guide cell motility. These relatively small protrusions are connected to the rest of the cell.

If a modeller is interested in understanding the dynamics in the small part of the computational domain (e.g. in a filopodium or an ion channel), then there is potential to decrease the computational cost of simulations by using a coarser model in the remainder of the computational domain. To do this, one has to have a good understanding of mathematical limits in which a fine description converges to a coarser one. This analysis can help us to develop so-called multi-resolution (or hybrid, or multiscale) models which use a detailed model only in a small part of the computational domain (where this level of detail is necessary) and a coarser modelling description in the remainder of the computational domain.

In this last chapter, we use two examples to show how models offering different level of detail can be used in the same dynamic simulation; each example illustrates one arrow in our diagram. Our first example is presented in Section 9.1 where Brownian dynamics of a simple diffusion system is used together with its compartment-based description. Our second example is presented in Section 9.2 where the hard sphere MD model (introduced in Section 8.4) is used in a multi-resolution approach with its coarser Langevin description (8.81)–(8.82). This is followed in Section 9.3 by an overview of related multi-resolution approaches in the literature.

9.1 Coupling SDE-Based and Compartment-Based Models

Consider N molecules diffusing in the one-dimensional interval $[-L, L]$ with reflective boundary conditions. A molecular-based model of a similar process was presented as the SSA (a8)–(c8). Denoting the positions of molecules as $X_i(t)$, $i = 1, 2, \ldots, N$, $t \geq 0$, the position of the ith molecule at time $t + \Delta t$ can be computed using (4.11), i.e.

$$X_i(t + \Delta t) = X_i(t) + \sqrt{2D \Delta t}\, \xi_i, \qquad (9.1)$$

where ξ_i is a normally distributed (with zero mean and unit variance) random number, Δt is the time step and D is the diffusion constant. In Exercise 9.1, you are asked to modify the SSA (a8)–(c8) so that it can simulate the above system with reflective boundary conditions at $x = -L$ and $x = L$. Such a simulation gives the time evolution of the positions of individual molecules, and it is the most detailed approach used in this section.

Our goal is to present multi-resolution methods which use a compartment-based model in some parts of the computational domain. In this section we only illustrate some of the challenges of such multiscale modelling. We use a simple test problem which is shown in Figure 9.1(a). We divide the computational

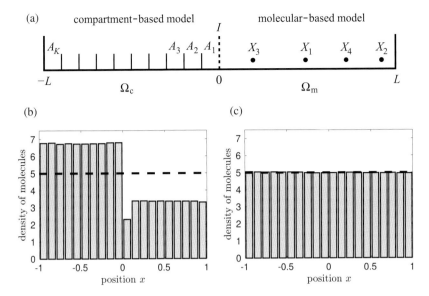

Figure 9.1 (a) *Schematic of the computational domain used in the SSA (a22)–(i22) for coupling a molecular-based model in* $\Omega_m = (0, L]$ *with a compartment-based model in* $\Omega_c = [-L, 0)$. (b) *Stationary distribution computed by the SSA (a22)–(i22), where* Φ *and* $f(x)$ *are given by (9.6) (grey histogram) compared with the exact solution given by (9.7) (dashed line).* (c) *Stationary distribution computed by the SSA (a22)–(i22), where* Φ *and* $f(x)$ *are given by (9.11) (grey histogram) compared with (9.7) (dashed line). The values of parameters are* $\Delta t = 10^{-4}$, $K = 10$, $h = 1/K = 1/10$, $L = 1$ *and* $D = 1$.

domain $[-L, L]$ into two parts $\Omega_c = [-L, 0)$ and $\Omega_m = (0, L]$ separated by the boundary (interface) $I = \{0\}$ (dashed line in Figure 9.1(a)). We use the compartment-based model in Ω_c and the molecular-based model in Ω_m. The compartment-based model can be formulated as the chain of "chemical reactions" (4.13). In our case, we divide the compartment-based subdomain Ω_c into K compartments of length $h = L/K$. We denote the number of molecules in the ith compartment $[-ih, -(i-1)h)$ by A_i, $i = 1, 2, \ldots, K$. Then the compartment-based model is given as the following chain of "chemical reactions":

$$A_K \underset{d}{\overset{d}{\rightleftarrows}} A_{K-1} \underset{d}{\overset{d}{\rightleftarrows}} \cdots \underset{d}{\overset{d}{\rightleftarrows}} A_3 \underset{d}{\overset{d}{\rightleftarrows}} A_2 \underset{d}{\overset{d}{\rightleftarrows}} A_1, \qquad (9.2)$$

where d is given by (4.14), i.e. $d = D/h^2$, and A_1 is the number of molecules in the compartment that is next to the interface I. In Section 4.2 we used the Gillespie SSA (a9)–(f9) to simulate the chemical system (9.2). To avoid technical details, we use a time-driven algorithm in this section instead of the Gillespie SSA. Then diffusive jumps in (9.2) can be simulated using the same finite time step Δt as we use in (9.1). A diffusive jump to the right from the ith compartment, $i = 2, 3, \ldots, K$, occurs with probability

$$d\,A_i(t)\,\Delta t = \frac{D}{h^2}\,A_i(t)\,\Delta t, \qquad (9.3)$$

and the same formula (9.3) can also be used for the jumps to the left. Next, we formulate the transition of molecules between Ω_c and Ω_m with the help of two parameters: a positive constant Φ, and a probability distribution function $f : [0, L] \to [0, \infty)$, where $\int_0^L f(x)\,dx = 1$. Possible choices of Φ and $f(x)$ are discussed below. We postulate that the transition rate for moving across the interface I from Ω_c to Ω_m is Φd, i.e. it is equal to the internal transition rate $d = D/h^2$ multiplied by the constant Φ. In particular, jumps from the first compartment to the right can also be written as the following first-order chemical reaction:

$$A_1 \xrightarrow{\Phi d} \text{transition to the molecular-based description in } \Omega_m. \qquad (9.4)$$

If a molecule jumps from the first compartment to the right, then a new molecule is initiated in Ω_m at a position which is sampled according to the distribution $f(x)$. On the other hand, if a molecule (which was in Ω_m at time t) crosses the interface during the time step $[t, t + \Delta t)$, then it will be incorporated into the first compartment in Ω_c.

Our choice of the same time step in both Ω_c and Ω_m (i.e. in both equations (9.3) and (9.1)) has been done to simplify the presentation of our illustrative multi-resolution algorithm, formulated on the next page as the SSA (a22)–(i22). In applications, it is more common to use a larger time step in a coarser model. In fact, we have already seen in Chapter 1 that the time-driven "naive" SSA (a1)–(b1) can be made more efficient by using the event-driven Gillespie SSA, which uses a larger time step than the SSA (a1)–(b1). To avoid errors, we could choose Δt so small in (9.3) that the probability of two events happening in the whole domain Ω_c during one time step $[t, t + \Delta t)$ is negligible. However, this could place significant restrictions on Δt. Alternatively, we could choose Δt to have $d\,A_i(t)\,\Delta t \ll 1$ in formula (9.3) to ensure that we are not missing any individual jumps. In this latter case, we could still have many jumps in Ω_c during one time step $[t, t + \Delta t)$ and we could obtain biases in our simulations depending on the order in which numbers A_i, $i = 1, 2, \ldots, K$, are updated. To avoid these problems, we store all changes to $A_i(t)$ in the variables δA_i, which are initialized as zero in the first step (a22). After adding all changes to δA_i, we update $A_i(t)$ to $A_i(t + \Delta t)$ in the last step (i22) using

$$A_i(t + \Delta t) = A_i(t) + \delta A_i, \qquad \text{for } i = 1, 2, \ldots, K. \tag{9.5}$$

We denote the positions of molecules in Ω_m by $X_i(t)$, $i = 1, 2, \ldots, N_m(t)$, where $N_m(t)$ is the number of molecules in subdomain Ω_m at time t. Then our rules for coupling Brownian dynamics with the compartment-based model can be formulated as the following multi-resolution SSA:

(a22) Set $\delta A_i = 0$ for $i = 1, 2, \ldots, K$.

(b22) Generate $K - 1$ random numbers r_i^L, $i = 1, 2, \ldots, K - 1$, uniformly distributed in $(0, 1)$.

    If $r_i^L < d\, A_i(t)\, \Delta t$, then $\delta A_i = \delta A_i - 1$ and $\delta A_{i+1} = \delta A_{i+1} + 1$.

(c22) Generate $K - 1$ random numbers r_i^R, $i = 2, 3, \ldots, K$, uniformly distributed in $(0, 1)$.

    If $r_i^R < d\, A_i(t)\, \Delta t$, then $\delta A_i = \delta A_i - 1$ and $\delta A_{i-1} = \delta A_{i-1} + 1$.

(d22) Compute the position of each molecule (which is in molecular-based subdomain Ω_m) at time $t + \Delta t$, i.e. compute $X_i(t + \Delta t)$ using (9.1) for $i = 1, 2, \ldots, N_m(t)$.

(e22) If $X_i(t + \Delta t)$, $i = 1, 2, \ldots, N_m(t)$, computed by (9.1) is greater than L, then use the reflective boundary condition

$$X_i(t + \Delta t) = 2L - X_i(t) - \sqrt{2D\, \Delta t}\, \xi_i.$$

(f22) If $X_i(t + \Delta t)$, $i = 1, 2, \ldots, N_m(t)$, computed by (9.1) is less than 0, then $\delta A_1 = \delta A_1 + 1$ and the trajectory of the ith molecule is terminated.

(g22) If $X_i(t + \Delta t)$, $i = 1, 2, \ldots, N_m(t)$, computed by (9.1) is greater than 0, then generate a random number r_1 uniformly distributed in $(0, 1)$. If $r_1 < \exp[-X_i(t + \Delta t)\, X_i(t)/(D\Delta t)]$, then $\delta A_1 = \delta A_1 + 1$ and the trajectory of the ith molecule is terminated.

(h22) Generate a random number r_2 uniformly distributed in $(0, 1)$. If $r_2 < \Phi\, d\, A_1(t)\, \Delta t$, then $\delta A_1 = \delta A_1 - 1$ and a new molecule is introduced at position $x \in \Omega_m$ sampled according to distribution $f(x)$.

(i22) Update numbers of molecules in all compartments using (9.5). Then continue with step (a22) for time $t + \Delta t$.

Steps (b22)–(c22) and (d22)–(e22) correspond to the simulations in Ω_c and Ω_m, respectively. In steps (b22)–(c22) we use the time-driven method to update the numbers of molecules in each compartment, i.e. we have implemented (9.2) using the probabilities (9.3). Steps (d22) and (e22) simulate trajectories of individual molecules in Ω_m and the reflective boundary condition at $x = L$, respectively.

Steps (f22)–(h22) implement the transfer of molecules across interface I. If a molecule (which was in Ω_m at time t) lands in Ω_c during time step $[t, t + \Delta t)$, then it is incorporated into the first compartment. This is implemented in steps (f22) and (g22), which can be justified using similar arguments to those used in Section 4.5. Clearly, if $X_i(t + \Delta t)$ computed by (9.1) is negative, a molecule needs to transfer from Ω_m to Ω_c – this is implemented in step (f22). However, as shown in (4.63), there is also a chance that a molecule crossed interface I during the finite time step $(t, t + \Delta t)$ even if $X_i(t + \Delta t)$ computed by (9.1) is positive; that is, during the time interval $(t, t + \Delta t)$ the molecule might have crossed interface I and then crossed back to X_i positive again. This probability is equal to (4.63). It turns out that the correct way to implement the interface conditions is to remove these molecules also and place them in the first compartment. This is implemented in step (g22). Finally, step (h22) simulates the transfer of molecules from Ω_c to Ω_m which is given by (9.4). Since our illustrative model only includes diffusion, we have a fixed number of molecules, N, in the whole domain $[-L, L]$, and this is also reflected in all steps of the SSA (a22)–(i22), which ensures the conservation of molecules given by $N = N_m(t) + \sum_{i=1}^{K} A_i(t)$ for all $t \geq 0$.

To use the SSA (a22)–(i22), we need to specify the value of Φ and distribution $f(x)$. Since the jump rate between compartments is equal to the same value, d, for all reactions in (9.2), it might look natural to impose the same transition rate for moving across the interface I from Ω_c to Ω_m, which would mean $\Phi = 1$. When a molecule moves from Ω_c to Ω_m, we sample its initial position in Ω_m according to distribution $f(x)$. One possibility is to assume that the initial position of each new molecule is estimated according to the uniform distribution in $[0, h]$. The main idea behind this ad hoc initialization is that the previous position of this molecule (at time t) was in the first compartment $[-h, 0]$. If (9.4) was interpreted as a jump to a neighbouring "compartment", then the molecule would land somewhere in $[0, h]$. Therefore we could simply postulate

$$\Phi = 1 \qquad \text{and} \qquad f(x) \equiv \begin{cases} 1/h & \text{for } x \in [0, h], \\ 0 & \text{otherwise.} \end{cases} \qquad (9.6)$$

This seems, on the face of it, reasonable. However, this ad hoc approach to multi-resolution modelling is not accurate and can lead to significant errors, as we show using a simple illustrative simulation in Figure 9.1(b).

To test the accuracy of the SSA (a22)–(i22) or any other multi-resolution approach, we need to first agree how the accuracy of a method is defined. In some applications, a hybrid model consists of a collection of computer codes, each of them based on a different mathematical technique and passing its output as an input of another computer code. The resulting hybrid model is then parametrized by fitting its results with available experimental data. However, even if we get a perfect fit of available experimental data, we cannot say that the hybrid approach is accurate and we have to be careful about interpreting any new predictions coming from the hybrid model. Such a data-driven approach hides that each individual computer code implements the corresponding mathematical method with a certain numerical error. If we obtain new phenomena from such hybrid models, we cannot be sure whether it is a numerical artefact or whether it could be observed in new experiments.

However, it is possible to study the accuracy and errors of the SSA (a22)–(i22) in a mathematically well-defined sense. Ideally, we would like the results of a multi-resolution approach be equal to the results that one would obtain by using the most detailed approach (if it was computationally feasible) in the whole domain. Thus, the error can be defined as a difference (or its suitable norm) between the computed spatio-temporal observable and the exact solution that would be obtained by the most detailed model. In Figure 9.1, this observable is the steady-state distribution, which can be analytically computed for our detailed model (see Exercise 9.2). It is a uniform density given by

$$n_s(x) = \frac{N}{2L}, \qquad \text{for } x \in [-L, L], \qquad (9.7)$$

which is plotted, for parameter values $L = 1$ and $N = 10$, as the dashed line in Figure 9.1(b) and (c). In Figure 9.1(b), we compare it with the result computed by using the SSA (a22)–(i22) where the coupling parameters are given by (9.6) and other simulation parameters are given as $\Delta t = 10^{-4}$, $K = 10$, $h = 1/K = 1/10$ and $D = 1$. To use (9.6), we have to sample from uniform distribution $f(x)$ in step (h22), which can be done by generating an additional random number r_0 uniformly distributed in $(0, 1)$ and introducing a new molecule at position $x = hr_0$ in Ω_m, i.e. we put $N_m(t + \Delta t) = N_m(t) + 1$ and $X_{N_m(t+\Delta t)}(t + \Delta t) = hr_0$.

In Figure 9.1(b), we clearly see that the SSA (a22)–(i22) with coupling parameters (9.6) fails to compute the exact solution. We obtain large errors because the algorithm does not correctly model the diffusive transport across the artificial interface I. In order to design a more accurate algorithm, we have to modify (9.6) by finding better values of Φ and $f(x)$. To do this, consider the SSA (a22)–(i22) for $N = 1$ molecule. We can equivalently describe this problem using mathematical equations, which can be analysed to get the correct values of Φ and $f(x)$. To simplify the resulting equations, we replace the finite domain, $[-L, L]$, by the infinite domain, $(-\infty, \infty)$. Then we do not need to worry about boundary conditions at $x = -L$ and $x = L$, which do not have to be considered to derive the correct values of coupling parameters Φ and $f(x)$.

Using $N = 1$, we denote by $p_i(t)$ the probability that the considered molecule is in the ith compartment, $i = 1, 2, 3, \ldots$. Let $p(x, t)$ be its probability density in $\Omega_m = (0, \infty)$. Using similar arguments as we used to derive the chemical master equation in Chapter 1, the random walk in the ith compartment can be described by

$$p_i(t + \Delta t) = \left(1 - 2\frac{D\Delta t}{h^2}\right) p_i(t) + \frac{D\Delta t}{h^2} (p_{i-1}(t) + p_{i+1}(t)), \quad \text{for } i \geq 2. \quad (9.8)$$

The same equation can also be derived from the diffusion master equation – see equation (4.24). Equation (9.8) describes steps (b22)–(c22) implementing (9.2) using the probabilities (9.3). The remaining steps of the SSA (a22)–(i22) modify the state of the first compartment and molecular-based subdomain Ω_m. They can be described using evolution equations for $p_1(t)$ and $p(x, t)$ in the following form (see Exercise 9.3):

$$p_1(t + \Delta t) = \left(1 - \frac{(1 + \Phi)D\Delta t}{h^2}\right) p_1(t) + \frac{D\Delta t}{h^2} p_2(t)$$

$$+ \frac{1}{h\sqrt{4\pi D\Delta t}} \int_0^\infty \int_0^\infty p(y, t) \exp\left(-\frac{(x + y)^2}{4D\Delta t}\right) dx\, dy \quad (9.9)$$

$$+ \frac{1}{h\sqrt{4\pi D\Delta t}} \int_0^\infty \int_0^\infty p(y, t) \exp\left(-\frac{xy}{D\Delta t}\right) \exp\left(-\frac{(x - y)^2}{4D\Delta t}\right) dx\, dy,$$

$$p(x, t + \Delta t) = \frac{1}{\sqrt{4\pi D\Delta t}} \int_0^\infty p(y, t) \left[\exp\left(-\frac{(x - y)^2}{4D\Delta t}\right) - \exp\left(-\frac{(x + y)^2}{4D\Delta t}\right)\right] dy$$

$$+ \frac{D\Delta t\, \Phi}{h} f(x)\, p_1(t). \quad (9.10)$$

Flegg et al. (2012) analysed equations (9.8)–(9.10) using the method of matched asymptotic expansions with the aim of finding the values of Φ and $f(x)$ that minimize the error at interface $I = \{0\}$. Such an analysis leads to the following formulae for Φ and $f(x)$:

$$\Phi = \frac{2h}{\sqrt{\pi D\Delta t}}, \qquad f(x) = \frac{\pi}{4D\Delta t} \operatorname{erfc}\left(\frac{x}{\sqrt{4D\Delta t}}\right), \qquad x \in \Omega_m, \quad (9.11)$$

where

$$\operatorname{erfc}(z) = \frac{2}{\sqrt{\pi}} \int_z^\infty \exp(-y^2)\, dy \quad (9.12)$$

is the complementary error function. This analysis also provides the justification for removing molecules that cross from the molecular domain Ω_m to the compartment domain Ω_c and back again during the interval $[t, t + \Delta t)$, as implemented in step (g22).

In Figure 9.1(c), we present illustrative results computed by the SSA (a22)–(i22), where Φ and $f(x)$ are given by (9.11) as the grey histogram. We get an excellent comparison with the exact result, the spatially homogeneous stationary distribution (9.7), plotted as the dashed line. We clearly see that the parameter choice (9.11) computes the stationary distribution more accurately than the parameter choice (9.6).

The accuracy of (9.11) is not limited to the computation of the stationary distribution of our simple diffusion model. The same approach can be applied to coupling Brownian dynamics models (studied in Section 6.2) with compartment-based stochastic reaction–diffusion models (from Section 6.1), as discussed by Flegg et al. (2012, 2014, 2015). Their analysis is based on the analysis of reactive (Robin) boundary conditions of stochastic diffusion models, which we introduced in Section 4.5. This is perhaps not surprising, because the interface between two subdomains (when different modelling regimes are used) can also be viewed as a reactive boundary of each subdomain.

One advantage of formula (9.11) is that it can be applied in simulations even by modellers who might not be interested in its mathematical derivation. All they need is a method to sample random numbers from probability distribution $f(x)$. In Figure 9.1(c), we have used an approach introduced in Exercises 9.4 and 9.5. These exercises present an algorithm to sample from the complementary error function distribution $\sqrt{\pi}\,\mathrm{erfc}(z)$, where $z \in (0, \infty)$. Once we sample such a random number, say ζ, we scale it as $\sqrt{4D\Delta t}\,\zeta$ to get a random number sampled from probability distribution $f(x)$ given by (9.11).

The main idea behind the algorithm in Exercise 9.4 is that the complementary error function distribution $\sqrt{\pi}\,\mathrm{erfc}(z)$ is relatively "close" to an exponential distribution for $z \in (0, \infty)$. Therefore we sample numbers from an exponential distribution which are sometimes accepted and sometimes rejected. We have already seen such an acceptance–rejection algorithm in Section 7.6, where the Metropolis–Hastings algorithm is discussed. The acceptance–rejection algorithm is efficient provided that we accept a significant number of random numbers. In Exercise 9.5, you are asked to optimize the parameters of the method from Exercise 9.4. They can be chosen so that 86% of proposed exponentially distributed random numbers are accepted. This acceptance–rejection method is also useful for simulations of the theoretical MD model introduced in Section 8.4, as we will show in the next section.

9.2 Coupling Molecular Dynamics with Langevin Dynamics

In the previous section, stochastic molecular-based modelling has been considered to be the most detailed approach, which we have coarse-grained in parts of the simulated domain by using the compartment-based approach. Here, we present a multi-resolution methodology where a stochastic approach is used as the macroscopic description of a more detailed MD model. Our illustrative MD example is an analytically tractable microscopic MD model of Brownian motion which was formulated in Section 8.4 in terms of a (heavy) diffusing particle (described as a ball B with radius R and mass m_p) colliding with a large number of (light) point solvent particles with masses $m_s \ll m_p$. Denoting the coordinates of the centre of the heavy particle by $\mathbf{X} = [X_1, X_2, X_3]$ and its velocity by $\mathbf{V} = [V_1, V_2, V_3]$, we showed that they converge to the solution of the Langevin equation (8.81)–(8.82) in the limit $m_p/m_s \to \infty$, provided that the density and velocity distributions of solvent particles are scaled accordingly.

In this section, we use this theoretical MD model to illustrate some multi-resolution approaches that have been designed in the literature to enable efficient simulations with an MD level of resolution. In Section 8.4, the MD model is formulated in the full three-dimensional space $\mathbb{R}^3$, i.e. we have infinitely many solvent particles. In order to simulate this model on a computer, we have to restrict to a finite computational domain. We consider that our computational domain is the cube $\Omega = [0, L] \times [0, L] \times [0, L]$, where $L > 2R$. Since the diffusing particle is the most important part of our simulation, we initially place it in the middle of our computational domain, i.e. $\mathbf{X}(0) = [L/2, L/2, L/2]$.

Assuming that Ω contains just one diffusing particle – the ball B – the number of solvent particles in Ω is distributed according to the Poisson distribution with mean $\lambda_\mu |\Omega \setminus B|$, where $|\Omega \setminus B|$ is the volume of the space available to solvent particles (i.e. the computational domain Ω without the ball B) and the constant λ_μ is given by equation (8.83).

To initialize the positions and velocities of the solvent molecules, we first sample a random number N_a from the Poisson distribution with mean $\lambda_\mu L^3$ and attempt to place N_a molecules at uniformly distributed positions inside the cube Ω. This can be done by sampling three independent uniformly distributed random numbers r_1, r_2 and r_3 in interval $[0, 1]$ for each solvent molecule and choosing its position in Ω as $\mathbf{x}^i = L[r_1, r_2, r_3]$. If the so-placed solvent molecule lies inside ball B, we remove it from our simulation. Then solvent molecules in $\Omega \setminus B$ are initially distributed according to the (spatial) Poisson distribution with density λ_μ. We also have to initialize their velocities according to the distribution (8.84)–(8.85). This can be done by sampling three independent normally distributed (with zero mean and unit variance) random numbers ξ_1, ξ_2 and ξ_3 for each solvent molecule and choosing its initial velocity as $\mathbf{v}^i = \sigma_\mu[\xi_1, \xi_2, \xi_3]$, where σ_μ is the constant given by (8.85).

Next, we have to decide whether we evolve time using a fixed time step, Δt, or whether we use an event-based algorithm. We choose the former approach. Then the time evolution of the positions of molecules during one time step can be updated by the "free-flight" equations

$$\mathbf{X}(t + \Delta t) = \mathbf{X}(t) + \mathbf{V}(t)\,\Delta t, \tag{9.13}$$

$$\mathbf{x}^i(t + \Delta t) = \mathbf{x}^i(t) + \mathbf{v}^i(t)\,\Delta t, \qquad \text{for } i = 1, 2, \ldots, N(t), \tag{9.14}$$

where $N(t)$ is the total number of solvent molecules at time t. If a solvent molecule does not collide with a large molecule, then the "free-flight" equation (9.14) provides the correct update rule for its position and its velocity does not change during the time step, i.e. $\mathbf{v}^i(t + \Delta t) = \mathbf{v}^i(t)$. Otherwise, we have to implement the elastic collision rule given by (8.78)–(8.79). The time evolution of the system can then be simulated by repeating the following three steps:

(a23) Calculate the "free-flight" positions of the heavy particle, $\mathbf{X}(t + \Delta t)$, and solvent molecules, $\mathbf{x}^i(t + \Delta t)$, using (9.13)–(9.14).

(b23) If there exists $j \in \{1, 2, \ldots, N(t)\}$ such that
$|\mathbf{X}(t + \Delta t) - \mathbf{x}^j(t + \Delta t)| < R$, then implement collisions of molecules according to (8.78)–(8.79).

(c23) Implement suitable boundary conditions on the boundary of Ω, with details given below.
Then continue with step (a23) for time $t + \Delta t$.

Note that we do not call algorithm (a23)–(c23) an SSA, because we have not yet introduced anything stochastic in steps (a23)–(c23). To simulate it on a computer, we have to provide more details of how steps (b23) and (c23) should be implemented. The key to accurate description of collisions in step (b23) is our choice of time step Δt. If it is chosen so small that only one collision happens in time interval $[t, t + \Delta t]$, then we can use the positions of colliding molecules at the beginning of the time step, $\mathbf{x}^j(t)$ and $\mathbf{X}(t)$, and their "free-flight" positions obtained by (9.13)–(9.14) to calculate both the time of the collision and the positions at the end of the time step, $\mathbf{x}^j(t + \Delta t)$ and $\mathbf{X}(t + \Delta t)$; see Exercise 9.6.

In step (c23), we have to implement suitable boundary conditions, because there will be some solvent molecules at time $t + \Delta t$ with their "free-flight" positions, computed in step (a23), outside Ω. One option is to use periodic boundary conditions as in Section 8.6, which are schematically shown (in two dimensions) in Figure 8.6(a). However, this would mean that the algorithm (a23)–(c23) would be a purely deterministic algorithm, which conserves energy at each step. In particular, periodic boundary conditions here would correspond to simulations with constant energy.

Since we want to study simulations with constant temperature, we consider our simulation box Ω as a part of a much larger domain. In fact, our MD model has been formulated in $\mathbb{R}^3$ in Section 8.4. Then solvent molecules that end up outside Ω are simply removed from our simulation in step (c23). In addition, we have to introduce new molecules that were outside Ω at time t, but are now inside Ω at time $t + \Delta t$.

How many new molecules should be introduced in step (c23)? First, consider the side of the cube Ω, which includes points with their first coordinate equal to zero. This part of the boundary consists of points of the form $(0, x_2, x_3)$, where $x_2 \in [0, L]$ and $x_3 \in [0, L]$. To simplify our presentation, we assume that we are far away from all other boundaries, so that we can effectively consider our MD model in the full space, $\mathbb{R}^3$, which is divided by a boundary plane with coordinates $(0, x_2, x_3)$, where $x_2 \in \mathbb{R}$ and $x_3 \in \mathbb{R}$. We are interested in solvent molecules that were outside of our computational domain at time t, but are inside it at time $t + \Delta t$.

Assume that heat bath particles are distributed according to the Poisson distribution with density λ_μ in the half space $(-\infty, 0) \times \mathbb{R}^2$ and their velocities are distributed according to (8.84)–(8.85) at time t. Next, compute the positions of these particles at time $t + \Delta t$ using "free-flight" translation, as in equation (9.14). Some of these particles will move to points in Ω, which we can identify in our simplified calculation with the half space $(0, \infty) \times \mathbb{R}^2$. A particle with velocity $\mathbf{v}$ which is at the point $\mathbf{x} \in \Omega$ at time $t + \Delta t$ was previously at the point $\mathbf{x} - \mathbf{v}\Delta t$ at time t. In particular, there will be new heat bath particles with velocity $\mathbf{v}$ at point $\mathbf{x}$ at time $t + \Delta t$ provided that $x_1 - v_1 \Delta t < 0$, which implies that the positions $\mathbf{x}$ and velocities $\mathbf{v}$ of these particles are distributed according to

$$H(-x_1 + v_1 \Delta t)\, \lambda_\mu\, f_\mu(\mathbf{v}), \qquad (9.15)$$

where $H(\cdot)$ is the Heaviside step function satisfying $H(z) \equiv 0$ for $z < 0$ and $H(z) \equiv 1$ for $z \geq 0$. Thus, in step (c23), we have to sample new particle positions and velocities using the distribution (9.15). One option would be to simply sample additional particles in Ω according to distribution $\lambda_\mu f_\mu(\mathbf{v})$, as we did when sampling the initial distribution, and then reject particles that do not satisfy condition $x_1 - v_1 \Delta t < 0$. However, this would be very inefficient – we would reject most of the molecules, because we have $x_1 \geq 0$ in our domain Ω and Δt is a small time step. Alternatively, we can integrate (9.15) to obtain the average number of new particles that move to Ω during one time step through the boundary at $x_1 = 0$ (see Exercise 9.7), which is given by

$$p_{new} = \frac{3\beta(\mu + 1)L^2\Delta t}{16\pi R^2}.$$ (9.16)

Assuming that Δt is chosen so small that (9.16) is much smaller than 1, we can interpret p_{new} as the probability of introducing one new heat bath particle through the boundary at $x_1 = 0$ during one time step. Thus, in step (c23), we can generate a random number r uniformly distributed in $(0, 1)$. If $r < p_{new}$, then we increase $N(t)$ by 1, and introduce a new heat bath particle in Ω. Alternatively, we can also generate a random number that is sampled from the Poisson distribution with mean p_{new} and introduce the corresponding number of particles. The two approaches are equivalent provided that Δt is chosen so small that the probability of introducing two particles during one time step is negligible. Due to symmetry, the same formula (9.16) is true for all other sides of cube Ω and we can implement the boundary condition in step (c23) using the same approach for all six sides of cube Ω. Once we know that a new heat bath particle should be introduced in Ω, we have to initialize its position and velocity. In Exercise 9.7, we derive the distribution of positions of new heat bath particles at point $\mathbf{x} \in \Omega$, provided that the heat bath particle enters domain Ω through the boundary at $x_1 = 0$. We obtain

$$\frac{\lambda_\mu}{2} \operatorname{erfc}\left(\frac{x_1}{\Delta t \, \sigma_\mu \sqrt{2}}\right), \qquad \text{for } x_1 \geq 0,$$ (9.17)

where erfc is the complementary error function given by (9.12). Thus we have to introduce the new heat bath particle at a position sampled according to the probability distribution proportional to (9.17). To sample from this distribution, we scale random number, ζ, sampled from the complementary error function distribution $\sqrt{\pi} \operatorname{erfc}(z)$, where $z \in (0, \infty)$; see Exercises 9.4 and 9.5 for an approach to sample from the complementary error function distribution. Once we sample such a random number, we initialize the position of the incoming solvent particle as $\mathbf{x}^i = [\Delta t \, \sigma_\mu \sqrt{2} \zeta, Lr_2, Lr_3]$, where r_2 and r_3 are uniformly distributed random numbers in (0.1).

The velocity of the incoming solvent particle is distributed according to the probability distribution proportional to $H(-x_1 + v_1\Delta t) f_\mu(\mathbf{v})$; see equation (9.15). This means that the second and third velocity coordinates, v_2 and v_3, are normally distributed, while the first velocity coordinate, v_1, comes from a truncated normal distribution, because we have to make sure that $v_1 > x_1/\Delta t$. One option to sample such a random number would be to generate normally distributed (with zero mean and unit variance) random number ξ_1 and attempt to choose $v_1 = \sigma_\mu \xi_1$. If $v_1 > x_1/\Delta t$, we would accept this value of initial veloc-ity. Otherwise, we would repeat the process. Clearly, the average probability of acceptance is 0.5 for $x_1 = 0$ and it is lower than 0.5 for $x_1 > 0$. In Exercise 9.8, we present an acceptance–rejection algorithm that has better acceptance prob-abilities. It is adapted from Robert (1995), where it is shown that its acceptance probability is 0.76 for $x_1 = 0$ and it converges to 1 as x_1 converges to infinity. A similar approach can also be used to initialize the positions and velocities of incoming solvent particles that enter through other sides of the cube.

In Figure 9.2(a), we present an illustrative trajectory of the diffusing ball obtained by the SSA (a23)–(c23). Since we have a stochastic implementation of boundary conditions, we now call our algorithm an SSA. In this simulation, we use dimensionless parameters $D = 1$, $\beta = 100$, $m_s = 1$, $m_p = 10^3$, $R = 1$, $L = 8$ and $\Delta t = 10^{-7}$. Consequently, equations (8.83), (8.85) and (9.16) imply that $\lambda_\mu \approx 47.3$, $\sigma_\mu \approx 316$ and $p_{new} \approx 3.82 \times 10^{-2}$. In particular, the simulation domain $\Omega \backslash B$ contains on average $\lambda_\mu |\Omega \backslash B| \approx 2.40 \times 10^4$ heat bath particles.

In Figure 9.2(a), we start our simulation with the diffusing molecule (yellow ball) in the middle of the domain, i.e. $\mathbf{X}(0) = [L/2, L/2, L/2]$, and we repeat steps (a23)–(c23) until the diffusing ball hits the boundary of cube Ω, which happened at time $t = 2.63$ of the presented realization of our algorithm. This corresponds to applying the SSA (a23)–(c23) over $2.63/\Delta t = 2.63 \times 10^7$ time steps during which we introduce, on average, $6p_{new} \times 2.63/\Delta t = 6.03 \times 10^6$ new heat bath particles to the cube Ω. In particular, at any given time point we are only keeping track of a relatively small fraction (0.4%) of all the heat bath molecules that passed through Ω during the simulation.

At the end of our simulation in Figure 9.2(a), the diffusing ball hits the boundary of the cube Ω. If we want to continue our simulation, we need to specify suitable boundary conditions for the ball. We have several possibili-ties, depending on the goals of our simulation. If the cube Ω is the subdomain in which we are interested in using the most detailed MD model, we can apply a multi-resolution approach by coupling the MD model in the cube Ω with a coarser model in the rest of the simulation domain. In our case, we want to use the Langevin description (8.81)–(8.82) for the heavy particle when it leaves the domain Ω. Then boundary conditions for heat bath particles can be imple-mented as in step (c23), while we have to carefully consider how we handle the transfer of the diffusing molecule between Ω and its surroundings.

Put differently, we are interested in simulating the behaviour of the diffus-ing molecule as though the MD model was applied in the whole space $\mathbb{R}^3$, but we want to save computational time by replacing solvent particles that are outside Ω by a coarse Langevin-type description. In Figure 9.2(b), we present a two-dimensional illustration of the heavy molecule leaving the domain Ω.

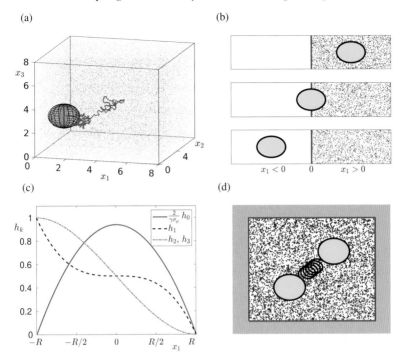

Figure 9.2 (a) *One realization of the SSA (a23)–(c23). The trajectory of the diffusing ball (red line) is calculated until the ball (which started in the middle of the domain) hits the boundary of Ω. It is plotted together with the final positions of solvent molecules (blue dots) and the ball (yellow). The parameters are $D = 1$, $\beta = 100$, $m_s = 1$, $m_p = 10^3$, $R = 1$, $L = 8$ and $\Delta t = 10^{-7}$. (b) A two-dimensional illustration of a diffusing molecule crossing the boundary $\{x_1 = 0\}$ between two subdomains where the Langevin description $(x_1 < 0)$ and MD simulations $(x_1 > 0)$ are used, respectively. (c) Functions $h_k \colon [-R, R] \to \mathbb{R}$, $k = 0, 1, 2, 3$, given by equations (9.22)–(9.23). (d) A schematic of a dimer molecule studied in Exercises 9.10, 9.11 and 9.12.*

The top panel shows the molecule inside Ω, the bottom panel shows it outside Ω, while the middle panel shows the heavy molecule when it intersects the side $\{x_1 = 0\}$. For the situation in the top panel we have formulated the SSA (a23)–(c23), while for that in the bottom panel we have the simplified Langevin description (8.81)–(8.82) (with the corresponding SSA developed in Exercise 4.7). However, we have yet to determine how to model the situation in the middle panel. In such a position the molecule is subject to explicitly simulated collisions with heat bath particles on the part of its surface that lies in Ω. This has to be compensated by adding a suitable random force (generated by those collisions not simulated on the other part of its surface), so that the overall model is equivalent to (8.81)–(8.82) in the Langevin limit. We finish this section by calculating this additional force. To simplify our presentation, we use the same time step in the whole domain, although efficient multi-resolution methods use different time steps in different parts of the computational domain.

The diffusing molecule is the ball with centre $\mathbf{X} = [X_1, X_2, X_3]$ with velocity $\mathbf{V} = [V_1, V_2, V_3]$ and radius R. It intersects interface $\{x_1 = 0\}$ if $X_1(t) \in (-R, R)$. Heat bath particles in Ω are simulated using the MD model. We do not explicitly simulate the heat bath particles outside Ω, but model their effect by an additional correction to the velocity of the diffusing molecule in the form

$$V_i(t + \Delta t) = \widetilde{V}_i(t + \Delta t) + f_i(t)\,\Delta t + g_i(t)\,\sqrt{\Delta t}\,\xi_i, \quad i = 1, 2, 3, \tag{9.18}$$

where Δt is a small time step and $\widetilde{\mathbf{V}}(t + \Delta t)$ is the post-collision velocity of the diffusing molecule at time $t + \Delta t$, which takes into account only collisions with the heat bath particles from inside Ω. Equation (9.18) is very similar to equation (8.86), with a drift term $f_i(t)\,\Delta t$ and a noise term $g_i(t)\,\sqrt{\Delta t}\,\xi_i$, where ξ_i is a normally distributed random number with zero mean and unit variance. In Section 8.4 we replaced all collisions with heat bath particles by Langevin-style drift and noise terms; this time we need to calculate effective drift and noise coefficients for those collisions with particles from outside Ω only.

Following the same approach as in Section 8.4 we find (see Exercise 9.9)

$$f_1(t) = -h_1(X_1(t))\,\beta\,V_1(t) + h_0(X_1(t)), \tag{9.19}$$

$$f_i(t) = -h_i(X_1(t))\,\beta\,V_i(t), \qquad \text{for } i = 2, 3, \tag{9.20}$$

$$g_i(t) = \sqrt{h_i(X_1(t))}\,\beta\,\sqrt{2D}, \qquad \text{for } i = 1, 2, 3, \tag{9.21}$$

where the functions $h_k : [-R, R] \to \mathbb{R}$ are given by

$$h_0(x_1) = \frac{3\beta\sigma_\mu\sqrt{\pi}}{8\sqrt{2}}\left(1 - \frac{x_1^2}{R^2}\right), \qquad h_1(x_1) = \frac{1}{2}\left(1 - \frac{x_1^3}{R^3}\right) \tag{9.22}$$

and

$$h_2(x_1) = h_3(x_1) = \frac{1}{4}\left(2 - 3\frac{x_1}{R} + \frac{x_1^3}{R^3}\right). \tag{9.23}$$

In Figure 9.2(c), we plot functions $h_k : [-R, R] \to \mathbb{R}$ for $k = 0, 1, 2, 3$. Substituting $x_1 = -R$ into equations (9.22)–(9.23), we obtain

$$h_0(-R) = 0, \quad h_1(-R) = h_2(-R) = h_3(-R) = 1.$$

Consequently, equations (9.19)–(9.21) give the same drift and diffusion coefficients as the Langevin model (8.81)–(8.82) for $x_1 = -R$. On the other hand, substituting $x_1 = R$, we obtain $h_k(R) = 0$ for $k = 0, 1, 2, 3$. Consequently, the multi-resolution model smoothly interpolates between the MD model with explicitly modelled heat bath (for $x_1 = R$) and the Langevin description (for $x_1 = -R$).

Equation (9.18) can be used to design a multi-resolution scheme which couples the presented MD model with its corresponding Langevin description. Considering the boundary $\{x_1 = 0\}$, the multi-resolution scheme simply uses the MD model for $X_1 \geq R$, the Langevin description for $X_1 \leq -R$ and equation (9.18) for $X_1 \in (-R, R)$. The other sides of the cube Ω are treated in a similar way. Technical details of such a multi-resolution scheme can be found in the paper by Erban (2014). We conclude this chapter with a discussion of other multi-resolution schemes developed in the literature.

9.3 Multi-Resolution Molecular and Brownian Dynamics

In Section 9.2, we have illustrated a multi-resolution approach using a theoretical MD model that converges to the Langevin description (8.81)–(8.82) in a certain limit. In particular, we can use this limiting process to replace the MD model by the corresponding Langevin description in a part of the simulated domain. To get closer to MD applications of multi-resolution modelling, we would need to replace our illustrative theoretical MD model with all-atom MD modelling, which has been discussed in Section 8.6. The heat bath would then be described by one of the many common molecular water models, for example, by the three-site SPC/E model schematically shown in Figure 8.6(e) or by other models developed in the literature to capture some of the relatively unusual properties of water. Indeed, water is such an important solvent that whole monographs have been written on its molecular-level modelling, dynamics and biological function; see, for example, books by Ben-Naim (2009) and Bagchi (2013). In addition to water, we would have to describe the diffusing biomolecule (which was simply a ball in Section 9.2) with the same level of atomistic resolution. This has become a relatively straightforward step in recent years, because there have been a number of online resources developed, where freely available data files can be downloaded, describing biological macromolecules (like proteins or nucleic acids) in a format that can be used in standard MD software packages (Berman et al., 2000). These databases continue to grow as new data are obtained in experiments.

Direct application of all-atom MD simulations is limited to modelling processes in relatively small domains over relatively short time intervals. Coarse-grained MD models have been developed to overcome this difficulty. Coarse-grained models represent groups of atoms by single sites or beads to reduce the number of degrees of freedom in MD simulations and allow for larger time steps, typically 10–20 femtoseconds, rather than 1–2 femtoseconds used in all-atom MD (Hadley and McCabe, 2012). Multi-resolution (hybrid) MD methods use all-atom MD and coarse-grained MD simulations in different parts of the simulation domain (Ensing et al., 2007, Praprotnik et al., 2008; Potestio et al., 2014). In these multi-resolution methodologies, the microscopic structure of interest is often placed in the centre of the simulation domain and it is solvated using a detailed atomistic MD water model in its immediate neighbourhood, which is coupled with a coarse-grained water description in the rest of the computational domain. Such multi-resolution techniques have recently enabled MD simulations of virus-like particles (Machado et al., 2017; Tarasova et al., 2017). Considering our illustrative MD model used in Section 9.2, we could design a similar multi-resolution scheme by assuming that our computational domain depends on time t as

$$\Omega \equiv \Omega(t) = \mathbf{X}(t) + \left[-\frac{L}{2}, \frac{L}{2}\right] \times \left[-\frac{L}{2}, \frac{L}{2}\right] \times \left[-\frac{L}{2}, \frac{L}{2}\right], \qquad (9.24)$$

where $\mathbf{X}(t) = [X_1, X_2, X_3]$ is the centre of the diffusing particle (ball with radius R) and $L > 2R$ is the length of the side of the co-moving cube (9.24). Gunaratne et al. (2019) generalized the approach of Section 9.2 to derive appropriate boundary conditions, which can be used in step (c23) in the case of time-dependent domain $\Omega(t)$. Then the diffusing molecule never escapes the subdomain where the MD model is used. Such an approach therefore enables long-time MD simulations, which can be used to estimate the details of the behaviour of the diffusing molecule.

Another type of multi-resolution methodology has been introduced for modelling large macromolecules, where a detailed model of an important part of a macromolecule is coupled with a coarser model of the rest of the macromolecule. For example, atomistic detail of the active part of an enzyme has been coupled with a coarser model of the rest of the protein (Fogarty et al., 2016), different resolutions have been used in the bead–spring modelling of DNA (Rolls et al., 2017) or for the modelling of polymer melts (Di Pasquale and Carbone, 2017). One important feature of these schemes (which has been left out from our discussion in Section 9.2) is the discretization of time. Although our illustrative simulations use the same time step for both the MD model and the Langevin description, this is not the most efficient or desirable strategy, because the (detailed) MD model requires a much smaller time step than the corresponding (coarser) Langevin equation. There is potential to design more efficient schemes by updating the coarser description only at certain multiples of the time step that is used in the most detailed model. This is also the case when a modeller further coarse-grains the Langevin description into a Brownian dynamics model which uses even large time steps. Indeed, we have already seen in Chapter 5 that one can design efficient and accurate SSAs by making use of the separation of time scales.

The MD model used in Section 9.2 converges to the Langevin description (8.81)–(8.82), while we have seen in Chapter 8 that more complicated MD models will better be described by the generalized Langevin equation (8.28)–(8.29). Equivalently, such systems could be described by introducing a few additional variables. For example, the one-particle solvent model (8.9)–(8.12) in Section 8.1 is an example of a stochastic coarse-grained model written with the help of additional variables and parameters, which can be estimated from detailed MD models of ions and can be used to design a multi-resolution approach (Erban, 2016). Additional variables can also be introduced to improve the time dependence of coarse-grained models estimated from detailed MD simulations (Davtyan et al., 2015).

To get further insights into multi-resolution methodologies at the MD level, we have included Exercises 9.10, 9.11 and 9.12 where the reader is asked to work with a dimer molecule, schematically depicted in Figure 9.2(d). The dimer molecule is modelled in each exercise as two monomers (balls) connected by a spring. We denote the positions of the centres of the first (resp. second) monomer by $\mathbf{X}_1 = [X_{1;1}, X_{1;2}, X_{1;3}]$ (resp. $\mathbf{X}_2 = [X_{2;1}, X_{2;2}, X_{2;3}]$). Each ball has the same mass, m_p. We denote by $\mathbf{r}$ the vector describing the separation between monomers, i.e. $\mathbf{r} = \mathbf{X}_2 - \mathbf{X}_1$. The interaction between monomers is given in terms of the potential $\Phi \equiv \Phi(r) : [0, \infty) \to \mathbb{R}$, where $r = |\mathbf{r}|$. This potential generates an equal and opposite force on each of the monomers with magnitude $\Phi'(r)$. Macroscopic equations describing the dimer are given as the following system of SDEs:

$$\mathbf{X}_i(t + dt) = \mathbf{X}_i(t) + \mathbf{V}_i(t)\, dt, \tag{9.25}$$

$$\mathbf{V}_i(t + dt) = \mathbf{V}_i(t) + (-1)^{i+1} \frac{\Phi'(r)}{m_p} \frac{\mathbf{r}}{r}\, dt - \beta\, \mathbf{V}_i(t)\, dt + \beta\, \sqrt{2D}\, d\mathbf{W}_i, \tag{9.26}$$

for $i = 1, 2$, where $\mathbf{V}_i = [V_{i;1}, V_{i;2}, V_{i;3}]$ are the velocities of the first and second monomer, respectively, $\mathbf{W}_i$ are three-dimensional vectors of independent Wiener processes, D is a diffusion coefficient and β is a friction coefficient. In Exercise 9.10, you are asked to study the macroscopic model (9.25)–(9.26) for different potentials Φ. In Exercise 9.11, the dimer's behaviour is analysed using the theoretical MD heat baths introduced in this book. Although some analytic progress can be made with such heat baths, they lack some properties of real water models and can only provide partial insights into the issues surrounding multi-resolution techniques. One problem is that the heat bath particles do not interact with each other. A more realistic theoretical model, which can be used to mimic hydrodynamic interactions, is given by a heat bath consisting of the Lennard-Jones fluid introduced in Section 8.6. We use it in our concluding Exercise 9.12.

The use of multiscale (hybrid) models is becoming more widespread, especially with the growing computational power that allows us to consider more complex systems in this manner. Considering Brownian dynamics modelling, we have presented an illustrative multi-resolution scheme in Section 9.1, where Brownian dynamics is coarse-grained using a compartment-based approach in parts of the computational domain. The decomposition of the computational domain into subdomains (i.e. the position of the interface I in Section 9.1) can also be made time-dependent. Robinson et al. (2014) use such an approach for simulating systems with travelling waves, where the subdomain requiring a detailed approach changes its location with time.

Another possible way to accelerate Brownian dynamics simulations for some systems is to consider event-based algorithms (van Zon and ten Wolde, 2005; Takahashi et al., 2010) or by using a different coarser model. For example, in Chapter 6, we have shown that the macroscopic mean-field description of a reaction–diffusion process can be written in terms of the corresponding reaction–diffusion PDEs for concentrations of the chemical species involved. Multi-resolution models that couple a stochastic reaction–diffusion approach with macroscopic PDEs have been studied by a number of authors; see, for example, the recent review by Smith and Yates (2018).

Another class of multiscale (hybrid) algorithms is used in spatio-temporal modelling of populations of cells and can be applied to models of chemotaxis (which we introduced in Section 7.3). For example, Dallon and Othmer (1997) developed a hybrid model for chemotaxis of slime mold *Dictyostelium discoideum* in which the cells are treated as individuals in a continuum field of the chemoattractant which evolves according to a reaction–diffusion PDE. These hybrid models couple individual-based descriptions of cells with PDE models of their environment; i.e. both an individual-based description and the macroscopic PDEs are used in the whole computational domain, but they describe different components of the studied system; see, for example, Franz and Erban (2012) for further discussion of such hybrid methods.

Exercises

9.1 Consider a system of $N = 10^3$ molecules diffusing in one-dimensional interval $[-L, L]$ with reflective boundary conditions, where the position of each molecule evolves according to (9.1). Modify the SSA (a8)–(c8) to write an SSA for simulation of this system. Implement it on a computer, calculate a trajectory of one molecule and the steady-state distribution of all molecules. Use $D = 1$ and an initial condition where all molecules are uniformly distributed in the left half of the domain, i.e. in interval $[-L, 0]$.

9.2 Consider the model from Exercise 9.1. Let $n(x, t)$ be the density of molecules, which are at point $x \in [-L, L]$ at time t. Write a PDE describing the time evolution of $n(x, t)$ and the corresponding boundary conditions. Solve the PDE for the steady-state solution to derive (9.7).

9.3 Consider $N = 1$ molecule that evolves according to the SSA (a22)–(i22) which does not include the step (e22), i.e. we do not use the reflective boundary condition at $x = L$ and we have $\Omega_m = (0, \infty)$. Denote by $p_1(t)$ (resp. by $p_2(t)$) the probability that the considered molecule is in the first (resp. in the second) compartment. Let $p(x, t)$ be its probability density in Ω_m. Derive equations (9.9)–(9.10) for the time evolution of $p_1(t)$ and $p(x, t)$ over one time step of the SSA.

9.4 Let a_1 and a_2 be positive constants such that inequality

$$z \leq a_2 \operatorname{erfc}(-a_1 \log(z)) \qquad \text{holds for all } z \in [0, 1]. \qquad (9.27)$$

Consider the following acceptance–rejection algorithm:

> - Generate a random number ζ_1 uniformly distributed in $(0,1)$.
> - Compute exponentially distributed random number ζ_2 by $\zeta_2 = -a_1 \log(\zeta_1)$.
> - Generate a random number ζ_3 uniformly distributed in $(0,1)$.
> - If $\zeta_1 \zeta_3 < a_2 \operatorname{erfc}(\zeta_2)$, then choose ζ_2 as a sample from the probability distribution $\sqrt{\pi} \operatorname{erfc}(z)$. Otherwise, repeat the algorithm.

Show that the sequence of random numbers generated by this algorithm are indeed random numbers that are distributed according to the complementary error function distribution $\sqrt{\pi} \operatorname{erfc}(z)$, where $z \in (0, \infty)$.

Show that the average probability to accept a generated exponentially distributed random number by this algorithm is

$$\frac{a_2}{a_1 \sqrt{\pi}}.\tag{9.28}$$

9.5   What is the best choice of parameters a_1 and a_2 of the acceptance–rejection algorithm in Exercise 9.4? Consider the values

$$a_1 = 0.532 \qquad \text{and} \qquad a_2 = 0.814.\tag{9.29}$$

Show that condition (9.27) is satisfied. Calculate the average acceptance probability of the acceptance–rejection algorithm in Exercise 9.4 by using (9.28).

9.6   Consider a heavy particle, modelled as a ball with radius R with its centre at position $\mathbf{X}(t)$, and a solvent molecule at $\mathbf{x}^j(t)$. Compute $\mathbf{X}(t + \Delta t)$ and $\mathbf{x}^j(t + \Delta t)$ according to the "free-flight" equations (9.13)–(9.14) and assume that they overlap, i.e. $|\mathbf{X}(t + \Delta t) - \mathbf{x}^j(t + \Delta t)| < R$. Calculate the time $t + \tau \in (t, t + \Delta t)$ when the solvent molecule collided with the heavy molecule. Write the equations for the correct positions of molecules at time $t + \Delta t$, modelling this collision according to equations (8.78)–(8.79).

9.7   Suppose that heat bath particles are distributed according to the (spatial) Poisson distribution with density λ_μ in half space $(-\infty, 0) \times \mathbb{R}^2$ and their velocities are distributed according to (8.84)–(8.85). Assume that there are no heat bath particles in $(0, \infty) \times \mathbb{R}^2$. Show that the average number of heat bath particles that move to domain $(0, \infty) \times (0, L)^2$ at time $t + \Delta t$ is equal to p_{new} given by equation (9.16), the positions and velocities of these particles are distributed according to distribution (9.15) and the positions of heat bath particles at point $\mathbf{x} \in (0, \infty) \times \mathbb{R}^2$ are distributed at time $t + \Delta t$ according to distribution (9.17).

Hint: Integrate (9.15) over all possible velocities $\mathbf{v} \in \mathbb{R}^3$ to obtain the distribution of positions of new heat bath particles at point $\mathbf{x} \in \Omega$ at time $t + \Delta t$, given by equation (9.17). Integrating this formula over $\mathbf{x}$ in domain $(0, \infty) \times (0, L)^2$, we obtain (9.16), the average number of particles that are in Ω at time $t + \Delta t$.

9.8   Let $A \in \mathbb{R}$. Consider the truncated normal distribution defined on the interval (A, ∞) by its probability distribution

$$p(z) = \frac{\sqrt{2}}{\sqrt{\pi}\,\text{erfc}(A/\sqrt{2})} \exp\left(-\frac{z^2}{2}\right), \qquad \text{for } z \in (A, \infty),\tag{9.30}$$

where we note that distribution (9.30) converges, in the limit $A \to -\infty$, to the standard normal distribution with zero mean and unit variance. Define the constant a by

$$a = \frac{A + \sqrt{A^2 + 4}}{2}$$

and consider the following acceptance–rejection algorithm:

- Generate a random number ζ_1 uniformly distributed in $(0,1)$.
- Compute translated exponentially distributed random number ζ_2 by $\zeta_2 = A - \log(\zeta_1)/a$.
- Generate a random number ζ_3 uniformly distributed in $(0,1)$.
- If $\zeta_3 < \exp(-(\zeta_2 - a)^2/2)$, then choose ζ_2 as a sample from truncated normal distribution (9.30). Otherwise, repeat the algorithm.

Show that the sequence of random numbers generated by this algorithm are indeed random numbers that are distributed according to the probability distribution (9.30).

9.9 Consider a heavy particle, modelled as a ball with radius R with its centre at position $\mathbf{X} = [X_1, X_2, X_3]$ where $X_1 \in [-R, R]$, in a heat bath of light solvent molecules, as in Sections 8.4 and 9.2. Suppose that solvent molecules in half space $(-\infty, 0) \times \mathbb{R}^2$ are not explicitly simulated, but replaced by the drift term $f_i(t) \Delta t$ and noise term $g_i(t) \sqrt{\Delta t} \, \xi_i$ in equation (9.18). Show that $f_i(t)$ and $g_i(t)$ are given by equations (9.19)–(9.21), where $h_k \colon [-R, R] \to \mathbb{R}$, $k = 0, 1, 2, 3$, are given by (9.22)–(9.23).

Hint: Adapt the approach used in Section 8.4. Integrate in equation (8.92) (resp. equation (8.94)) over the part of the surface that is inside $(-\infty, 0) \times \mathbb{R}^2$, i.e. inside the region where we want to replace collisions with solvent molecules by an SDE with drift coefficient $f_i(t)$ (resp. noise coefficient $g_i(t)$).

9.10 Consider a dimer molecule, modelled as two balls connected by a spring, as shown schematically in Figure 9.2(d) and described by macroscopic equations (9.25)–(9.26). Use the harmonic potential $\Phi_{\mathrm{ha}}(r)$ introduced in (8.100) and derive a formula for the expected length of the dimer, the diffusion constant of its centre of mass $(\mathbf{X}_1 + \mathbf{X}_2)/2$ and the autocorrelation function of its velocity $(\mathbf{V}_1 + \mathbf{V}_2)/2$. How do the results change if the potential Φ_{ha} is replaced by a harmonic spring with rest length $\ell_0 > 0$, i.e. by $\Phi(r) = k(r - \ell_0)^2/2$? What can you say about the case when, in addition to the spring force, each monomer is a hard sphere with radius R, i.e. we have $r \geq 2R$ which is enforced by elastic collisions?

9.11 Consider a dimer molecule that is modelled as two hard balls of radius R and mass m_p connected by a spring with potential $\Phi(r) = k(r - \ell_0)^2/2$, where ℓ_0 is the rest length of the spring. Replace the terms with diffusion and friction constants, D and β, in the macroscopic equations (9.25)–(9.26) by the following explicitly modelled heat baths:

(a) The heat bath introduced in Section 8.4, given by (light) point solvent particles with masses $m_s \ll m_p$ which collide with both monomers.

(b) The heat bath introduced in Section 8.1, given by one solvent molecule per monomer.

Implement each model on a computer. Discuss the differences between the two microscopic models and the results obtained by using macroscopic equations (9.25)–(9.26) in Exercise 9.10.

In each case (a) and (b), design a multi-resolution scheme where the first monomer has its heat bath interactions replaced by suitable macroscopic equations, while the second monomer is modelled using the corresponding microscopic model. In the case (a), use a co-moving cube, $\Omega(t)$, given by (9.24), where $\mathbf{X}(t) = \mathbf{X}_2(t)$.

In case (a), design a different multi-resolution scheme, where the domain Ω is fixed and the dimer can move between Ω, where the MD model is used, and $\mathbb{R}^2 \setminus \Omega$, where the Langevin description is used (the region denoted by grey in Figure 9.2(d)). Implement equation (9.18) when the dimer crosses the boundary between subdomains (denoted as the red interface in Figure 9.2(d)).

9.12 Consider a dimer molecule that is modelled as two hard balls of radius R and mass m_p connected by a spring with potential $\Phi(r) = k(r - \ell_0)^2/2$, where ℓ_0 is the rest length of the spring. Suppose that the dimer is in the Lennard-Jones fluid introduced in Section 8.6. Write a computer code that studies the dimer's length, diffusion constant and velocity autocorrelation function. Discuss similarities and differences with the results obtained in Exercises 9.10 and 9.11.

Appendix

A: Deterministic Modelling of Chemical Reactions

As explained in Chapter 1, a well-stirred system of chemical reactions can be
simulated by the Gillespie SSA provided that we know the propensity function
of each reaction. In Tables 1.1 and 1.2, the propensity functions of chemical
reactions up to the third order are given. In a number of places, we compare
the results of the Gillespie SSA with the corresponding deterministic model
which is written in terms of ODEs for concentrations of the chemical species
involved. The examples include the ODE (1.51) for the chemical system (1.27)
and the ODE system (1.58)–(1.59) for the chemical system (1.55)–(1.56).
Here, we briefly summarize how the deterministic models are constructed.

First, let us consider a well-stirred chemical system. Let $a(t)$ be the concentration of the chemical species A; the units of $a(t)$ are usually moles (or number
of molecules) per unit of volume. Let us assume that the chemical species A
is produced or destroyed according to k chemical reactions, which are labelled
from 1 to k. Then the time evolution of concentration $a(t)$ is given by the ODE

$$\frac{\mathrm{d}a}{\mathrm{d}t} = \sum_{i=1}^{k} c_i \, r_i,$$

where r_i is the rate of the ith reaction and c_i is the change in the number of
molecules of A corresponding to the occurrence of one ith reaction, i.e. it is the
difference between the number (stoichiometric coefficient) in front of A on the
right-hand side of the reaction and the corresponding stoichiometric coefficient
on the left-hand side. For example, $c_i = -2$ for the reaction $A + A \rightarrow B$; $c_i = -1$
for the reaction $A + B \rightarrow C$; $c_i = 1$ for the reaction $A + B \rightarrow 2A$; and $c_i = 2$ for
the reaction $B \rightarrow A + A$. The rate $r_i \equiv r_i(t)$ is computed as a product of the rate
constant and the concentrations of the reactants. The rates r_i for the chemical
reactions up to the third order are given in Table A.1.

Table A.1 *Rates $r_i(t)$ in deterministic modelling of chemical reactions up to the third order.*

chemical reaction	order	rate $r_i(t)$	units of k
$\emptyset \xrightarrow{k} A$	zeroth-order	k	$m^{-3}\,sec^{-1}$
$A \xrightarrow{k} \emptyset$	first-order	$a(t)k$	sec^{-1}
$A + B \xrightarrow{k} \emptyset$	second-order	$a(t)b(t)k$	$m^3\,sec^{-1}$
$A + A \xrightarrow{k} \emptyset$	second-order	$a^2(t)k$	$m^3\,sec^{-1}$
$A + B + C \xrightarrow{k} \emptyset$	third-order	$a(t)b(t)c(t)k$	$m^6\,sec^{-1}$
$2A + B \xrightarrow{k} \emptyset$	third-order	$a^2(t)b(t)k$	$m^6\,sec^{-1}$
$3A \xrightarrow{k} \emptyset$	third-order	$a^3(t)k$	$m^6\,sec^{-1}$

If the chemical system is not well stirred and molecules are only transported by diffusion, then the deterministic description is given in terms of a spatially varying concentration $a(\mathbf{x}, t) \equiv a(x_1, x_2, x_3, t)$ which evolves according to the PDE

$$\frac{\partial a}{\partial t} = D\triangle a + \sum_{i=1}^{k} c_i\, r_i,$$

where c_i and r_i are given as before (see Table A.1), D is the diffusion constant and $\triangle \equiv \nabla^2$ is the Laplace operator

$$\triangle = \sum_{j=1}^{3} \frac{\partial^2}{\partial x_j^2}.$$

It is worth noting that the same definitions of c_i and r_i, $i = 1, 2, \ldots, k$, are used in both ODE-based and PDE-based models.

B: Discrete Probability Distributions

A random variable is called discrete if it can achieve only a finite or count-ably infinite number of values. It can be characterized by the *probability mass function* which gives a probability that a discrete random variable is equal to a given value. In this book, we used the following discrete probability distributions.

Poisson distribution: Its support is the set of non-negative integers. The probability mass function is

$$p(k) = \frac{\mu^k}{k!} \exp[-\mu], \qquad \text{for} \quad k = 0, 1, 2, \ldots,$$

where $\mu > 0$ is its mean. Its variance is equal to the mean μ.

Binomial distribution: Its support is the finite set $\{0, 1, 2, \ldots, n\}$ where the integer n is one of two parameters of this distribution. The second parameter is $\lambda \in [0, 1]$. The probability mass function is

$$p(k) = \binom{n}{k} \lambda^k (1 - \lambda)^{n-k}, \qquad \text{where} \quad k \in \{0, 1, 2, \ldots, n\}.$$

Its mean is $n\lambda$ and variance is $n\lambda(1 - \lambda)$.

Discrete uniform distribution: Its support is the finite set $\{1, 2, \ldots, n\}$ where n is a positive integer. The probability mass function is

$$p(k) = \frac{1}{n}, \qquad \text{for} \quad k \in \{1, 2, \ldots, n\}.$$

Its mean is $(n + 1)/2$ and variance is $(n^2 - 1)/12$.

Multinomial distribution: It is a generalization of the binomial distribution to the case where we simultaneously generate m random numbers $k_1, k_2, \ldots, k_m$. Its support is the finite set

$$\left\{ [k_1, k_2, \ldots, k_m] \in \{0, 1, 2, \ldots, n\}^m \text{ such that } k_1 + k_2 + \cdots + k_m = n \right\},$$

where the integer n is one of the parameters of this distribution. The other parameters are $\lambda_1, \lambda_2, \ldots, \lambda_m \in [0, 1]$ satisfying $\lambda_1 + \lambda_2 + \cdots + \lambda_m = 1$. The joint probability mass function is

$$p(k_1, k_2, \ldots, k_m) = \frac{n!}{k_1! k_2! \cdots k_m!} \lambda_1^{k_1} \lambda_2^{k_2} \cdots \lambda_m^{k_m}.$$

The mean of k_i is $n\lambda_i$ and its variance is $n\lambda_i(1 - \lambda_i)$. The covariance of the ith and jth random number, k_i and k_j for $i \neq j$, is equal to $-n\lambda_i\lambda_j$.

C: Continuous Probability Distributions

A continuous random variable can take a continuous range of values from the set Ω. It is characterized by the probability density function $p: \Omega \to [0, \infty)$. The probability that the continuous random variable takes a value in the subset $A \subset \Omega$ is equal to $\int_A p(x) \, dx$. Examples of continuous probability distributions used in this book are as follows.

Normal distribution: Its support is $\Omega = \mathbb{R}$. The probability density function is

$$p(x) = \frac{1}{\sigma \sqrt{2\pi}} \exp\left[-\frac{(x - \mu)^2}{2\sigma^2}\right],$$

where $\mu \in \mathbb{R}$ is the mean and σ^2 is the variance ($\sigma > 0$ is the standard deviation). An algorithm to sample random numbers from the normal distribution with zero mean and unit variance is given in Exercise 3.1. An algorithm to sample random numbers from the truncated normal distribution, defined by equation (9.30), is presented in Exercise 9.8.

Exponential distribution: Its support is $\Omega = [0, \infty)$. The probability density function is

$$p(x) = \frac{1}{\mu} \exp\left[-\frac{x}{\mu}\right],$$

where $\mu > 0$ is the mean. Its variance is equal to μ^2.

Continuous uniform distribution: Its support is a finite interval (a, b) where $a, b \in \mathbb{R}$, $a < b$. The probability density function is

$$p(x) = \frac{1}{b - a}, \qquad \text{for} \quad x \in (a, b).$$

Its mean is $(a + b)/2$ and variance is equal to $(b - a)^2/12$.

Complementary error function distribution: Its support is $\Omega = (0, \infty)$. The probability density function is

$$p(x) = \sqrt{\pi} \, \text{erfc}(x), \qquad \text{for} \quad x \in (0, \infty),$$

where the complementary error function is given by (9.12). Its mean is $\sqrt{\pi}/4$ and variance is equal to $(1/3 - \pi/16)$. An algorithm to sample random numbers from this distribution is given in Exercises 9.4 and 9.5.

References

Agbanusi, I. and Isaacson, S. (2014), 'A comparison of bimolecular reaction models for stochastic reaction–diffusion systems', *Bulletin of Mathematical Biology* **76**(4), 922–946.

Alen, M. and Tildesley, D. (2017), *Computer Simulation of Liquids*, 2nd ed., Oxford University Press.

Andersen, H. (1983), 'Rattle: a "velocity" version of the Shake algorithm for molecular dynamics calculations', *Journal of Computational Physics* **52**, 24–34.

Andrews, S. and Bray, D. (2004), 'Stochastic simulation of chemical reactions with spatial resolution and single molecule detail', *Physical Biology* **1**, 137–151.

Andrews, S., Addy, N., Brent, R. and Arkin, A. (2010), 'Detailed simulations of cell biology with Smoldyn 2.1', *PLOS Computational Biology* **6**(3), e1000705.

Bagchi, B. (2013), *Water in Biological and Chemical Processes: from Structure and Dynamics to Function*, Cambridge University Press.

Barkai, N. and Leibler, S. (1997), 'Robustness in simple biochemical networks', *Nature* **387**, 913–917.

Ben-Naim, A. (2009), *Molecular Theory of Water and Aqueous Solutions*, World Scientific.

Berendsen, H., Grigera, J. and Straatsma, T. (1987), 'The missing term in effective pair potentials', *Journal of Physical Chemistry* **91**(24), 6169–6271.

Berg, H. (1983), *Random Walks in Biology*, Princeton University Press.

Berg, H. and Purcell, E. (1977), 'Physics of chemoreception', *Biophysical Journal* **20**, 193–219.

Berman, H., Westbrook, J., Feng, Z., Gilliland, G., Bhat, T., Weissig, H., Shindyalov, I. and Bourne, P. (2000), 'The Protein Data Bank', *Nucleic Acids Research* **28**, 235–242.

Berneche, S. and Roux, B. (2003), 'A microscopic view of ion conduction through the K^+ channel', *Proceedings of the National Academy of Sciences USA* **100**, 8644–8648.

Biancalani, T., Fanelli, D. and Di Patti, F. (2010), 'Stochastic Turing patterns in a Brusselator model', *Physical Review E* **81**, 046215.

Bruna, M. and Chapman, S. (2014), 'Diffusion of finite-size particles in confined geometries', *Bulletin of Mathematical Biology* **76**(4), 947–982.

Buhl, J., Sumpter, D., Couzin, I., Hale, J., Despland, E., Miller, E. and Simpson, S. (2006), 'From disorder to order in marching locusts', *Science* **312**, 1402–1406.

Bussi, G. and Parrinello, M. (2007), 'Accurate sampling using Langevin dynamics', *Physical Review E* **75**, 056707.

Cao, Y. and Erban, R. (2014), 'Stochastic Turing patterns: analysis of compartment-based approaches', *Bulletin of Mathematical Biology* **76**(12), 3051–3069.

Cao, Y., Li, H. and Petzold, L. (2004), 'Efficient formulation of the stochastic simulation algorithm for chemically reacting systems', *Journal of Chemical Physics* **121**(9), 4059–4067.

Cao, Y., Gillespie, D. and Petzold, L. (2005*a*), 'Multiscale stochastic simulation algorithm with stochastic partial equilibrium assumption for chemically reacting systems', *Journal of Computational Physics* **206**, 395–411.

Cao, Y., Gillespie, D. and Petzold, L. (2005*b*), 'The slow-scale stochastic simulation algorithm', *Journal of Chemical Physics* **122**(1), 14116.

Cao, Y., Gillespie, D. and Petzold, L. (2006), 'Efficient step size selection for the tau-leaping simulation method', *Journal of Chemical Physics* **124**, 044109.

Chandrasekhar, S. (1943), 'Stochastic problems in physics and astronomy', *Reviews of Modern Physics* **15**, 2–89.

Chapman, J., Erban, R. and Isaacson, S. (2016), 'Reactive boundary conditions as limits of interaction potentials for Brownian and Langevin dynamics', *SIAM Journal on Applied Mathematics* **76**(1), 368–390.

Chen, W., Erban, R. and Chapman, S. (2014), 'From Brownian dynamics to Markov chain: an ion channel example', *SIAM Journal on Applied Mathematics* **74**(1), 208–235.

Chib, S. and Greenberg, E. (1995), 'Understanding the Metropolis–Hastings algorithm', *American Statistician* **49**(4), 327–335.

Corry, B., Kuyucak, S. and Chung, S. (2000), 'Test of continuum theories as models of ion channels. II. Poisson–Nernst–Planck theory versus Brownian dynamics', *Biophysical Journal* **78**, 2364–2381.

Cotter, S., Zygalakis, K., Kevrekidis, I. and Erban, R. (2011), 'A constrained approach to multiscale stochastic simulation of chemically reacting systems', *Journal of Chemical Physics* **135**, 094102.

Cotter, S., Vejchodsky, T. and Erban, R. (2013), 'Adaptive finite element method assisted by stochastic simulation of chemical systems', *SIAM Journal on Scientific Computing* **35**(1), B107–B131.

Dallon, J. and Othmer, H. (1997), 'A discrete cell model with adaptive signalling for aggregation of *Dictyostelium discoideum*', *Philosophical Transactions of the Royal Society B: Biological Sciences* **352**(1351), 391–417.

Dama, J., Sinitskiy, A., McCullagh, M., Weare, J., Roux, B., Dinner, A. and Voth, G. (2013), 'The theory of ultra-coarse-graining. 1. General principles', *Journal of Chemical Theory and Computation* **9**, 2466–2480.

Davtyan, A., Dama, J., Voth, G. and Andersen, H. (2015), 'Dynamic force matching: a method for constructing dynamical coarse-grained models with realistic time dependence', *Journal of Chemical Physics* **142**, 154104.

DeVille, R., Muratov, C. and Vanden-Eijnden, E. (2006), 'Non-meanfield deterministic limits in chemical reaction kinetics', *Journal of Chemical Physics* **124**, 231102.

Di Pasquale, N. and Carbone, P. (2017), 'Local and global dynamics of multi-resolved polymer chains: effects of the interactions atoms–beads on the dynamic of the chains', *Journal of Chemical Physics* **146**(8), 084905.

Dobramysl, U., Rüdiger, S. and Erban, R. (2016), 'Particle-based multiscale modeling of calcium puff dynamics', *Multiscale Modelling and Simulation* **14**(3), 997–1016.

Duncan, A., Liao, S., Vejchodský, T., Erban, R. and Grima, R. (2015), 'Noise-induced multistability in chemical systems: discrete versus continuum modeling', *Physical Review E* **91**, 042111.

Dürr, D., Goldstein, S. and Lebowitz, J. (1981), 'A mechanical model of Brownian motion', *Communications in Mathematical Physics* **78**, 507–530.

E, W., Liu, D. and Vanden-Eijnden, E. (2005), 'Nested stochastic simulation algorithm for chemical kinetic systems with disparate rates', *Journal of Chemical Physics* **123**, 194107.

Earnest, T., Lai, J., Chen, K., Hallock, M., Williamson, J. and Luthey-Schulten, Z. (2015), 'Toward a whole-cell model of ribosome biogenesis: kinetic modeling of SSU assembly', *Biophysical Journal* **109**(6), 1117–1135.

Einstein, A. (1905), 'Über die von der molekularkinetischen Theorie der Wärme geforderte Bewegung von in ruhenden Flüssigkeiten suspendierten Teilchen', *Annalen der Physik* **17**, 549–560.

Elowitz, M., Levine, A., Siggia, E. and Swain, P. (2002), 'Stochastic gene expression in a single cell', *Science* **297**, 1183–1186.

Engblom, S., Ferm, L., Hellander, A. and Lötstedt, P. (2009), 'Simulation of stochastic reaction–diffusion processes on unstructured meshes', *SIAM Journal on Scientific Computing* **31**, 1774–1797.

Ensing, B., Nielsen, S., Moore, P., Klein, M. and Parrinello, M. (2007), 'Energy conservation in adaptive hybrid atomistic/coarse-grain molecular dynamics', *Journal of Chemical Theory and Computation* **3**, 1100–1105.

Erban, R. (2014), 'From molecular dynamics to Brownian dynamics', *Proceedings of the Royal Society A* **470**, 20140036.

Erban, R. (2016), 'Coupling all-atom molecular dynamics simulations of ions in water with Brownian dynamics', *Proceedings of the Royal Society A* **472**, 20150556.

Erban, R. and Chapman, S. J. (2007*a*), 'Reactive boundary conditions for stochastic simulations of reaction–diffusion processes', *Physical Biology* **4**(1), 16–28.

Erban, R. and Chapman, S. J. (2007*b*), 'Time scale of random sequential adsorption', *Physical Review E* **75**(4), 041116.

Erban, R. and Chapman, S. J. (2009), 'Stochastic modelling of reaction–diffusion processes: algorithms for bimolecular reactions', *Physical Biology* **6**(4), 046001.

Erban, R. and Haskovec, J. (2012), 'From individual to collective behaviour of coupled velocity jump processes: a locust example', *Kinetic and Related Models* **5**(4), 817–842.

Erban, R. and Othmer, H. (2004), 'From individual to collective behaviour in bacterial chemotaxis', *SIAM Journal on Applied Mathematics* **65**(2), 361–391.

Erban, R. and Othmer, H. (2005), 'From signal transduction to spatial pattern formation in E. coli: a paradigm for multi-scale modeling in biology', *Multiscale Modeling and Simulation* **3**(2), 362–394.

Erban, R., Kevrekidis, I., Adalsteinsson, D. and Elston, T. (2006), 'Gene regulatory networks: a coarse-grained, equation-free approach to multiscale computation', *Journal of Chemical Physics* **124**, 084106.

Erban, R., Chapman, S. J. and Maini, P. (2007), 'A practical guide to stochastic simulations of reaction–diffusion processes', 35 pages. arXiv:http://arxiv.org/abs/0704.1908.

Erban, R., Chapman, S. J., Kevrekidis, I. and Vejchodsky, T. (2009), 'Analysis of a stochastic chemical system close to a SNIPER bifurcation of its mean-field model', *SIAM Journal on Applied Mathematics* **70**(3), 984–1016.

Fange, D. and Elf, J. (2006), 'Noise-induced Min phenotypes in E. coli', *PLoS Computational Biology* **2**(6), 637–648.

Flegg, M. (2016), 'Smoluchowski reaction kinetics for reactions of any order', *SIAM Journal on Applied Mathematics* **76**(4), 1403–1432.

Flegg, M., Chapman, J. and Erban, R. (2012), 'The two-regime method for optimizing stochastic reaction–diffusion simulations', *Journal of the Royal Society Interface* **9**(70), 859–868.

Flegg, M., Chapman, J., Zheng, L. and Erban, R. (2014), 'Analysis of the two-regime method on square meshes', *SIAM Journal on Scientific Computing* **36**(3), B561–B588.

Flegg, M., Hellander, S. and Erban, R. (2015), 'Convergence of methods for coupling of microscopic and mesoscopic reaction–diffusion simulations', *Journal of Computational Physics* **289**, 1–17.

Fogarty, A., Potestio, R. and Kremer, K. (2016), 'A multi-resolution model to capture both global fluctuations of an enzyme and molecular recognition in the ligand-binding site', *Proteins* **84**(12), 1902–1913.

Franz, B. and Erban, R. (2012), 'Hybrid modelling of individual movement and collective behaviour', *in* M. Lewis, P. Maini and S. Petrovskii, eds, *Dispersal, Individual Movement and Spatial Ecology: A Mathematical Perspective*, Springer.

Frenkel, D. and Smit, B. (2002), *Understanding Molecular Simulation, from Algorithms to Applications*, 2nd ed., Academic Press, Elsevier.

Gadgil, C., Lee, C. and Othmer, H. (2005), 'A stochastic analysis of first-order reaction networks', *Bulletin of Mathematical Biology* **67**, 901–946.

Gibson, M. and Bruck, J. (2000), 'Efficient exact stochastic simulation of chemical systems with many species and many channels', *Journal of Physical Chemistry A* **104**, 1876–1889.

Gierer, A. and Meinhardt, H. (1972), 'A theory of biological pattern formation', *Kybernetik* **12**, 30–39.

Gillespie, D. (1976), 'A general method for numerically simulating the stochastic time evolution of coupled chemical reactions', *Journal of Computational Physics* **22**, 403–434.

Gillespie, D. (1977), 'Exact stochastic simulation of coupled chemical reactions', *Journal of Physical Chemistry* **81**(25), 2340–2361.

Gillespie, D. (2000), 'The chemical Langevin equation', *Journal of Chemical Physics* **113**(1), 297–306.

Gillespie, D. (2001), 'Approximate accelerated stochastic simulation of chemically reacting systems', *Journal of Chemical Physics* **115**(4), 1716–1733.

Gottwald, F., Karsten, S., Ivanov, S. and Kühn, O. (2015), 'Parametrizing linear generalized Langevin dynamics from explicit molecular dynamics simulations', *Journal of Chemical Physics* **142**, 244110.

Grima, R. (2011), 'Construction and accuracy of partial differential equation approximations to the chemical master equation', *Physical Review E* **84**, 056109.

Grima, R., Thomas, P. and Straube, A. (2011), 'How accurate are the chemical Langevin and Fokker–Planck equations?', *Journal of Chemical Physics* **135**, 084103.

Gunaratne, R., Wilson, D., Flegg, M. and Erban, R. (2019), 'Multi-resolution dimer models in heat baths with short-range and long-range interactions', *Interface Focus* **9**(3), 20180070.

Hadley, K. and McCabe, C. (2012), 'Coarse-grained molecular models of water: a review', *Molecular Simulation* **38**, 671–681.

Hänggi, P., Talkner, P. and Borkovec, M. (1990), 'Reaction-rate theory: fifty years after Kramers', *Reviews of Modern Physics* **62**, 251–341.

Hattne, J., Fange, D. and Elf, J. (2005), 'Stochastic reaction–diffusion simulation with MesoRD', *Bioinformatics* **21**(12), 2923–2924.

Hellander, S. and Petzold, L. (2016), 'Reaction rates for a generalized reaction–diffusion master equation', *Physical Review E* **93**, 013307.

Hellander, S., Hellander, A. and Petzold, L. (2012), 'Reaction–diffusion master equation in the microscopic limit', *Physical Review E* **85**, 042901.

Hellander, S., Hellander, A. and Petzold, L. (2015), 'Reaction rates for mesoscopic reaction–diffusion kinetics', *Physical Review E* **91**, 023312.

Hillen, T. (2001), 'Transport equations and chemosensitive movement', Habilitation thesis, University of Tuebingen, Germany.

Hillen, T. and Stevens, A. (2000), 'Hyperbolic models for chemotaxis in 1-D', *Nonlinear Analysis: Real World Applications* **1**, 409–433.

Hoover, W. (1985), 'Canonical dynamics: equilibrium phase-space distributions', *Physical Review E* **31**(3), 1695–1697.

Howard, M. (2012), 'How to build a robust intracellular concentration gradient', *Trends in Cell Biology* **22**(6), 311–317.

Huggins, D. (2012), 'Correlations in liquid water for the TIP3P-Ewald, TIP4P-2005, TIP5P-Ewald, and SWM4-NDP models', *Journal of Chemical Physics* **136**(6), 064518.

Hunenberger, P. and Reif, M. (2011), *Single-Ion Solvation: Experimental and Theoretical Approaches to Elusive Thermodynamic Quantities*, RSC Publishing.

Isaacson, S. (2009), 'The reaction–diffusion master equation as an asymptotic approximation of diffusion to a small target', *SIAM Journal on Applied Mathematics* **70**(1), 77–111.

Isaacson, S. (2013), 'A convergent reaction–diffusion master equation', *Journal of Chemical Physics* **139**(5), 054101.

Isaacson, S. and Zhang, Y. (2018), 'An unstructured mesh convergent reaction–diffusion master equation for reversible reactions', *Journal of Computational Physics* **374**, 954–983.

Israelachvili, J. (2011), *Intermolecular and Surface Forces*, 3rd ed., Academic Press, Elsevier.

Jahnke, T. and Huisinga, W. (2007), 'Solving the chemical master equation for monomolecular reaction systems analytically', *Journal of Mathematical Biology* **54**(1), 1–26.

Jung, G., Hanke, M. and Schmid, F. (2017), 'Iterative reconstruction of memory kernels', *Journal of Chemical Theory and Computation* **13**, 2481–2488.

Kac, M. (1974), 'A stochastic model related to the telegrapher's equation', *Rocky Mountain Journal of Mathematics* **4**(3), 497–509.

Kang, H., Zheng, L. and Othmer, H. (2012), 'A new method for choosing the computational cell in stochastic reaction–diffusion systems', *Journal of Mathematical Biology* **65**(6-7), 1017–1099.

Karch, G. (2000), 'Selfsimilar profiles in large time asymptotics of solutions to damped wave equations', *Studia Mathematica* **143**, 175–197.

Kazeev, V., Khammash, M., Nip, M. and Schwab, C. (2014), 'Direct solution of the chemical master equation using quantized tensor trains', *PLOS Computational Biology* **10**(3), 1–19.

Keller, E. and Segel, L. (1971a), 'Model for chemotaxis', *Journal of Theoretical Biology* **30**, 225–234.

Keller, E. and Segel, L. (1971*b*), 'Traveling bands of chemotactic bacteria: a theoretical analysis', *Journal of Theoretical Biology* **30**, 235–248.

Kepler, T. and Elston, T. (2001), 'Stochasticity in transcriptional regulation: origins, consequences and mathematical representations', *Biophysical Journal* **81**, 3116–3136.

Koneshan, S., Rasaiah, J., Lynden-Bell, M. and Lee, S. (1998), 'Solvent structure, dynamics and ion mobility in aqueous solutions at 25°C', *Journal of Physical Chemistry B* **102**, 4193–4204.

Lee, S. and Rasaiah, J. (1996), 'Molecular dynamics simulation of ion mobility. 2. Alkali metal and halide ions using the SPC/E model for water at 25°C', *Journal of Physical Chemistry* **100**, 1420–1425.

Leimkuhler, B. and Matthews, C. (2013), 'Rational construction of stochastic numerical methods for molecular sampling', *Applied Mathematics Research Express* **1**, 34–56.

Leimkuhler, B. and Matthews, C. (2015), *Molecular Dynamics*. Vol. 39 of *Interdisciplinary Applied Mathematics*, Springer.

Lennard-Jones, J. (1924), 'On the determination of molecular fields. II. From the equation of state of a gas', *Proceedings of the Royal Society of London Series A* **106**, 463–477.

Lennard-Jones, J. (1931), 'Cohesion', *Proceedings of the Physical Society* **43**, 461–482.

Lewars, E. (2016), *Computational Chemistry: Introduction to the Theory and Applications of Molecular and Quantum Mechanics*, 3rd ed., Springer.

Liao, S., Vejchodský, T. and Erban, R. (2015), 'Tensor methods for parameter estimation and bifurcation analysis of stochastic reaction networks', *Journal of the Royal Society Interface* **12**(108), 20150233.

Lipkova, J., Zygalakis, K., Chapman, J. and Erban, R. (2011), 'Analysis of Brownian dynamics simulations of reversible bimolecular reactions', *SIAM Journal on Applied Mathematics* **71**(3), 714–730.

Lipkow, K., Andrews, S. and Bray, D. (2005), 'Simulated diffusion of phosphorylated CheY through the cytoplasm of *Escherichia coli*', *Journal of Bacteriology* **187**(1), 45–53.

Liu, Z., Pu, Y., Li, F., Shaffer, C., Hoops, S., Tyson, J. and Cao, Y. (2012), 'Hybrid modeling and simulation of stochastic effects on progression through the eukaryotic cell cycle', *Journal of Chemical Physics* **136**, 034105.

Machado, M., Gonzáles, H. and Pantano, S. (2017), 'MD simulations of viruslike particles with supra CG solvation affordable to desktop computers', *Journal of Chemical Theory and Computation* **13**(10), 5106–5116.

Mann, R., Perna, A., Strömbom, D., Garnett, R., Herbert-Read, J., Sumpter, D. and Ward, A. (2013), 'Multi-scale inference of interaction rules in animal groups using Bayesian model selection', *PLOS Computational Biology* **9**(3), 1–13.

Mark, P. and Nilsson, L. (2001), 'Structure and dynamics of the TIP3P, SPC, and SPC/E water models at 298 K', *Journal of Physical Chemistry A* **105**(43), 9954–9960.

Metropolis, N., Rosenbluth, A., Rosenbluth, M., Teller, A. and Teller, E. (1953), 'Equation of state calculations by fast computing machines', *Journal of Chemical Physics* **21**, 1087–1092.

Munsky, B. and Khammash, M. (2006), 'The finite state projection algorithm for the solution of the chemical master equation', *Journal of Chemical Physics* **124**(4), 044104.

Muratov, C., Vanden-Eijnden, E. and E, W. (2005), 'Self-induced stochastic resonance in excitable systems', *Physica D* **210**, 227–240.

Murray, J. (2002), *Mathematical Biology*, Springer.

Nosé, S. (1984), 'A molecular dynamics method for simulations in the canonical ensemble', *Molecular Physics* **52**(2), 255–268.

Nymand, T. and Linse, P. (2000), 'Ewald summation and reaction-field methods for potentials with atomic charges, dipoles and polarizabilities', *Journal of Chemical Physics* **112**(14), 6152–6160.

Othmer, H. and Schaap, P. (1998), 'Oscillatory cAMP signaling in the development of *Dictyostelium discoideum*', *Comments on Theoretical Biology* **5**, 175–282.

Othmer, H., Dunbar, S. and Alt, W. (1988), 'Models of dispersal in biological systems', *Journal of Mathematical Biology* **26**, 263–298.

Paulsson, J., Berg, O. and Ehrenberg, M. (2000), 'Stochastic focusing: fluctuation-enhanced sensitivity of intracellular regulation', *Proceedings of the National Academy of Sciences USA* **97**(13), 7148–7153.

Perera, L., Essmann, U. and Berkowitz, M. (1995), 'Effect of the treatment of long-range forces on the dynamics of ions in aqueous solutions', *Journal of Chemical Physics* **102**(1), 450–456.

Plesa, T., Vejchodský, T. and Erban, R. (2016), 'Chemical reaction systems with a homoclinic bifurcation: an inverse problem', *Journal of Mathematical Chemistry* **54**(10), 1884–1915.

Plesa, T., Zygalakis, K., Anderson, D. and Erban, R. (2018), 'Noise control for molecular computing', *Journal of the Royal Society Interface* **15**(144), 20180199.

Potestio, R., Peter, C. and Kremer, K. (2014), 'Computer simulations of soft matter: linking the scales', *Entropy* **16**(8), 4199–4245.

Praprotnik, M., Delle Site, L. and Kremer, K. (2008), 'Multiscale simulation of soft matter: from scale bridging to adaptive resolution', *Annual Review of Physical Chemistry* **59**, 545–571.

Rao, C., Wolf, D. and Arkin, A. (2002), 'Control, exploitation and tolerance of intracellular noise', *Nature* **420**, 231–237.

Reeves, G., Kalifa, R., Klein, D., Lemmon, M. and Shvartsmann, S. (2005), 'Computational analysis of EGFR inhibition by Argos', *Developmental Biology* **284**, 523–535.

Risken, H. (1989), *The Fokker–Planck Equation, Methods of Solution and Applications*, Springer.

Robert, C. (1995), 'Simulation of truncated normal variables', *Statistics and Computing* **5**(2), 121–125.

Roberts, E., Stone, J. and Luthey-Schulten, Z. (2013), 'Lattice Microbes: high-performance stochastic simulation method for the reaction–diffusion master equation', *Journal of Computational Chemistry* **34**(3), 245–255.

Robinson, M., Flegg, M. and Erban, R. (2014), 'Adaptive two-regime method: application to front propagation', *Journal of Chemical Physics* **140**(12), 124109.

Robinson, M., Andrews, S. and Erban, R. (2015), 'Multiscale reaction–diffusion simulations with Smoldyn', *Bioinformatics* **31**(14), 2406–2408.

Rogers, K. and Schier, A. (2011), 'Morphogen gradients: from generation to interpretation', *Annual Review of Cell and Developmental Biology* **27**(1), 377–407.

Rolls, E., Togashi, Y. and Erban, R. (2017), 'Varying the resolution of the Rouse model on temporal and spatial scales: application to multiscale modelling of DNA dynamics', *Multiscale Modeling and Simulation* **15**(4), 1672–1693.

Roux, B., Allen, T., Berneche, S. and Im, W. (2004), 'Theoretical and computational models of biological ion channels', *Quarterly Reviews of Biophysics* **37**(1), 15–103.

Rowlinson, J. (2002), *Cohesion: A Scientific History of Intermolecular Forces*, Cambridge University Press.

Saunders, T., Pan, K., Angel, A., Guan, Y., Shah, J., Howard, M. and Chang, F. (2012), 'Noise reduction in the intracellular pom1p gradient by a dynamic clustering mechanism', *Developmental Cell* **22**(3), 558–572.

Schlögl, F. (1972), 'Chemical reaction models for non-equilibrium phase transitions', *Zeitschrift für Physik* **253**(2), 147–161.

Schnakenberg, J. (1979), 'Simple chemical reaction systems with limit cycle behaviour', *Journal of Theoretical Biology* **81**, 389–400.

Schnoerr, D., Sanguinetti, G. and Grima, R. (2014), 'The complex chemical Langevin equation', *Journal of Chemical Physics* **141**, 024103.

Shimmi, O., Umulis, D., Othmer, H. and Connor, M. (2005), 'Facilitated transport of a Dpp/Scw heterodimer by Sog/Tsg leads to robust patterning of the *Drosophila* blastoderm embryo', *Cell* **120**(6), 873–886.

Shin, H., Kim, C., Talkner, P. and Lee, E. (2010), 'Brownian motion from molecular dynamics', *Chemical Physics* **375**, 316–326.

Shoup, D. and Szabo, A. (1982), 'Role of diffusion in ligand binding to macromolecules and cell-bound receptors', *Biophysical Journal* **40**, 33–39.

Sick, S., Reinker, S., Timmer, J. and Schlake, T. (2006), 'WNT and DKK determine hair follicle spacing through a reaction–diffusion mechanism', *Science* **314**, 1447–1450.

Sjöberg, P., Lötstedt, P. and Elf, J. (2009), 'Fokker–Planck approximation of the master equation in molecular biology', *Computing and Visualization in Science* **12**, 37–50.

Smith, C. and Yates, C. (2018), 'Spatially-extended hybrid methods: a review', *Journal of the Royal Society Interface* **15**(139), 20170931.

Smoluchowski, M. (1917), 'Versuch einer mathematischen Theorie der Koagulationskinetik kolloider Lösungen', *Zeitschrift für physikalische Chemie* **92**, 129–168.

Soloveichik, D., Seelig, G. and Winfree, E. (2010), 'DNA as a universal substrate for chemical kinetics', *Proceedings of the National Academy of Sciences USA* **107**(12), 5393–5398.

Spiro, P., Parkinson, J. and Othmer, H. (1997), 'A model of excitation and adaptation in bacterial chemotaxis', *Proceedings of the National Academy of Sciences USA* **94**, 7263–7268.

Srinivas, N., Parkin, J., Seelig, G., Winfree, E. and Soloveichik, D. (2017), 'Enzyme-free nucleic acid dynamical systems', *Science* **358**, eaal2052.

Stiles, J. and Bartol, T. (2001), 'Monte Carlo methods for simulating realistic synaptic microphysiology using MCell', *in* E. Schutter, ed., *Computational Neuroscience: Realistic Modeling for Experimentalists*, CRC Press, pp. 87–127.

Sumpter, D. (2010), *Collective Animal Behavior*, Princeton University Press.

Takahashi, K., Tanase-Nicola, S. and ten Wolde, P. (2010), 'Spatio-temporal correlations can drastically change the response of a MAPK pathway', *Proceedings of the National Academy of Sciences USA* **107**, 19820–19825.

Tarasova, E., Farafonov, V., Khayat, R., Okimoto, N., Komatsu, T., Taiji, M. and Nerukh, D. (2017), 'All-atom molecular dynamics simulations of entire virus capsid reveal the role of ion distribution in capsid', *Journal of Physical Chemistry Letters* **8**, 779–784.

Tostevin, F., ten Wolde, P. and Howard, M. (2007), 'Fundamental limits to position determination by concentration gradients', *PLOS Computational Biology* **3**, 763–771.

Turing, A. (1952), 'The chemical basis of morphogenesis', *Philosophical Transactions of the Royal Society of London* **237**, 37–72.

van Zon, J. and ten Wolde, P. (2005), 'Green's-function reaction dynamics: a particle-based approach for simulating biochemical networks in time and space', *Journal of Chemical Physics* **123**, 234910.

Vicsek, T., Czirók, A., Ben-Jacob, E., Cohen, I. and Shochet, O. (1995), 'Novel type of phase transition in a system of self-driven particles', *Physical Review Letters* **75**(6), 1226–1229.

Wils, S. and De Schutter, E. (2009), 'STEPS: modeling and simulating complex reaction–diffusion systems with Python', *Frontiers in Neuroinformatics* **3**(15), 1–8.

Wolpert, L., Beddington, R., Jessel, T., Lawrence, P., Meyerowitz, E. and Smith, J. (2002), *Principles of Development*, Oxford University Press.

Wooldridge, M. (2002), *An Introduction to Multi-agent Systems*, John Wiley & Sons.

Zauderer, E. (1983), *Partial Differential Equations of Applied Mathematics*, John Wiley & Sons.

Zhuravlev, P. and Papoian, G. (2009), 'Molecular noise of capping protein binding induces macroscopic instability in filopodial dynamics', *Proceedings of the National Academy of Sciences USA* **106**(28), 11570–11575.

Index